BASIC GUIDE
TO PESTICIDES

"It is the public that is being asked to assume the risks that the insect controllers calculate. The public must decide whether it wishes to continue on the present road, and it can do so only when in full possession of the facts. In the words of Jean Rostand, 'The obligation to endure gives us the right to know.' "

Rachel Carson, *Silent Spring*

BASIC GUIDE TO PESTICIDES

Their Characteristics and Hazards

Shirley A. Briggs and the staff of
RACHEL CARSON COUNCIL

CRC Press
Taylor & Francis Group
Boca Raton London New York

CRC Press is an imprint of the
Taylor & Francis Group, an **informa** business
A TAYLOR & FRANCIS BOOK

BASIC GUIDE TO PESTICIDES: Their Characteristics and Hazards

First published 1992 by Taylor & Francis

Published 2019 by CRC Press
Taylor & Francis Group
6000 Broken Sound Parkway NW, Suite 300
Boca Raton, FL 33487-2742

CRC Press is an imprint of Taylor & Francis Group, an Informa business

First issued in paperback 2019

No claim to original U.S. Government works

ISBN 13: 978-0-367-45023-6 (hbk)
ISBN 13: 978-1-56032-253-5 (pbk)

Visit the Taylor & Francis Web site at
http://www.taylorandfrancis.com

and the CRC Press Web site at
http://www.crcpress.com

DISCLAIMER CLAUSE or STATEMENT OF WARRANTY

The author and publisher are in no way responsible for the application or use of the chemicals mentioned or described herein. They make no warranties, expressed or implied, as to the accuracy or adequacy of any of the information presented in this book, nor do they guarantee the current status of registered uses of any of the chemicals with the U.S. Environmental Protection Agency. Also, by the omission, either unintentionally or from lack of space, of certain trade names and of some of the formulated products available to the public, the author is not endorsing those companies whose brand names or products are listed.

A CIP catalog record for this book is available from the British Library.

Library of Congress Cataloging-in-Publication Data

Briggs, Shirley A. (Shirley Ann), date.
    Basic guide to pesticides: their characteristics and hazards /
Shirley A. Briggs.
        p. cm.
    Includes bibliographical references and index.

    1. Pesticides—Toxicology.   I. Title.
RA1270.P4B73   1992
363.17'92—dc20   92-7024
                                                            CIP

    ISBN 1-56032-253-5

Publisher's Note
The publisher has gone to great lengths to ensure the quality of this reprint but points out that some imperfections in the original may be apparent.

Contents

Foreword

    Basic Guide to Pesticides is a fitting tribute to the memory of Rachel Carson. It covers all that people need to know about some 700 pesticides and their contaminants. This book is important in dealing with environmental problems both in general and in individual cases.

    Rachel Carson's gifts as both poet and scientist turned *Silent Spring* into an eloquent book. Because of her, we undertook landmark hearings in the U. S. Senate that aroused Congress and the nation to the dangers she described. Her purpose, she told me before she died, was to call attention to the ever-increasing contamination on the balance of nature, global in scope and detrimental to mankind.

    This present book is a guide for humanity as a whole. Ultimately, if we fail to use chemicals properly, we will injure deeply all nature and mankind.

Senator Abraham Ribicoff

Preface

THE PURPOSE OF THIS GUIDE

This book had its origin in the publication of *Silent Spring* in 1962, and more directly in the situation in which Rachel Carson found herself thereafter. She was overwhelmed by requests for information that reached her from people everywhere, and realized that no individual could deal with the amount and scope of the need. She spoke of establishing an organization that would keep abreast of new research, and would respond to requests from individuals, organizations, and governments with problems in use and control of pesticides. After her death in 1964, friends and colleagues with whom she had discussed this hope established what is now the Rachel Carson Council, an information center on chemical toxins, especially pesticides. As an independent, objective source, the Rachel Carson Council has continued to seek all sound information available, and to respond to requests from all over the world. Her book caused such a universal increase in concern with and comprehension of problems of pesticide contamination that it led to a steady growth in scientific study in the field and of government requirements for better testing. When we began, there were very few manuals and references available, and these covered limited aspects of the subject. With the help of those among our directors and consulting experts who represent pesticide toxicology, medicine, ecology, fish and wildlife, agriculture, and related subjects, we have gathered an extensive library and files over the years, assessed the reliability of the data, and continued to share the information with the public. Our library and files are available to those who wish to delve further.

Our years of dealing with the concerned public have given us insight into the problems people encounter, the kind of information they need, and the form in which it has proved most useful. Our exploration of the technical literature and attention to the ways of government regulatory agencies have expanded our understanding of the technical issues involved, and the ways in which economic and social pressures affect the way regulations are actually enforced. (Should not the information we have amassed be easily available to all from government? In an ideal world, perhaps, but much of our task has been running interference for the public against controlling groups, both inside government and out.)

Basic Guide to Pesticides is the product of all these years of gathering data and explaining them. Here we tell either the beginner or the specialist what they *need* to know, not just what is readily at hand. If key facts are not yet known, we make this clear so that caution is indicated. The final task of updating all our files for this book took place in 1990 and 1991. While we have included new information as much as possible, the formal cutoff date was September 1991.

With a wide range of potential users in mind, we have tried to arrange the facts so that a reader can quickly find just what is sought, without having to read through lengthy text material to sift out a few pertinent facts. Tabular presentations, with definitions of the categories and kinds of information given, have proved the most useful. They also show clearly where gaps in our present knowledge occur. A blank space in a column with a question mark in it shows that the trait or problem may exist, but as yet we cannot tell.

Many people will want more details than we can give in a necessarily terse presentation. Through our lists of recommended general sources and our specific references we point out further research routes. In the supplementary material in the appendices, experts in important aspects of pesticides summarize what should be understood by everyone in our increasingly chemical world.

DIMENSIONS OF THE SUBJECT

Everyone, whether consciously or not, is exposed to a large number of pesticides through many routes. Residues occur in our food, drinking water, air, clothing, and household furnishings. We may encounter more concentrated amounts in schools, churches, offices, apartment buildings, factories, golf courses, or from spraying of or run-off from agricultural lands or our neighbors' gardens. Communities have widespread spraying programs attempting to deal with nuisance insects or pests of city trees. It is often difficult to identify even the apparent sprays and dusts to which we are exposed, and the total array that may reach an individual in a short space of time is impossible to distinguish by either kind or quantity. It is this total, pervasive burden of toxic materials that we must consider when we have a decision to make about using a pesticide ourselves, or when involuntary exposure causes problems. It may not be just the latest exposure to chemicals that can have adverse effects on us, or on exposed animals and plants, but the final combination. In a world that has absorbed ever-increasing amounts of pesticides in the past 46 years—many of them synthetic toxins never before found in nature—both the immediate and the long-term reactions can be serious.

Pesticides, with few exceptions, are very biologically active substances. They can have profound effects on living matter in various ways, and are designed to kill at least certain forms. They may have different effects on different organisms, doing one thing to plants, another to birds, or poisoning the target pest by a different physiologic reaction than that caused in other forms of life. There are, however, basic similarities in the ways that cells function, whether in plants or animals, and it can be assumed that a substance that can kill one organism may have a marked effect on many others. In a few cases we do indeed have materials that affect only a narrow range of plants or animals, and these are the most desirable pesticides. It is more profitable to manufacture products with many uses, so the pesticides in common use are usually "broad spectrum," which means that they can damage plants and animals that the user may not expect or wish to harm.

Pesticides include broadly toxic substances that are released into our environment and may have effects far from the point of application, both in space and time. To gauge the whole impact of any one would require a knowledge of the intricate operation of many ecosystems far beyond our present information. It is unlikely that we shall ever have a sufficient grasp of all of these factors and their interactions to make an adequate assessment. Because Rachel Carson made the elements of such understanding clear to the public in *Silent Spring* in order to explain the scope of the danger from uncontrolled use of pesticides and the vulnerability of our living environment, she has been called the mother of the environmental movement.

AMOUNTS IN USE

Since Rachel Carson first described the problem in 1962, pesticide production and use in the United States and around the world has vastly increased. Whereas she was concerned about a U.S. total of 637,666,000 pounds a year in 1960, we now stand at 1.1 billion, and if all materials correctly designated as pesticides are included, at 2.1 billion pounds.

(Originally, the figures omitted wood preservatives, disinfectants, and sulfur.) These figures are for active ingredients only, and come from the latest report from the Environmental Protection Agency for 1989 (Economic Analysis Branch, Office of Pesticide Programs). The United States produces 1.3 billion pounds, imports 200 million, and exports 400 million to reach the 1.9 billion pounds of "conventional" pesticides used. The expenditure for this use was $7.615 billion. Herbicides have become the most-used kind of pesticides, at 61%, with insecticides at 21%, fungicides at 10%, and all others at 7%. In the May 1991 *EPA Journal* summarizing pesticide programs, a graph shows the amounts for the top 10 pesticides, with a total of 44,020,000 pounds per year. Two, carbaryl and malathion, are insecticides; the rest are herbicides. They account for 40% of all U.S. usage.

alachlor   100 million pounds
atrazine   100 million pounds
2,4-D   52.67 million pounds
butylate   44.58 million pounds
metolachlor   44.55 million pounds
trifluralin   30.35 million pounds
cyanazine   20.25 million pounds
carbaryl   12.25 million pounds
malathion   15.20 million pounds
metribuzin   13.17 million pounds

Since the United States accounts for one-third of the world figure, by multiplication we now exist on an earth where 6.3 billion pounds* of these toxic materials are added every year, to join the continuing residues that make their way, like the air and ocean currents, all over the globe.

To live on such an earth, clear understanding of these materials is essential for everyone. To this end, we offer our *Basic Guide to Pesticides*.

Shirley A. Briggs

*Most estimates of world consumption are based on the shorter list of "conventional" pesticides. Data are elusive, but it is reasonable to assume that the United States uses of the three additional types in the full 2.1 billion total are proportional to world usage, thus the 6.3 billion figure.

Illustrations

To contend with the larger issues of pest control is to become enmeshed in many aspects of our attitude toward the natural world. Because she dealt with these aspects clearly and convincingly, Rachel Carson has been credited with launching what is now called the *environmental movement,* successor to previous periods of concern for our habitat called *conservation.* We must balance short-term against long-term effects and our self-centered aims against broader needs of other forms of life, and gain a concept of the dynamics of ecology—a term the book *Silent Spring* first made common currency. In selecting the illustrations for this guide, which deals mostly with the tools for pest control from which we must choose, we wish to suggest attitudes that ei-

ther focus on wiping out immediate annoyances and threats or seek to promote a continuing healthy environment.

When the Rachel Carson Council was established, the now-deceased Mauritz Escher gave his support by granting permission to use his drawings in our publications. They express so well the unity of nature, the beauty of creatures that some find alien, and the sense of proportion and humor that were also fundamental to Rachel Carson's world view. Cartoons also can express these concerns pointedly, with a look at both the surface hilarity of human quirks and blindnesses and the underlying import of our behavior. Cartoons from the British magazine *Punch* are used here by permission.

"This is the dog that bit the cat that killed the rat that ate the malt that came from the grain that Jack sprayed."

PUNCH, March 6, 1963

"Spring is here—listen, the first crop-spraying helicopter!"

"Come on, son, eat up your Warfarin or you'll never grow up big and immune like your dad."

"Tin of Killo, packet of Pestdoom and a quart of Liquideath."

Acknowledgments

The plan and supporting files for the *Basic Guide to Pesticides* have been developed over the past quarter century, and everyone who has been on our staff over these years has had a part in preparing for the final publication. To give special credit to those whose work pertained closely to the guide, while recognizing that we could not have done it without much help with office routine from many others, called for some close decisions. First, we must thank all of those who have served as Council officers and directors, supporting the effort with expert information as well as by keeping the organization alive.

Next are those who made the final all-out effort to bring all the data up to date within the year, and into consistent form for publication: Nathan Erwin, Theresa Laranang, Taher Husain, DaVisa Hughes, and Howard and Jane Whitlow. Dr. William Lijinsky also gave us much scientific guidance. Those who spent considerable time and skill on earlier preparations include Rubin Borasky, Martha Damon, Cynthia French, Susan Garabrant, Edwin J. Jolly, Lisa Lefferts, Margaret Quarles, Merry Rabb, Ellen Rainer, Donald Weber, Robin West, and Feseha Woldu. Others whose contribution was shorter, but appreciated, include Charlotte Aggerholm, Leith Bernard, Leah Devlin, Christina Edwards, Kathleen Lucatorto, Barbara Pitkin, Marjorie Van Nostrand, and Ann Vogel. We are especially grateful to the National Coalition Against Misuse of Pesticides for taking over responses to routine inquiries on pesticides, and for sharing their files with us. As the one who got all of these people, and other staff and volunteers, into this, I most appreciate the dedication and cheerful spirit with which the work has been sustained.

We are greatly indebted to the experts in the field who reviewed the whole manuscript: Dr. William A. Butler, Dr. John L. George, Dr. Marion Moses, Dr. David Pimentel, Dr. Frederick W. Plapp, Dr. Robert L. Rudd, Dr. Marvin Schneiderman, and Dr. Thomas G. Scott.

Financial Support

Since 1965, Rachel Carson Council has been sustained by contributions from many individuals who approve of our work and have been helped by the information we dispense. Proposals to foundations for special grants have brought in additional funds, but until the last few years these were not sufficient to raise staff and resources to the level needed for the final concentrated effort to bring the guide to publication. Credit for the recent attainment of these resources goes first to the late Louise Tomkins Smith, long one of our mainstays, who gave a matching grant that began to attract other funds, and then to Director Nancy Greenspan and Treasurer David McGrath, who launched a major campaign to build on this start. Those who have given substantial help over the years include these foundations and individuals, some of whom gave through family foundations not also listed.

Foundations: Geraldine R. Dodge Foundation, Henry Doubleday Research Association, The Charles Engelhard Foundation, The Henry P. Kendall Foundation, National Fish and Wildlife Foundation, Marjorie Mosher Schmidt Foundation, The Florence and John Schumann Foundation, The Sears Family Foundation, Towncreek Foundations, Inc., Wallace Genetic Foundation, Inc., The Waletzky Lead Trust, Westinghouse Foundation, and The Wildcat Foundation.

Individuals: Marjorie Arundel, Mr. and Mrs. Stuart Avery, Lewis H. Batts, Paul Brooks, Mrs. C. C. Buckland, Mrs. John Buckland, Mrs. John Burgher, Cornelia Cameron, Ann Catts, David Challinor, Mary Cooper, Joan H. Griswell, Mr. and Mrs. Edwun Crocker, Maria Ealand, Mr. and Mrs. Theodore Edison, Richard H. Goodwin, Russell Hall, John W. Hanes, Jr., Mr. and Mrs. H. V. Harsha, John B. Hatcher, Mrs. Yarnall Jacobs, Mrs. Louis I. Jaffe, Raymond E. Johnson, Mrs. Paul Knight, Catherine Le Maistre, William Lijinsky, Mrs. V. S. Littauer, Mrs. J. A. Llewellyn, David McCalmont, Mr. and Mrs. James McCarthy, Cynthia K. McGrath, Catherine P. Middlebush, Northern Stone Supply, Inc., Ann Notnes, Louise D. Peck, Esther Peterson, also contributions in memory of Oliver Peterson, Margaret D. Potvin, Mr. and Mrs. Nelson Poynter, Mrs. Chandler Robbins II, Mrs. David Rockefeller, Sylvia Rosenheck, Mr. J. C. Rotunda, Slade Gorton and Company, Inc., Jocelyn Arundel Sladen, Mr. and Mrs. Stanley Smigel, Mrs. Robert P. Solem, Arthur Strauss, Ruth Strosnider, Mr. and Mrs. David Swetland, Mr. and Mrs. Lee M. Talbot, Norma Terris, Pearl Thussen, Ruth Thompson, Margaret van den Bosch, Joseph K. Wagoner, Mrs. Thomas M. Waller, and Mrs. Alexander Wetmore.

PLAN AND SOURCES

Chapter One

Plan and Sources

Selection of Pesticide Materials Included

ACTIVE INGREDIENTS

Commercial pesticide products combine several kinds of ingredients, but testing for kinds of immediate and long-term toxicity is done on substances called active ingredients, which are those with pesticide action against target pests. Commercial products may include a number of very similar formulations marketed by different producers and their various components are not tested for their separate or combined effects, except as these are active ingredients. The large number of formulations marketed would present an impossible amount of testing, and the task of adequate testing of just the authorized active ingredients has been many years in reaching the present partial percentage.

A recent estimate of currently registered active ingredients is 650, though no one at the Environmental Protection Agency seems very sure of this rapidly shifting number. At one time there were about 1400, but many have been withdrawn for excessive toxic hazard or lack of use and are thus of no economic value. New testing requirements and a fee for continuing registration have contributed to major deletions in the last couple of years. New materials, perhaps 15 or 20, are added each year. The former estimate was that about 600 active ingredients were used enough to matter, and of these, perhaps 120 are major constituents of most-used products. We have chosen those most used or of special hazard, either because of toxicity or those whose persistence in our environment means that they will be with us for many years to come. We include several not permitted in the United States, but used widely in other countries, since this book is designed for readers worldwide. Where feasible, we also include common names used elsewhere. Products used only in veterinary medicine, especially internally, are omitted.

INERT INGREDIENTS

Materials in pesticide formulations that are called inert are not so classified because they are inactive, but only because they have no pesticidal effect on the target organisms. They may be solvents, propellants, surfactants, emulsifiers, wetting agents, carriers, or diluents. They may, in fact, be very active from a biological standpoint, and are sometimes the most generally toxic portion of a pesticide product. Hundreds of these are in current use, and have been considered trade secrets by the producers and therefore have not been listed on the label. EPA has recently given them more of the attention they deserve and has selected the most toxic for scrutiny, identifying first 50 substances of special concern. Almost all of these have now been removed from products by the registrants, while those still in use must be identified on labels. EPA policy now calls for using the least toxic inerts available. A second group of 65 potentially too hazardous inerts has been selected for study and testing. Uncertainty about the danger from many inerts comes from the lack of testing. Very few commercial chemical products are tested for immediate or long-term toxicity, certainly not to the extent now required for pesticides.

SYNERGISTS

These ingredients, which may not have pesticidal action by themselves, are added to heighten the effects of the active ingredients, especially when these are expensive materials. By enhancing the combined toxicity, the effect on a target pest may be increased several hundred times. Piperonyl butoxide is a member of a commonly used class of synergists, the methylenedioxyphenyls (MDPs), which are added to pyrethrum and pyrethroids commonly, and also have a strong effect on the toxicity of carbamates. They act by inhibiting the target pest's ability to detoxify the primary poison. They can also make the pesticide far more toxic to humans and other nontarget creatures by the same process. Their effects must be carefully considered in the choice of a pesticide, or in deciding whether the use of a chemical compound is justified. We have included the most commonly used of these ingredients.

EFFECTS OF COMBINED INGREDIENTS

With the numerous formulations on the market, many similar to each other, it is neither practical nor possible to list the comparative hazards of each one. Nor is this known, since most testing is by active ingredients separately, and not by the combination in a single product. In many cases, a fairly good estimate of the total effects of a product can be made by adding the known qualities of individual constituents. In many cases, however, a combination creates a synergistic effect and the resulting product may be many times more toxic than would be expected by the known toxicity of the several parts. Two chemicals of a low or medium range of toxicity may combine to make something that ranks as very toxic. This is true of a number of mixtures with malathion, for instance. It can also occur when a person, plant, animal, or other exposed organism is or has also been exposed to a substance that interacts with the pesticide. Contact with malathion after being exposed to parathion, for example, can cause a severe reaction, because the parathion can exhaust the body's supply of a detoxifying enzyme for the time being, and the malathion has no opposition. Many pesticides should not be used by anyone taking certain drugs or drinking alcoholic beverages. The familiar danger of combining exposures of barbituates and alcohol is an example of the kind of thing that can happen with many substances to which we may be exposed, voluntarily or involuntarily.

The wary user of pesticides should allow a wide margin of safety when there is any question of potentiation of combined toxicants either in the product or available to react with it.

PESTICIDES NOT INCLUDED

A number of pesticides are not studied in detail in this guide either because of lack of information or minor use. We maintain active files on many of these and welcome more information. Those who cannot find the pesticide they seek here may inquire directly of the Rachel Carson Council for data.

Sources of Information on Pesticides

This guide is a compilation of the best factual material that we have been able to assemble since 1965. The data base is far from ideal: we have consulted the relevant manuals, computer listings, technical journals, and experts in the field over the years, and gradually built our supporting files. The major part of the testing and other research on pesticide toxicity has been done by or for the pesticide manufacturers to provide the data required for government registration of products allowed on the market. Pesticide manufacturers have done much research themselves, hired commercial testing firms to do it, or provided grants for study in academic institutions. In some cases the possible bias suggested by this process has been found as laboratories or researchers slanted results to achieve what the producers hoped to find. The case of the Industrial Bio-Test Laboratories was most notorious in this respect. Reputable manufacturers realize the perils of incorrect testing, of course, and strive for reports that can stand close scrutiny. This testing is very expensive, justified if a company can expect to make sufficient profits from sale of the product, but beyond the capacity of most independent researchers, or even of most other national governments. The result has been a wide dependence on results obtained in the United States, so we have a worldwide responsibility to be accurate and to consider all important aspects.

A variety of manuals and directories have been published to serve the pesticide industry, agricultural users, research chemists, or the medical profession. None includes all of the kinds of information needed by the person applying the pesticide or the person who may be exposed to it. In no single source could we find all of the pesticide ingredients listed here, or all of their characteristics that should be known.

We cannot vouch for the accuracy of some of these sources. Often, manuals and compilations do not indicate their sources and many seem to have been copied from each other in long succession. Sometimes the findings of one scientist or organization contradict the conclusions of another. Research methods, if they are known at all, are not always known in enough detail to assess the validity of a study. Replicate studies may not be available to verify original experimental results, especially with the high cost of much of this testing. Little original, independent research may be done on most pesticides to give us needed comparisons. Once the evidence is provided to the Environmental Protection Agency, the federal bureau responsible for registering pesticides for use and enforcing the control rules, it has in the past remained buried in their files, much of it classified as a "trade secret" by the producer. For most of the years that we have pursued this information, it took lengthy negotiations through the Freedom of Information Act to gain access to industry test material, and then access was given only if the company in question agreed. Though the law governing pesticide regulation, the Federal Insecticide, Fungicide, and Rodenticide Act (FIFRA), says that toxicology and environmental fate data should be open to the public, it took a Supreme Court decision in 1981 to confirm that this should indeed be so. It can still take months to obtain a desired document, however, since EPA still requires the Freedom of Information process.

The U.S. General Accounting Office, a congressional agency, studies ways in which laws are carried out, and makes other valuable studies of government performance. Their investigations into pesticide regulation over the years have been commendable. Two recent studies are especially valuable for the average concerned person: *Nonagricultural Pesticides; Risks and Regulation*, GAO/RCED-86-97, issued in April 1986, and *Lawn Care Pesticides; Risks Remain Uncertain While Prohibited Safety Claims Continue*, GAO/RCED-90-134, March 1990. Up to five copies of each GAO report are free on request.

The 1972 revision of FIFRA required for the first time that pesticides allowed on the market be tested for a wide range of effects, short- and long-term. Before, while this regulation was under the U.S. Department of Agriculture from 1947 until the EPA was established in 1970, only effectiveness against the target pests and simple immediate toxicity tests were done. No matter what hazards were found, no product was denied registration under USDA auspices. The 1972 law required that all active ingredients be more thoroughly tested and the risks found be balanced against the estimated benefits from use of the product. Each commercial formulation is thus not tested more than originally, as far as combined effects are concerned. EPA set about deciding which tests should be made, and on which pesticides.

Years passed while this was being decided, and the guidelines for testing were set up. The 1977 deadline, by which all registered pesticides were to be retested on the new rules, passed. Under new administrations since 1980, emphasis on limiting regulation has prevailed, and the guidelines were revised to be less stringent. A FIFRA revision in 1988 finally called EPA to stricter account, requiring that

all of this reregistration of old products remaining on the market, as well as registration of new ones, be completed by 1997. So far, few products have gone through even the reduced testing now required. Pesticides used on food have priority in these rules and nonagricultural uses have much more limited requirements.

For those who wish to explore just how much testing is required, and for which effects, see Section 158 of the Pesticides Registration, Data Requirements, issued by EPA with guidelines for testing, evaluation procedures, and laboratory practices, revised from time to time. In the summary section 158, charts list which tests are required for each kind of pesticide, and which are optional. Noting how many are optional, the reader then finds that a later clause lets the EPA administrator waive any required test.

Consulting the detailed volumes of the complete guidelines on all kinds of testing reveals many curious gaps. Tests for products that may cause birth defects, for example, will only be done on pesticides to which *human* females are likely to be exposed extensively in places where large numbers of them are expected to be found. No spill accidents in other places are considered and neither are venturesome women who strike out to less crowded places. Nor is it recognized that exposure of men to teratogens, substances that can cause birth defects, can be equally damaging.

When test guidelines were sent out for comment in 1982, Rachel Carson Council noticed that no provision was made to check the plant-killing potential of pesticides not registered for use as herbicides. We found 98 other pesticides known to damage plants too. These hazards were usually discovered by experience, not by comprehensive tests. This and other gaps in the requirements explain the lack of needed information in several areas, since comments from us and others did not change EPA policy.

Study of such EPA documents and submission of comments when requested have given us experience in the amount and effectiveness of current pesticide testing. In the book *Toxicity Testing*, published by the National Academy of Sciences, the estimate of the proportion of pesticides for which testing was adequate to make human health hazard assessment was only 10%. For 38% of pesticides nothing useful was known, and the rest fell somewhere in between.

Despite these discouraging conditions, it is crucial that people have the best estimate possible of the hazards of pesticides. They should also realize that the law regulating pesticides differs from those laws aiming to achieve clean air or water. FIFRA is a law to enable sale of pesticides through a balancing act between the claimed benefits (mostly to one group of people) against the known risks (usually to a completely different group.) The law specifies that no pesticide be labeled "safe," "non-toxic," "safe if used as directed," or "approved" by EPA. All pesticides exist because they are toxic to something, and EPA just registers by a marketing formula rather than approval.

FINDING AND EVALUATING DATA

If you note our list of major sources, you will find several that are compendiums of information on a wide range of pesticides and other commercial products. When we entered the final phase of compilation of this guide, we set a schedule to review every file and chart to bring each up to date within the year. Beyond the data and references we already had, we consulted such large, inclusive sources as the *Registry of Toxic Effects of Chemical Substances,* issued periodically by the National Institute of Occupational Safety and Health. Updates are available quarterly on microfiche, and all chemicals in commercial use are supposed to be listed, along with a terse summary of known toxic effects, citing the study quoted. They do not vouch for the reliability of the studies; this falls to the reader. Many times, the only study on a key point appears in a foreign journal, sometimes obscure. We then have to see whether this can be obtained from the National Library of Medicine, the USDA Library of Agriculture, or university libraries in the area. We have also gone through computer listings for Medline, Toxline, and Agricola services of libraries to ferret out journal or book articles we may have missed elsewhere. For each useful article found, we go through its list of references to be sure that we have the essential primary study for each key point on hand.

Another major source comes from EPA studies. When they single out a pesticide for special review because of its priority on their reregistration list, they issue a registration standard, showing what they know about it and where the gaps that must be filled by further testing exist. If they then decide that action should be taken to restrict or eliminate some or all uses of this pesticide, they issue position documents in a series documenting their findings and recommendations as study continues.

A final decision either to restrict, cancel, or reregister summarizes the supporting data. From all of

these sources we can normally pin down specific ratings for the criteria considered, and we can usually determine their primary source. If the registration standard is not clearly documented we seek the key primary studies and cite them. We may have to cite the registration standard where it is not fully documented, if they have not provided the original source in response to inquiry or have not yet given it to us through a Freedom of Information request.

With the key studies on hand, we apply accepted rules for assessing their thoroughness, methods used, and overall credibility. Conflicting evidence is resolved by asking experts in the field, with the council's professionally expert Directors and Consulting Experts called upon first.

In these ways we have done the best we could with a large but various body of information with our first consideration the hazards to an exposed person or other creature and the surrounding environment on which we depend.

In the explanation of the pesticide charts, we give our criteria and standards. We have explored as much of the literature on these pesticides as possible, judged it by standards we can support, and consulted objective scientists.

The lists of exact references for certain points should answer the needs of most people seeking more detailed information. For those who need to have a list of references for each point on a chart, this can be provided on request for a modest handling fee. These lists give the principal, most current sources.

Our complete files for each pesticide may contain a succession of studies going back to early inquiries, all of which comprise our supporting data. Our files and library are available to anyone who needs to go into the subject at this length. We have reviewed information on all pesticides in the guide through September 1991.

LISTS OF REFERENCES

The three lists found in Chapter 6 of the guide cover our principal sources, some specific references for details on the charts, and a final list of general background material. Some of these should be available in libraries.

HOW TO USE THIS GUIDE

Chapter Two

How To Use This Guide

THE CHARTS

To find the information for a specific pesticide ingredient, first take the name or names on the label or other description, and by use of the index of names, find the official common name under which the chart is headed, listed in alphabetical order.

Use of the Index of Names

Each pesticide active ingredient can have four kinds of names: the officially designated common name, various trade names for commercial formulations, the chemical name or names, and a CAS number. These names are all listed in alphabetical order, with the CAS numbers in numerical order. Each of these then gives the common names under which the chart appears. In the charts, the common name recognized in the United States is given first, with perhaps one used in another country. Next come trade names for formulations in which it is the principal active ingredient. These are capitalized. Chemical names can be numerous and confusing, since the chemical formula may be translated into words in various ways, some of which do not look much alike. We give those most commonly used, especially on labels. Where no common name exists, the chart is headed by a prominent trade name.

Trade names of products with several active ingredients, some present in small amounts, are not given, for reasons of space and clarity. To estimate the hazards and characteristics of such a product, check each of the active ingredients listed on the label. This will give a general idea, though it does not allow for interactions among the ingredients that may give unexpected effects or more toxicity.

If the only identification known is a trade name, and you cannot consult a label for a list of ingredients, start with this trade name in our index. If you cannot find it there, try to determine how old the product is. Trade names are sometimes changed or superseded, and may describe very different formulations from time to time. If you can find a corresponding edition of *Farm Chemicals Handbook,* it may have the old name. A current package of the product, or the *FCH,* may give you the name of the producer to whom you can write. You may also ask the Registration Division of the Environmental Protection Agency.

In some cases, so many brands use a particular ingredient that listing all or most brand names is impossible. These very commonly used pesticides are usually clearly listed as ingredients.

To find a chemical name in an alphabetical list requires following certain special rules. Some beginning letters or terms are not used in alphabetizing: these are in italics and include letter locants (*O-, m-, p-, sec-, tert-, N-, O-, S-,* etc.) and stereochemical descriptors (*cis-, trans-, (R)-, (S)-, (E)-, (Z)-, endo-,* etc.), and Greek letters. These are in italics, so looking past them for the key letter is not difficult. If a name begins with numerals, and there are more than one of the same name but different numerals, these will be in numerical order. Whether a Greek letter prefix is given as such or is written out (alpha, beta, gamma, etc.) the same order applies. You may find them either way, elsewhere. Greek characters used in chemical names correspond to their word designations, as written out below. Few in this guide go beyond the first six.

α	alpha	ν	nu
β	beta	ξ	xi
γ	gamma	o	omicron
δ	delta	π	pi
ϵ	epsilon	ρ	rho
ζ	zeta	ς	sigma
η	eta	τ	tau
θ	theta	υ	upsilon
ι	iota	ϕ	phi
κ	kappa	χ	khi
λ	lambda	ψ	psi
μ	mu	ω	omega

An example: if you look for chlordane, a common name, you will find it directly in the index as CHLORDANE, and the capitalization tells you that this is the name of the chart. Or you may find *Kypchlor: see chlordane,* or *octochloro-4,7-methanotetra-hydroindane: see chlordane.*

CAS numbers are assigned by the Chemical Abstract Service, at Ohio State University, Columbus, Ohio. This is the most generally accepted system for sure, concise identification of chemicals in commerce. They try to give each distinct chemical substance a number so that it can be identified through whatever confusion of trade and chemical names and their many versions it may have. Some pesticide ingredients may have more than one number if it is necessary to identify their various isomers, salts, esters, or other aspects. A few may not yet have received a number. In a few cases we found a conflict in the numbers given in equally authoritative sources, so even this system for dispelling confusion may occasionally falter. In case of two possibilities, we give the one with the best authority first, and the other second. We have no way of knowing from what source our readers may start to trace a material. Even a number or a chemical name that is technically incorrect may be included if it is in common usage.

CAS numbers begin with those with the fewest number of digits, in numerical order, then to the next number of digits, etc.

Nonchemical pesticides, such as bacteria, viruses, botanical materials, and so on, do not come within the CAS system. Their scientific names may be italicized, but they are listed in clear alphabetical order—*Bacillus thuringiensis*, for example.

**Transformation Products,
Contaminants, Components**

These related compounds that are given on the charts are treated in the index as are the alternative names: listed alphabetically, with the chart name on which they are found—*malaoxon: see malathion.*

EXPLANATION OF TERMS ON THE CHARTS

The various names described above are found in the first column, along with subsections for transformation products, contaminants, and other components of toxicological concern.

**Transformation Products,
Contaminants, Components**

Transformation products include the many results of introducing a pesticide into the environment, even in storage. Almost all break down eventually into constituent parts, are altered by contact with light, water, and other chemicals in soil or plants, or are metabolized by animals that absorb them. Entirely new chemical compounds may result. It is not enough to test for residues of the original

pesticide to know what has become of it or what further hazards it may create; the several transformation products must be known and sought also. Some of these may be more persistent than the original pesticide, and some are more toxic. We have listed those products that have had known problems result from their presence, and those with effects differing from those of the parent compound. Other names for these compounds are degradation products, derivatives, or metabolites.

Nitrosamines are a special class of transformation products, frequently found to be carcinogenic, in some cases among the most potent carcinogens known. They form when an amine in a compound comes into contact with a nitrosating agent such as nitrous acid in the saliva and stomach of ingesting animals, in soil, water, and air, and in certain industrial processes.

Production methods may create unintended *contaminants* that çan have the same range of effects. These are thus included. Other unexpected toxic effects may come from constituents in the formulation or the technical grade of the active ingredient that we call *components*, also included where known to have adverse effects.

Classes of Pesticides

Column 2 in the charts tells to which general group of pesticides this ingredient belongs. In certain cases this may be almost all that we know about a certain material. If we can learn the essential characteristics of this family or class of pesticides, we can have some idea of the probable behavior of a member of the group. Some classes are well established and studied, others have not been defined. Where we find no consistent precedent, we have made categories on the basis of the toxic action of the class, since that is the primary concern of those who will use this guide. The class name in column 2 refers to the later section of the guide called Toxic Characteristics of Classes of Pesticides, where we give the following information for each group:

All pesticides in the guide that fall into this class
The way in which these toxicants work, where known
Immediate toxicity, symptoms and effects, since this may be too long to include on the original chart
Long-term toxicity, kinds and eventual effects
Environmental effects, where known

Sources come from tests on appropriate test animals, and also on evidence of human effects where this is known. We cannot test people as we do laboratory animals, but we can compile the evidence of many medical records.

This is all given in necessarily condensed form. For more details see the list of principal references.

In too many cases, we still do not know all that we should about the mode of action of certain classes, in different species, or what can be done to treat adverse reactions. For some, no antidote is known. Long-term effects especially need far more study.

For the intensity of immediate toxicity, and the main kinds of long-term toxicity, see columns 4 and 5. These columns give the most important warnings, while the section on Toxic Characteristics of Classes of Pesticides tells how the poison works and what the danger signals may be, as well as known precise results of exposure.

Chief Pesticide Use and Status

The overall term *pesticide* has several subdivisions indicating the pest that is its chief target and for which it is sold. This does not mean that its effect is limited to one class of pests alone. Many herbicides, for example, are especially toxic to mammals or insects, and some pesticides are so broadly lethal that they are called *biocides*. The kinds listed under column 3 include

Acaricides, which kill mites and spiders (include miticides)
Algacides, which kill algae
Antibiotics, which kill bacteria and viruses (include bactericides and disinfectants)
Avicides, which kill birds
Desiccants, which dry up animals and plants, either to kill or permit early harvesting
Fungicides, which kill fungi
Herbicides, which kill plants
Insecticides, which kill insects
Molluscicides, which kill molluscs
Nematocides, which kill nematodes
Piscicides, which kill fish
Plant Regulators, which retard or speed the growth of plants
Repellents, which drive pests away
Rodenticides, which kill rodents
Sterilants, which stop reproduction

Wood preservatives are sometimes given as a class of pesticides. They include insecticides and fungicides to delay rotting and tunneling in wood.

All of these are correctly grouped under the general term *pesticides.* It is both incorrect and confusing to use the phrase "pesticides and herbicides." It implies that herbicides are not pesticides, or are toxic only to plants, while others are more widely dangerous. Note the biocidal range of some herbicides to overcome this idea.

Column 3 also gives information on the legal status of a pesticide: if it has been banned or restricted in use in the United States, in individual states, or in other countries. EPA is directed to ban a pesticide from some or all uses when it determines that such use presents an imminent hazard. The process by which this is done can take many months or years, but in case of a severe emergency, the pesticide can be suspended—taken off the market immediately—before the whole legal process is completed. If EPA finds a pesticide to be especially hazardous but still so economically valuable that a ban is not in order, it can be labeled "restricted," which means that it may only be applied by "certified applicators." These are people trained under state auspices, in compliance with EPA rules. Even so, qualifications for certified applicators vary, and the actual application may be done by uncertified people under the supervision of a certified applicator, who may not always be on the spot. This program, designed to protect the general public from exposure to the most toxic pesticides, has some curious aspects. A certified applicator may put one of these products on a home ground, or public place, where it can remain for some time with a potential for exposing vulnerable people or animals. A pesticide known to be restricted should be avoided anywhere the public can encounter it, and especially where the most vulnerable people are found: the very young, the old, and those with special susceptibilities or illnesses that reduce resistance.

When the registration of a pesticide has been suspended, cancelled, or restricted by the full legal process, there is a conclusive designation listed on our charts. We do not include other uses of these categories employed by EPA as punitive measures for failure to provide required data, pay fees, or respond to other EPA rules, since these may be temporary and reflect on the registrants rather than on the product.

EPA registration of a pesticide for sale in the United States must balance the risk of use perceived against the estimated benefits to ensue. Risk often affects a different group of people than those who will benefit. Immediate financial gain is thus balanced against health and environmental damage that may have continuing effects. The law under which EPA regulates pesticides does not put first priority on health or environment and is the only environmental law in the United States that does not permit citizens to go to court to insist on better enforcement. Our few comments on the status of individual pesticides give only a limited review of these legal aspects, which continue to change. (See Appendix 5 on U.S. pesticide regulation.)

Statements in Quotation Marks

Where a statement in quotation marks appears on a chart instead of the standard wording, it means that no quantitative data was found to fit our rating system, but such a statement exists in a usually trustworthy source. We use it in quotation marks to show that it is not comparable, but does give a clue.

Question Marks

Where only a question mark appears in a column, we have no reliable information. Where it accompanies a word or statement, it means that this is the case in our best judgment, but there is some inadequacy in the source material.

Persistence

Persistence is the length of time that a pesticide remains in the environment, whether it stays where it was put or moves through air, soil, water, or living organisms. It is not always clear whether references to a pesticide's persistence apply only to the original formulation, or to this and its transformation products. A pesticide product, which includes the so-called inert ingredients as well as the active ones, usually moves or changes under environmental impacts. What remains after a certain span of time is the residue. As Robert Rudd defines this, "The residue itself may contain reduced portions of the original toxic ingredient, metabolic derivatives of this chemical, physically transformed derivatives of quite different chemical structure, and surviving portions of the solvent and diluent carriers of the original material. The very wide differences in chemical responses to the even wider variables of nature preclude any precise definition of the word 'residue' " (Pesticides and the Living Landscape). To consider the real effect of applying a pesticide we should be able to trace these stages of change, and identify and define the toxicity of the various transformation products and their persistence. Seldom is the information available to do this, especially since many inert ingredients are not required to be identified. Menzie's Metabolism of Pesticides covered much of the field until the series was discontinued with the author's retirement from the post of Fish and Wildlife Service pesticide toxicologist.

Persistence times usually given for pesticides seem to apply only to the original active ingredient as far as its pesticidal effectiveness lasts. These figures give us a general idea of the life of a product once it is released from the applicator's hands. Since chemicals may react differently in differing climates, soils, kinds of surfaces, and accompanying chemicals, any rating must be very general. The water solubility of a chemical can affect this—will it dissolve and run off quickly and move to other areas? If it is oil-soluble, it can be stored in the fatty tissues of animals and accumulate as exposures continue. If it

is volatile, it may quickly evaporate into the air and move widely. Other factors can come from the method of application: by aerial, ground, broadcast, or precise hand application, or by its form, whether a liquid, emulsion, dust, or granules.

Taking all of this into account, and using admittedly inadequate data in many cases, we adopted a scale for rating persistence in four stages devised for the Council on Environmental Quality, of the Executive Office of the President of the United States, in their first annual report in 1970. Individual cases may not conform to these stages precisely, of course, but they give the best estimate that can be made now. This is for outdoor conditions only; indoor persistence is likely to be considerably longer but testing is not required.

Non-persistent (non-pers): effectiveness lasts from a few hours to several days, rarely more than 12 weeks

Moderately-persistent (mod-pers): from 1 to 18 months

Persistent (pers): retains toxicity for years, perhaps as many as 50 to 100

Permanent (perm): non-degradable to non-toxic materials in the environment; this includes elements like mercury

TOXICITY

Four principal questions should be answered about any toxic material to which people and the environment are exposed:

1. How does it affect mammals, the group to which humans belong?
2. What is its immediate toxicity?
3. What are its long-term effects, either from one exposure or from repeated exposures over a period of time?
4. How does it affect other non-target species, and the whole environment into which it is introduced?

Answers to these questions are presented in columns 5, 6, and 7.

Effects on Mammals

In many cases we have evidence of the effects only on certain species of mammals other than humans, while rarely do we have only human evidence. Only on laboratory animals whose reactions are known to be similar to those of humans can we conduct carefully controlled experiments to measure kind and amount of exposure and results. Only where a specific group of humans is known to have had a certain exposure that has produced consistent results can we estimate our susceptibility. But when suitable test animals have been well tested, and their reactions are known to relate to human reactions, we get as clear a warning as we are apt to have of the hazard to our species.

Immediate Toxicity

The technical term long used by toxicologists for the immediate effects of exposure to a poison is *acute*. This means what happens to the exposed creature immediately, or shortly after, contact with the poison. Since the word *acute* has other meanings in common use, and might be interpreted to mean severe or critical we use the more clearly descriptive term *immediate* with *acute* in parentheses.

Ratings are given for the amount of active ingredients involved in relation to the body weight of the exposed individual. This is done on the metric scale of milligrams of the substance in relation to kilograms of weight of the individual: mg/kg. Thus a rating can apply equally to a small test animal or an animal many times its size. A milligram (1/1000 of a gram) may be very hazardous to a mouse but not very dangerous to a 200-pound (90.72 kg) human.

Immediate toxicity is commonly measured by a test called Lethal Dose 50, (LD_{50}) or in case of creatures exposed through air or water, Lethal Concentration 50 (LC_{50}). This test, devised in 1927, tries to set the amount that will kill half of the test animals in a specified time, presumably an average. Sometimes the LDlo (low) is given, the dose at which the first animals died. The lethal dose given for humans is based on medical records. The LD_{50} test is a crude measurement, affected by many conditions: species of test animals, and their age, weight, sex, genetic strain, health, diet, temperature, housing conditions, season, and probably other environmental conditions at the time of the test. The method of administration also matters whether by various means of feeding, or by injection, exposure through skin, or inhalation. Many of the LD_{50} ratings given are based on tests done a long time ago, under less rigorous requirements than now exist.[1]

At best, the LD_{50} rating for the relative degree of toxicity of a material has many inadequacies, and can be used only as a very rough estimate. Unfortunately, it is usually all that we have. It was originally designed to check lethality of very toxic medicines,

but was adopted for testing all manner of toxic materials for which it may be less appropriate. It gives an illusion of being a precise numerical rating beyond its capability, but this is often ignored by those who want a simple answer. More important, it deals only with the death of the test animals, not the immediate or lasting impairment they may suffer.

We use the LD_{50} on the scale adopted by the Environmental Protection Agency, ranging from Very High, High, Medium to Low, based on the mg/kg amount. This can tell us something about the danger of immediate exposure to a substance, but not what kind of damage it may cause short of death. (For this information, see Chapter 5 on classes of pesticide ingredients.)

Until the 1972 law required wider testing of effects of pesticides only the immediate toxicity was determined by the LD_{50} route. The test is fairly easy and inexpensive to carry out, and the U.S. Department of Agriculture, which was responsible for regulation before EPA was established, cared mainly about the ability of a pesticide to kill the pest. Whatever the immediate toxicity tests showed, USDA had never denied registration to a product because of toxicity to non-pests.

Better tests for immediate toxicity are being developed, internationally, so the LD_{50} may be superseded in the near future. It is unlikely that all toxics will be retested promptly, however.

Means of exposure may be as important as the degree of toxicity. Much illness and death from pesticide poisoning occurs because the victims did not realize that many compounds are just as or more poisonous if they touch the skin or are inhaled than if they are swallowed. Dermal toxicity refers to the ability of many toxics to penetrate intact skin. Others are especially dangerous if they touch a break in the skin. Will children or animals touch the plants or soil in your garden after you have used one of these pesticides? Will you have your hands in the soil before the end of the pesticide's period of persistence? Are you using a form of application like spraying or fogging that makes inhalation or skin contact very hard to avoid? Is the pesticide a volatile substance that will continue to give off poisonous vapors long after it has been applied? This can be especially dangerous indoors or with aerial drift that can reach unintended targets.

Under the **Immediate Toxicity** column, three categories are given: **oral,** for pesticides that are swallowed, **dermal,** for those that penetrate the skin, and **inhalation,** for those that are breathed in.

Oral exposure routes the toxic material through the digestive tract, and to the liver and kidneys that provide the principal detoxifying process. Dermal and inhalation exposure may be especially dangerous

---

[1]G. Zhinden and M. Flury-Roversi. 1981. Significance of the LD_{50} test for the toxicological evaluation of chemical substances. *Archives of Toxicology* 47:77–99.

because the poison goes directly into the bloodstream and may reach crucial organs before the liver and kidneys have a chance to do any detoxifying. Poisons that enter the body through the skin may irritate the skin in the process, but this is not necessarily so. Likewise, those that go through the digestive tract or those that go through the lungs may have adverse effects on those organs, but they may also produce serious effects in other parts of the body.

Immediate Toxicity Ratings

Oral and dermal ratings are expressed in terms of LD_{50}; inhalation in terms of LC_{50}.

Our Rating	EPA Rating	Type of Exposure	Amount of Exposure	Probable Lethal Dose for 150-pound Human
Very high	I	Oral	0–50 mg/kg	0–1 teaspoon
		Dermal	0–200 mg/kg	
		Inhalation	0–0.2 mg/l	
High	II	Oral	50–500 mg/kg	1 teaspoon–
		Dermal	200–2000 mg/kg	1 ounce
		Inhalation	0.2–2 mg/l	
Medium	III	Oral	500–5000 mg/kg	1 ounce–1 pint
		Dermal	2000–20,000 mg/kg	(or 1 pound)
		Inhalation	2–20 mg/l	
Low	IV	Oral	over 5000 mg/kg	over 1 pint or
		Dermal	over 20,000 mg/kg	1 pound
		Inhalation	over 20 mg/l	

LD_{50} = lethal dose that kills 50% of test animals in a given time.

LC_{50} = lethal concentration in air or water in which test animals live that kills 50% in a given time.

mg/l = milligrams per liter. A milligram is 1/1000 of a gram. This measurement is comparable to parts per million (ppm).

mg/kg = milligrams (of a toxin) per kilogram (of body weight of animal). This measurement is comparable to parts per million (ppm).

Label warning statements on U.S.-registered products are geared to these classes of immediate toxicity, based on the oral rating. Very High requires "Danger," "Poison," skull and crossbones symbol, description of treatment and antidote, "Call physician immediately," and "Keep out of the reach of children." High requires "Warning," and "Keep out of the reach of children." Medium requires "Caution" and "Keep out of the reach of children." Low requires no warning statement beyond "Keep out of the reach of children." Labels may not include misleading claims of safety.

Long-Term Toxicity

Toxicologists use the term **chronic toxicity** to refer to various kinds of long-lasting reactions to poi-

sons. The word *chronic* may not have all of the same connotations to many laymen, so we chose the alternate *long-term*. This includes long-lasting or permanent damage from one exposure, continuing exposure, gradual accumulation in the body, or effects that may appear long after the original exposure that began the process. Some effects may appear only in later generations of an animal; some may affect the next generation only. Some can be cured partially or altogether; some are irreversible. Long-term effects are more difficult to assess and understand than the immediate results of a poisoning incident. They may be far more serious, especially when the results appear long after anything can be done to prevent them. We also know much less about long-term effects than immediate reactions. Most can only be determined with any precision by laboratory tests with suitable test animals over a considerable length of time. Most must cover the animals' lifetimes, or go through successive generations. This is an expensive process, requiring very skilled technicians and pathologists, with precise equipment. Results of these tests are our best warning that similar effects may be caused in humans. Direct human evidence cannot be obtained by controlled tests, of course, but may be found where a limited group of people who are all exposed to a particular poison develop consistent symptoms. Individual medical records of exposed people provide additional evidence.

Some defenders of the uncritical use of pesticides make extraordinary claims—that any chemical if used in large enough doses will be highly toxic, or cause cancer. This contradicts scientific fact. They may also misrepresent the meaning of careful laboratory tests, saying that because test animals were necessarily given high exposures and humans might receive less, we need not worry. The tests are to determine whether a kind of reaction occurs, and this is not invalidated by the quantity of exposure of other victims. A succession of small exposures may, in fact, be more damaging than one large one.

Terms used in column 6 tell the kinds of long-term damage that may occur. These are discussed in the order used.

Cumulative: If an animal absorbs a toxic substance rapidly and excretes it much more slowly, intermittent or constant exposure will result in an accumulation in the animal's body. This occurs at varying rates with many factors involved, so we cannot set up a tidy scale to define the rate at which this may happen with a given substance or particular animal. If we say that a substance is cumulative, however, the difference between absorption and excretion is so marked that the levels that can build up are a matter for concern. A series of small exposures may bring the body level to the lethal limit, and the animal will die just as surely as if it were given one massive dose. The organochlorines are a family of pesticides that are cumulative because they are oil soluble and persistent. If a substance is quickly metabolized into low-toxic components in the body, of course it does not accumulate in its original form, though some metabolities may not be quickly excreted. With the organochlorines and similar-acting materials, they dissolve in the body lipids, or oils, and become fixed in the fatty tissue of the body. They are not easily metabolized or excreted unless the animal draws rapidly on its fat reserves. The residues may then enter the bloodstream in concentrations that may be seriously toxic, acting on the brain and nervous system or other vulnerable organs. This happens, for instance, with migrating birds, who draw down their fat reserves in migration, and may suddenly suffer the accumulated effects of all the contaminated insects and other food they have eaten in the past season. Human symptoms can result from sudden weight loss or stress that starts the same process.

When a substance is also carcinogenic, as are many of the organochlorines, accumulation can be especially serious. Retention in the body means that the adjacent cells are under steady exposure to a potentially damaging agent. The crucial interaction can occur at any time, or repeatedly. The cancer-causing potential is thus much greater than from a substance that passes quickly through the body. Other long-term toxic effects can also be increased by such continuous exposure.

Biomagnification, or the build-up in food chains, is another result of this cumulative characteristic. A chemical present in very small quantities in soil, air, or water may be taken up by plants and small animals that are then eaten by larger animals, and so to the top of the food chain. Amounts accumulate in each stage, so the small amounts present in the numerous prey species add together until the top animal receives the grand total, perhaps 1000 times as much as in the surrounding environment. Man is one of the most vulnerable animals at the top of food chains. This biomagnification is a major reason for avoiding the use of persistent poisons that resist transformation into less damaging materials. Their persistence also enables them to disperse widely through the environment over time.

Carcinogen, a substance that can cause cancer, differs in important ways from other kinds of poison. Exposure to a carcinogen can begin a process that can grow over many years. No minimum amount of a carcinogen has been found below which it has no effect. Theoretically, one molecule of a carcinogen affecting one susceptible cell can start the proliferation of cancerous cells that can lead in time to as lethal a cancer as might come from much more exposure. Since there are many kinds of cancer and many kinds of carcinogens, only rarely can one kind of exposure be targeted exactly as the cause of a particular cancer. The important fact is the total burden of carcinogen exposures for each individual. Many creatures are subject to cancer, and we have all evolved in the presence of natural carcinogens. Defenses against prevailing exposures have evolved also. In this century we have added large amounts of new synthetic carcinogens to our environment and there has been a corresponding increase in cancer rates. A number of these synthetic carcinogens are pesticides. Because these are man-made substances the production and use of which we could control or eliminate, a first defence against rising cancer rates is the study and control of these products. In most cases, exposed people cannot know about or avoid exposures, and our fellow creatures are indeed helpless. Where we do have control over our exposure, as with pesticides used in our homes, gardens, businesses, and communities, we can exercise particular caution.

Our designations of certain pesticides as carcinogens may not coincide with the verdicts of other rating systems. Agencies such as EPA and the U.S. Food and Drug Administration (FDA) are under legal requirements to be more specific about the nature and potency of carcinogens that they are required to regulate than can really be supported, scientifically, with the data we now have. They are forced to rely on formulas devised to produce very exact-sounding conclusions. In the risk/benefit balancing that EPA must do, these contribute to such judgments as the estimated number of cancers per million people as opposed to the estimated economic benefits from the continued use of the pesticide. We question both the underlying assumption that such a calculation can be made so accurately and some of the factors that they include in their equations. In Appendix 3 Dr. William Lijinsky explains the most important of these. The EPA's methods of extrapolating test results from animals to humans are questioned, and their assumption that a cancer that requires the lifetime of a short-lived test animal to develop will require the lifetime of a human given the same dose also is without foundation.

These are our criteria:

1. A substance that has been shown in well-conducted tests with suitable test animals to cause cancer is a carcinogen.
2. Exposure to any carcinogen increases the risk of cancer to any species, but this cannot be quantified with any precision.
3. The risk to an individual animal depends

most on the total number of exposures to carcinogens, and the animal's own susceptibility.

4. Ratings that focus only on estimates of the danger to humans cannot be justified scientifically. Rachel Carson Council is equally concerned about the danger to all species, which can also not be extrapolated exactly from tests on other animals. A given species may be far more susceptible than the test animals.

5. In some cases, where a carcinogen produces a distinctive kind of cancer, and a given population has had known exposure and develops that cancer, we can make the connection. Human epidemiology can show cause and effect in such cases where a group of people has in fact been used as test animals.

6. The distinction often made between "benign" and "malignant" tumors can give false assurance. Tumors can progress from the "benign" stage to the "malignant," which can metastasize (spread through the body). In some cases it does not matter whether the cancer spreads; a growing brain tumor can be lethal in itself.

7. Carcinogenicity is a rare property of chemicals. A positive test, showing that a substance does possess this capability, is a warning to all. There may be other tests that do not show the trait, which reflects the inexact nature of such testing but does not cancel the warning of positive tests (see Appendix 3).

8. Testing methods must be understood. A misleading claim is often made by apologists for a carcinogenic product that the test animals were given high doses, while their product provides just a small dose that is not dangerous. There is no connection between the significance of the dose given a test animal, which must be enough to show a significant effect in some of a small group of animals who live only two years, and that absorbed by a human. Considerable evidence indicates that it can be more dangerous to have a series of small exposures than to have one larger exposure. (Medicines are not required to be non-carcinogenic. Accompanying data sheets sometimes make the unjustified claim that a specified dose is all right because it is lower than one causing cancer in animals.)

Suspect Carcinogen: when the tests indicating carcinogenicity are old, and so were not conducted by current standards, or the evidence is not as conclusive as the rules now require, but is still sufficient to serve as a warning.

Mutagen: causes mutation in cells, in test animals (in vivo testing).

Suspect mutagen: causes mutations when tested on various kinds of cells separately, outside living animals or plants (in vitro). Any mutations give a sound warning of similar reactions in mammals, but are not as specifically conclusive.

Teratogen: can cause birth defects when it reaches egg or sperm cells of the parents, or the developing fetus.

Suspect Teratogen: the same qualifications as for suspect carcinogen and making a distinction between tests on animals and tests on cell cultures: in vivo versus in vitro.

Fetotoxin: can poison the fetus, in contrast to deforming it by a teratogen. An *embryotoxin* affects an early stage of the fetus.

Organ damage: includes various kinds and degrees of damage to the designated organ, whether liver, kidneys, nervous system, blood, testes, ovary, thyroid, or endocrine system. Liver damage, for example, can range from increase in size to severe necrosis: death of tissue.

Neurotoxin: damages nerves.

Delayed toxicity: effects may become evident some time after exposure, which may delay treatment until it is less effective or useless. Delayed neurotoxicity for example, means damage to the nervous system that may be very long-lasting but where the original exposure did not produce usual signs of poisoning.

Immunotoxin: damages the immune system.

Viral enhancer: increases the toxicity of viruses that an exposed person or animal also encounters. Reye Syndrome in children, associated with aspirin and virus exposure, is an example. Some pesticides are also suspect in this case.

Adverse Effects on other Non-Target Species: Physical Properties

We assume that most species covered under Effects on Mammals (except rodents such as rats, mice, and ground squirrels, and miscellaneous animals sometimes considered pests) are not the targets of pesticides used and mammalian effects are not repeated. If a pesticide is classified as sold to kill any kind of creature, we do not include it in column 7. The reader can assume that if a pesticide is an avicide, birds are at particular risk, as are many kinds of fish from piscicides. Bees are an exception to this rule. Bees have an essential role in the natural environment, which depends on them for pollination of crops and many other desired plants. Studies have been done therefore to find the effects on bees of insecticides and other kinds of pesticides. Since this information is often available, and bees are rarely intended targets of pesticides, we include them.

A **biocide** is a broad-spectrum poison that is harmful or lethal to a wide range of animals and often plants. They have dominated the pesticide market for economic reasons, since mass production can bring down production costs on a product that can then be advertised for use against a large number of pest species. Because of their wide potential damage to non-target species and the environment as a whole, they should be avoided unless the manner of application can be selective down to the exact pest targeted, and their subsequent escape into the environment is prevented.

Immediate toxicity for species other than mammals is given, followed by any known **long-term toxicity**. We know little about species that may be damaged inadvertently. Most of the early research on this aspect was done by federal and state fish and wildlife experts, with limited resources. From this research we received the first danger signals from the new synthetic pesticides after World War II. Evidence is still meager. We must judge the reactions of all fish from perhaps two or three species tested, or other species exposed to accidental spills into waters. In only a few cases has anyone checked the reactions of reptiles and amphibians, molluscs, worms, and such arthropods as shrimp, fish food organisms, and desirable insects. Only when there is a commercial angle are these tests apt to be made, but all forms of life are integral parts of a healthy ecosystem. We especially need data on effects on insect predators and parasites critical to sound biological control of pests.

Mutations recorded for plants are a warning for all forms of life. Research on mutagenicity is often done first on plants, because cells of plant and animal tissues may have similar reactions. Warning results can then direct attention to effects on animals.

Physical Properties

Physical properties determine the way a pesticide may travel through soil, air, water, and organisms. They also relate to hazards from use, transport, and storage of these products.

Water solubility can be important in determining the course of a pesticide through the environment. If it is water soluble it is not apt to remain in one place or one organism, but will percolate through living and inert materials into ground and surface water, and will travel as far as its persistence in original form or transformation products will allow.

Oil solubility means that a substance can be readily absorbed by an animal's fatty tissues, and with persistence this can lead to biomagnification. An oil-soluble material released into a body of water may be so quickly taken up by small organisms that no trace can be found in the water shortly after.

Scales for rating solubility of toxic materials vary from those that simply determine ordinary rules for a saturate solution. The key must be how much must be taken up by the solvent to be hazardous to creatures exposed to it. Our scale for rating solubility is shown in comparison to the usual scale for pharmaceutical solutions. Sometimes even a scale for toxic solubility is not adequate to deal with extremely toxic pesticides. The EPA rating for diflubenzuron (Dimilin) for example is "soluble" although only a very few parts per trillion may be dissolved in water. This is all it takes to be lethal to crustaceans and other chitinous creatures, however.

Our Rating	Solubility at Room Temperature (20 to 30°C, 68 to 86°F)	U.S. Pharmacopeia Rating
Insoluble	Less than 1 part per million (ppm)	Insoluble
Slightly soluble	1 to 10 ppm 10 to 100 ppm	
Soluble	100 to 1000 ppm	Very slightly soluble
	1000 to 10,000 ppm (.1 to 1%)	Slightly soluble
Very soluble	10,000 to 33,333 ppm	Sparingly soluble
	333,333 to 100,000 ppm	Soluble
	100,000 to 1,000,000 ppm (10% to 100%)	Freely soluble
	Over 1,000,000 (over 100%) but not miscible	Very soluble
	Miscible	Infinitely soluble

Volatility is the capacity of a substance to evaporate, thus moving into the air, being easily inhaled, and moving widely as its persistence permits. Our scale for rating this trait is:

Volatility Rating	Vapor Pressure at 20 to 30°C
Non-volatile	Less than 1×10^{-7} mmHg (.0000001 millimeters of mercury)
Slightly volatile	10^{-7} to 10^{-4} mmHg (.0000001 to .0001 mmHg)
Volatile	10^{-4} to 10^{-2} mmHg (.0001 to .01 mmHg)
Highly volatile	Greater than .01 mmHg

Fire hazard of a chemical or formulation is crucial safety information in handling, storing, or transporting. Tests for this quality depend on many variable factors, including temperature, humidity, containment, available oxygen, and volatility, so that any number can be only an approximation. The key figure is the flash point temperature. Flash point is the temperature at which vapor from a substance will ignite in air, and different tests may produce a range of values for one substance. As a very general scale for warning of fire hazard, a division is made between those substances whose flash point lies above 140°F (45.7°C), normally above the temperatures at which pesticides would be used, or below, into usual temperature levels. The word **combustible** is used for those substances that only burn above 140° and are less hazardous, **flammable** for those with a flash point below 140°, more likely to ignite in use or storage, and **explosive** is used for those that can burn so fast and fiercely that they detonate.

These ratings apply only to the pure active ingredient or other component of the chart. The formulated products actually used may contain other components that alter the fire hazard. Some use petroleum products as solvents or carriers, for example. For a more detailed rating of a particular substance, if you can determine the significant ingredients, see the *Fire Protection Guide on Hazardous Materials,* 9th ed. (1986), or later versions, from the National Fire Protective Association.

Environmental Effects

Information in columns 5,6, and 7 can give an estimate of the overall effects of a pesticide on the natural functioning of the area it reaches. From impacts on specific organisms, and some physical properties, clues exist for those who understand the ecosystem. Dr. Charles R. Walker considers these aspects in Appendix 4.

VARYING VULNERABILITY TO PESTICIDE POISONING

The toxicity ratings we give indicate what may happen after exposure to one of these products. Individual cases will differ with the susceptibility of the individual exposed, and some of the circumstances of the incident. Age, either especially young or old, previous or current illnesses, concurrent exposures, or the cumulative effects of other exposures, can make the difference.

Animals defend themselves against toxic substances through various organs and processes. In humans, the detoxifying organs include the liver, kidneys, and respiratory epithelium (lining of the nose). Too massive an exposure or too many repeated episodes can damage these organs, some-times permanently. The victim then may have little ability to cope with the later exposures to a wide range of toxic chemicals (Lijinsky 1989).

A condition known as multiple chemical sensitivity syndrome has been defined by Dr. Mark Cullen, head of Occupational Medicine at Yale, "This syndrome is described as an acquired intolerance of common environmental chemicals with symptoms involving multiple organ systems. The onset of illness follows a toxic exposure, often after organophosphate poisoning" (Cullen 1989).

The chemicals at issue are often synthetics to which life forms have not had time to adjust. Some confusion may arise between the detoxifying system for chemical toxins and that for microbial or fungus infections, commonly called the immune system. The immune system has branches, dealing respectively with parasites, bacteria, viruses, and fungi. While it does not cope directly with chemical toxics, there can be a reverse effect with chemicals damaging the immune system itself and so interfering with its ability to counter infections (Meggs 1991; Olson 1986).

Confusion also comes from imprecise use of the term *allergy.* Many people use the word very generally to describe the reaction of a susceptible person to a substance that normally causes no adverse effect on most people. In technical usage, an allergy is an altered reactivity that provokes an immune response. The body's defenses overreact to an introduced substance and the inflammation and congestion, even anaphylactic shock, create the problem. Genuine allergies are often treated by a series of shots introducing small amounts of the offending substance to cause the body to adjust its defenses more normally. This is the reason that confusing allergy with reaction to an actual poison can have serious consequences. A person already damaged by poisoning needs to be withdrawn from further exposure as completely as possible to allow the detoxifying organs a chance to recuperate as much as possible. To be subjected to deliberate injections of the key poison aggravates the original injury. Some pesticides do induce genuine allergic reactions as well as toxic ones. Perhaps avoiding contact is a preferred treatment here for both responses.

People damaged by pesticide exposure may have difficulty finding a doctor who understands the situation. Even now, standard medical education includes relatively little toxicology. Effects of all of the new synthetic toxic chemicals in the past 46 years have introduced new and still not fully understood aspects of toxicology. Injured people should not be discouraged in their search for help by uncomprehending doctors who may dismiss the symptoms as just psychological.

REFERENCES

1. Cullen, M. 1989. The worker with multiple chemical sensitivities. *Occupational Medicine: State of the Art Reviews* 2(4):655–662.
2. Lijinsky, W. 1989. Occupational and environmental exposures to N-nitroso compounds. *Environmental and occupational cancer, scientific update.* Princeton, NJ: Princeton University Press.
3. Meggs, W. 1991. Personal Communication.
4. Olson, L. J. 1986. The immune system and pesticides. *Journal of Pesticide Reform* 6:20–25.

INDEX OF PESTICIDE NAMES

Chapter Three

Index of Pesticide Names

This is a cross-index of common, trade, and chemical names as well as CAS numbers. Common names under which the charts are alphabetized are listed in all capitals.

A 1093: see ametryn
A 363: see ametryn
A363: see aminocarb
A7-Vapam: see metam-sodium
A-820: see butralin
Aadibroom: see ethylene dibromide
AA Outdoor Dog Repellent: see safrole
AAtack: see thiram
AAtrex: see atrazine
2-AB: see 2-butanamine
ABAMECTIN
Abar: see leptophos
Abat: see temephos
Abate: see temephos
Abathion: see temephos
Abazan: see temephos, trichlofon
AC 222,705: see flucythrinate
AC 26691: see cythioate
AC 4049: see malathion
AC 4124: see dicapthon
AC 52160: see temephos
AC 5223: see dodine
AC 84777 Finaven: see difenzoquat
AC 92100: see terbufos
AC 92553: see pendimethalin
Acacide: see Aramite
Acaraben: see chlorobenzilate
Acaralate: see chloropropylate
Acarelte: see dinobuton
Acarelte Forte: see dinobuton
Acarfor: see dicofol, ethion
Acarie: see mancozeb
Acarin: see dicofol
Acaristop: see clofentezine
Acarithion: see carbonphenothion
Acarol: see bromopropylate
Acaron: see chlordimeform
Acarthane: see dicofol, dinocap
Acavers 35: see dicofol, methomyl
Accelerate: see endothall
Accelerate: see endrine
Accelerator Thiuram: see thiram
Acclaim: see fenoxaprop-ethyl
Accord: see glyphosate
Accotab: see pendimethalin
Accothion: see fenitrothion
Acenit: see acetochlor
ACEPHATE

ACEPHATE-MET
acetamide: see methomyl
ACETOCHLOR
acetonitrile: see methomyl
3-(α-acetonylbenzyl)-4-hydroxycoumarin: see warfarin
3-(α-(2-acetonylfurfuryl)-4-hydroxycoumarin: see coumafuryl
acetoxy-triphenylstanname: see triphenyltin acetate
acid lead o-arsenate: see lead arsenate
acid lead arsonate: see lead arsenate
ACIFLUORFEN
Acifon: see azinphosethyl
ACL 56: see sodium dichloroisocyanurate dihydrate
ACL 59: see potassium dichloroisocyanurate
ACL 66: see penta-s-triazinetrione
ACL 70: see dichloroisocyanuric acid
ACL 90: see trichloroisocyanuric acid
Acme Garden Fungicide: see captan
Acme MCPA Amine 4: see MCPA
ACP-63303: see ioxynil
ACPM-673-A: see fenac
Acquinite: see chloropicrin
Acrex: see dinobuton
Acricide: see binapacryl
Acritet: see acrylonitrile
ACROLEIN
acrylaldehyde: see acrolein
acrylic aldehyde: see acrolein
Acrylofume: see acrylonitrile
Acrylon: see acrylonitrile
ACRYLONITRILE
Actamer: see bithionol
Actellic: see pirimiphos-methyl
Actellifog: see pirimiphos-methyl
Acti-Aid: see cycloheximide
Acti-Dione: see cycloheximide
Acti-Dione BR Concentration: see cycloheximide
Acti-Dione PM: see cycloheximide
Acti-Dione RZ: see cycloheximide, pentachloronitrobenzene
Acti-Dione TGF: see cycloheximide
Acti-Dione Thiram: see cycloheximide, thiram
Actispray: see cycloheximide
activated 7-dehydro-colesterol: see cholecalciferol
Actor: see diquat, paraquat
Actosin C: see chlorophacinone
Actril: see ioxynil

Actril DS: see 2,4-D, ioxynil
Actril S: see bromoxynil, ioxynil, dichlorprop, MCPA
Acutox: see pentachlorophenol
Admiral: see fonofos
A-Dust: see calcium cyanide
Afalon: see linuron
Afesin: see monolinuron
Affirm: see abamectin
Aficide: see lindane
Afnor: see chlorophacinone
AG 500: see diazinon
Agriben: see borax, bromacil
Agricide Maggot Killer: see toxaphene
Agricorn D: see 2,4-D
Agridip: see coumaphos
Agrimek: see abamectin
Agrine: see endrin
Agritol: see Bacillus thuringiensis var. kurstaki
Agrocide: see lindane
Agrogas: see ethylene dibromide
Agrotect: see 2,4-D
Agrothion: see fenitrothion
Agroxone: see MCPA
AI3-29 117: see kadethrin
AI3 29391: see flucythrinate
Akar: see chlorobenzilate
AKTON
AL-50: see dichloran
ALACHLOR
Alar: see daminozide
Alar-85: see daminozide
ALDICARB
aldicarb sulfone: see aldicarb
aldicarb sulfoxide: see aldicarb
Aldrex: see aldrin
ALDRIN
aldrine: see aldrin
Aldrite: see aldrin
Aldrosol: see aldrin
Alfa-tox: see diazinon
Algae-Rhap CU: see copper tea complex
Algistat: see dichlone
Aliette: see fosetyl-al
Aliette Extra: see captan, fosetyl-al, thiabendazole
Alirox: see EPTC
Alkron: see parathion
alkyl dimethyl benzylammonium choride: see benzalkonium chloride
Alleron: see parathion
ALLETHRIN
d-cis trans-allethrin: see allethrin
d-trans-allethrin: see bioallethrin
(+)-trans-allethrin: see bioallethrin
Allethrin concentrate MGK: see allethrin
allethrin stereoisomer: see allethrin, bioallethrin, esbiothrin
d-allethronyl d-trans-allethrin: see S-bioallethrin

(+)-allethronyl (+)-trans-allethrin: see S-bioallethrin
(RS)-allethronyl (1R,cis,trans) chrysanthemate: see allethrin
(RS)-allethronyl [1R,trans]-chrysanthemate: see bioallethrin
(RS)-allethronyl [1R,trans]-chrysanthemate: see esbiothrin
Alleviate: see allethrin
allidochlor: see CDAA
Allisan: see dichloran
Alltox: see toxaphene
allycinerin: see allethrin
ALLYL ALCOHOL
allyl aldehyde: see acrolein
4-allyl-1,2-methylenedioxybenzene: see safrole
(RS)-3-allyl-2-methyl-4-oxocyclopent-2-enyl(1RS-cis,trans-chrysanthemate: see allethrin
Alrato: see antu
Altacide Extra: see atrazine, sodium chlorate
Altavar: see atrazine, 2,4-D, sodium chlorate
Altazin: see amitrole, atrazine
Altosand: see methoprene
Altosid: see methoprene
Altosid Briquet: see methoprene
Altox: see aldrin
ALUMINUM PHOSPHIDE
aluminum tris(o-ethylphosphonate): see fosetyl-al
Alvit: see dieldrin
AMA: see ammonium methanearsonate
Amaze: see isofenphos
Ambox: see binapacryl
Ambush: see permethrin
Ambushat: see permethrin
Ambush C: see cypermethrin
Ambushfog: see permethrin
Amchem 70-25: see butralin
Amchem Garden Weeder: see chloramben
Amcide: see AMS
Amdon: see picloram
Amdro: see hydramethylnon
American Cyanamid 28023: see famphur
American Cyanamid 4124: see dicapthon
American Cyanamid C1 26691: see cythioate
Amerol: see amitrole
Ametrex: see ametryn
AMETRYN
ametryne: see ametryn
Amex 820: see butralin
Amiben: see chloramben
Amidox: see 2,4-D
Amilon-WP: see chloramben
amine methanearsonate: see ammonium methanearsonate
(S)-4-[[3-amino-5-[(aminoiminomethyl)methylamino]-1-oxopentyl]amino]-1-(4-amino-2-oxo-1-(2H)-pryimidinyl)-1,2,3,4-tetradeoxy-d-erythro-hex-2-enopyranuronic acid: see blasticidin S
2-aminobutane: see 2-butanamine

4-amino-6-*tert*-butyl-3-3ethylthio-1,2,4-triazin-5(*4H*)-
    one: see ethiozin
4-amino-6-*tert*-butyl-3-(methylthio)-*as*-triazin-5(*4H*)-
    one: see metribuzin
AMINOCARB
3-amino-2,5-dichlorobenzoicacid: see chloramben
4-amino-6-(1,1-dimethylethyl)-3-(ethylthio)-1,2,4-
    triazin 5(*4H*)-one: see ethiozin
4-aminopyridine: see Avitrol 200
4-aminopyridine hydrochloride: see Avitrol 200
Amino Triazole: see amitrole
aminotriazole: see amitrole
3-amino-1,2,4-triazole: see amitrole
3-amino-*s*-triazole: see amitrole
4-amino-3,5,6-trichloropyridine-2-carboxylic acid:
    see picloram
Aminozide: see daminozide
Amiral: see triadimefon
AMITRAZ
Amitrile T.L.: see amitrole, ammonium thiocyanate
AMITROLE
Amitrole-T: see amitrole, ammonium thiocyanate
Amizine: see amitrole, simazine
Amizol: see amitrole
Ammat: see AMS
Ammate X: see AMS
Ammate X-NI: see AMS
Ammo: see cypermethrin
ammonium amidosulphate: see AMS
ammonium arsenate: see arsenic
ammoniumdimethy-dithiocarbamate: see diram
ammonium ethyl carbamoylphosphoneat: see
    fosamine ammonium
ammonium methanearsonate: see arsenic
ammonium rhodanide: see ammonium thiocyanate
ammonium salt of imazapyr: see imazapyr
ammonium sulfamate: see AMS
ammonium sulfocyanate: see ammonium
    thiocyanate
ammonium sulfocyanide: see ammonium
    thiocyanate
ammonium sulphamidate: see AMS
AMMONIUM THIOCYANATE
AMOBAM
Amobam: see amobam
amorphous (fumed) silica dust: see silica aerogel
Amoxone: see 2,4-D
AMS
Amthio: see ammonium thiocyanate
Anclirox: see EPTC
ANILAZINE
Aniten: see flurecol-butyl, MCPA
Ankilostin: see tetrachloroethylene
Anofex: see DDT
Anone: see cyclohexanone
Anprolene: see ethylene oxide
Ansar 157: see ammonium methanearsonate
Antabuse: see disulfiram

Anter: see dieldrin
Anthon: see trichlorfon
Anthracene: see anthraquinone
9,10-anthracenedione: see anthraquinone
anthradione: see anthraquinone
ANTHRAQUINONE
9,10-anthraquinone: see anthraquinone
Anti-Carie: see hexachlorobenzene
Anticarie: see hexachlorobenzene
Antimicrobial: see pentachlorophenol
Antimilace: see metaldehyde
ANTIMYCIN A
antimycin A1: see antimycin A
antimycin A3: see antimycin A
Antinonnin: see dinitrocresol
antiphen: see dichlorophen
antipiricullin: see antimycin A
Antor: see diethatyl
Antu: see antu
Anturat: see antu
Apachlor: see chlorfenvinphos
Apavap: see dichlorvos
Apavinfos: see mevinphos
Apex 5E: see methoprene
Apistan: see γ-fluvalinate
Apl-Luster: see thiabendazole
Apobas: see bromadiolone
Apollo: see clofentezine
Apolo: see clofentezine
aprocarb: see propoxur
Apron: see metalaxyl
Aqua Kleen: see 2,4-D
Aqua-Vex: see silvex
Aquacide: see diquat
Aqualin: see acrolein
Aqualin Biocide: see acrolein
Aqualin Slimicide: see acrolein
Aquathol: see endothall
Aquazine: see simazine
Aquinite: see chloropicrin
Arab Rat Deth: see warfarin
Aracide: see Aramite
Aragran: see terbufos
ARAMITE
Aramite-15W: see Aramite
Athane: see dinocap
Athron: see Aramite
Arbotect: see thiabendazole
Aresin: see monolinuron
Argold: see cinmethyline
Arinosu-Korori: see hydramethylnon
Ariotox: see metaldehyde
Arkotine: see DDT
Arnold Weed-O-Spray: see 2,4-D
AROSURF
Arotex-extra: see chlormequat chloride
arprocarb: see propoxur
Arresin: see monolinuron

Arrest 75W: see carboxin
Arrhenal: see DSMA
Arrivo: see cypermethrin
Arsenal: see imazapyr
arsenate of lead: see lead arsenate
ARSENIC
arsenic acid: see arsenic
o-arsenic acid: see arsenic acid
arsenic acid anhydride: see arsenic pentoxide
arsenic acid, calcium salt: see calcium arsenate
arsenic acid, disodium salt, heptahydrate: see
    sodium arsenate [III]
arsenic acid, lead salt: see lead arsenate
arsenic anhydride: see arsenic pentoxide
arsenic oxide: see arsenic pentoxide,arsenic
    trioxide
arsenic (III) oxide: see arsenic trioxide
arsenic [V] oxide: see arsenic pentoxide
arsenic penthoxide: see arsenic pentoxide
arsenic pentoxide: see arsenic
arsenic sesquioxide: see arsenic trioxide
arsenious acid: see arsenic trioxide, sodium
    arsenite
arsenious acid copper (2+) salt: see cupric arsenite
arsenious acid, monosodium salt: see sodium
    arsenite
arsenious acid, sodium salt: see sodium arsenite
    arsenic trioxide: see arsenic
arsenolite: see arsenic trioxide
Arsenolite: see arsenic trioxide
arsenious acid: see arsenic trioxide
arsenious acid, anhydride: see arsenic trioxide
arsenious acid, sodium salt: see sodium arsenite
arsenious oxide: see arsenic trioxide
arsenious oxide, anhydride: see arsenic trioxide
arsentrioxide: see arsenic trioxide
Arsinyl: see DSMA
Arsodent: see arsenic trioxide
arsonic acid copper (2+) salt: see cupric arsenite
Arvest: see ethephon
Arylam: see carbaryl
Ashlade D-Moss: see chloroxuron, ferrous sulfate
Asilan: see asulam
ASP-47: see sulfoTEPPE
Asparasin: see lindane
Aspon-chlordane: see chlordane
Aspor: see zineb
Asporum: see zineb
Assault: see imazapyr
Assert: see imazamethabenz
Asset: see benazolin
ASULAM
Asulfox F: see asulam
Asulox: see asulam
Asulox 40: see asulam
Asuntol: see coumaphos
asymmetrical dimethylhydrazine: see
    unsymmetrical 1,1-dimethylhydrazine

AT: see amitrole
3,A-T: see amitrole
ATA: see amitrole
Atacamite: see copper oxychloride
Aten: see niclosamide
Atenase: see niclosamide
Atgard: see dichlorvos
Atlas 'A': see sodium arsenite
Atlazin: see amitrole, atrazine
Atoxan: see carbaryl
Atradex: see atrazine
Atradex 50: see atrazine
Atranex: see atrazine
Atratol: see atrazine, prometon
ATRAZINE
Atrimmec: see dikegulac sodium
Atrinal: see dikegulac sodium
Atroban: see permethrin
Attack: see *Bacillus thuringiensis* var. *kurstaki*
Attack: see permethrin, pirimiphos-methyl
Aules Chipco Thiram 75: see thiram
AURAMINE
Avadex: see diallate
Avadex BW: see triallate
Avenge: see difenzoquat
Avermectin B_1: see abamectin
Avicol: see PCNB
Avid: see abamectin
AVITROL 100
AVITROL 200
Avomec: see abamectin
Axall: see bromoxynil, ioxynil, mecoprop
Axiom: see Akton
Aygard V: see dichlorvos
AZ 500: see isoxaben
Azac: see terbucarb
azacholesterol: see azacosterol
AZACOSTEROL
azacosterol hydrochloride: see azacosterol
Azadieno: see amitraz
Azaform: see amitraz
Azak: see terbucarb
Azar: see terbucarb
Azasterol: see azacosterol
2-azido-4-isopropylamino-6-methylthio-*s*-triazine:
    see aziprotryne
4-azido-*N*-(1-methylethyl)-6-(methylthio)-1,3,5-triazin-
    2-amine: see aziprotryne
Azinos: see azinphosethyl
AZINPHOSETHYL
azinphos-ethyl: see azinphosethyl
AZINPHOSMETHYL
azinphos-methyl: see azinphosmethyl
aziprotryn: see aziprotryne
AZIPROTRYNE
AZOBENZENE
azobenzenoxide: see azoxybenzene
azobenzide: see azobenzene

azobenzol: see azobenzene
Azodrin: see monocrotophos
Azolan: see amitrole
Azole: see amitrole
Azomyte: see azoxybenzene
azosydibenzene: see azoxybenzene
azoxybenzene: see azobenzene
azoxybenzide: see azoxybenzene
B-3015: see benthiocarb
B-622: see anilazine
B-995: see daminozide
BAAM: see amitraz
Bacillus popillae: see milky spoire
Bacillus lentimorbus: see milky spore
BACILLUS THURINGIENSIS (BERLINER)
Bacillus thuringiensis var. aizawei: see *Bacillus thuringiensis* (Berliner)
Bacillus thuringiensis var. *israelensis*: see *Bacillus thuringiensis* (Berliner)
Bacillus thuringiensis var. *kurstaki*: see *Bacillus thuringiensis* (Berliner)
Bacillus thuringiensis var. *san diego*: see *Bacillus thuringiensis* (Berliner)
Bacillus thuringiensis var. *tenebrionis*: see *Bacillus thuringiensis* (Berliner)
Bactimos: see *Bacillus thuringiensis* var. *israelensis*
Bactospeine: see *Bacillus thuringiensis* var. *kurstaki*
Bactur: see *Bacillus thuringiensis* var. *kurstaki*
Bakthane: see *Bacillus thuringiensis* var. *kurstaki*
Balan: see benefin
Balcom Butyl 6 Ester Weed Killer: see 2,4-D
Baldafume: see sulfoTEPP
Balfin: see benefin
BAM: see 2,4-dichlorobenzamide
Bandock: see dicamba, mecoprop, 2,4,5-T
Ban-Dook: see dicamba, mecoprop, 2,4,5-T
Banex: see dicamba
Banlene-Plus: see dicamba, MCPA, mecoprop
Banlene Solo: see dicamba, dichlorporp, ioxynil
Banol: see carbanolate
Bantrol: see ioxynil
Bantu: see antu
Banvel D: see dicamba
Banvel K: see 2,4-D, dicamba
Banvel M: see dicamba, MCPA
Banvell II: see dicamba
Banzsalox: see benazolin
Baran: see fluoroacetamide
barbamate: see barban
BARBAN
barbane: see barban
Barbodan: see carbofuran
BARIUM METABORATE
BARIUM POLYSULFIDE
Barleyquat-B: see chlormequat chloride
Barnon Plus: see flamprop-isopropyl
Baron: see erbon
Barquat MB-50: see benzalkonium chloride

Barquat MB-80: see benzalkoinum chloride
Barricade: see cypermethrin
Barrier 2G: see dichlobenil
Barrier 50W: see dichlobenil
Barrix: see diethatyl, ethofumesate
BARTHRIN
BAS-3520: see vinclozolin
Basagran DP: see bentazon, dichlorprop
Basagran Ultra: see bentazon, dichlorprop, ioxynil
Basagran-M60: see bentazon
Basalin: see fluchloralin
Basamid: see dazomet
Basfapon: see dalapon
BASF-Grunkupfer: see copper oxychloride
BAS-392-H: see fluchloralin
basic copper carbonate: see copper
basic copper chloride: see copper
basic copper chloride: see copper oxychloride
basic copper sulfate: see copper
basic copper sulfate: see copper sulfate
basic cupric acetate: see copper
basic cupric chloride: see copper oxychloride
Basudin: see diazinon
Batasan: see triphenyltin acetate
Batazina: see simazine
Battal: see carbendazim
Bavel K: see 2,4-D, dicamba
Bavistin: see carbendazim, maneb
Bavistin M: see carbendazim, maneb
BAY 10756: see demeton
BAY 15203: see demeton-methyl
BAY 16259: see azinphosethyl
BAY 17147: see azinphosmethyl
BAY 19639: see disulfoton
BAY 21/116: see demeton-methyl
BAY 21/199: see coumaphos
BAY 22555: see fenaminosulf
BAY 25141: see fensulfothion
BAY 29493: see fenthion
BAY 37344: see methiocarb
BAY 39007: see propoxur
BAY 41831: see fenitrothion
BAY 44646: see aminocarb
BAY 45432: see omethoate
BAY 47531: see dichlofluanid
BAY 5072: see fenaminosulf
BAY 60618: see benzthiazuron
BAY 6159H: see metribuzin
BAY 68138: see fenamiphos
BAY 71628: see acephate-met
BAY 78418: see edifenphos
BAY 78537: see carbofuron
BAY 9214: see isofenphos
BAY 94337: see metribuzin
BAY H-321: see methiocarb
BAY L 13/59: see trichlorfon
BAY ME B6447: see triadimefon
BAY SMY 1500: see ethiozin

BAY SRA 12869: see isofenphos
BAY SRA 7747: see chlorphoxim
Baycid: see fenthion
BAY-E-393: see sulfoTEPP
Bayer 4895: see fenthion
Baygon Spray: see cyfluthrin, dichlorvos, propoxur
Bayleton: see triadimefon
Bayleto AN: see cymoxanil, triadimefon
Bayleton Total: see carbendazim, triadimefon
Bayleton Triple: see captafol, carbendazim,
    triadimefon
Bayluscid: see niclosamide
Bayluscide: see niclosamide
Bayluscit: see niclosamide
Baymix: see coumaphos
Baytex: see fenthion
Baythion C: see chlorphoxim
Baythroid F: see cyfluthrin, omethoate
Baythroid TM: see acephate-met, cyfluthrin
BBC 12: see DBCP
BBH: see lindane
BCM: see carbendazim
BCPE: see chlorfenethol
beachwood creosote: see creosote (wood tar)
Bedfume: see methyl bromide
Beet Kleen: see chlorpropham, fenuron, propham
Bellater: see atrazine, cyanazine
Belmark: see fenvalerate
Belt: see chlordane
BENALAXYL
BENAZOLIN
Benazolox: see benazolin, clopyralid
Bencornox: see benazolin
Bendex: see fenbutatin-oxide
BENDIOCARB
bendioxide: see bentazon
Benefex: see benefin
BENEFIN
benfluralin: see benefin
Benfos: see dichlorvos
Benlate: see benomyl
BENOMYL
Benopan: see benazolin
Bensecal: see benazolin
BENSULIDE
BENTAZON
bentazone: see bentazon
BENTHIOCARB
Bentrol: see ioxynil
Benzahex: see benzene hexachloride
BENZALKONIUM CHLORIDE
Benzar: see benazolin
2-benzen-4-chlorophenol: see *O*-benzyl-*p*-
    chlorophenol
BENZENE
BENZENE HEXACHLORIDE
S-2-benzenesulfonamidoethyl *O,O*-di-
    isopropylphophorodithioate: see bensulide

1,2,4-benzenetriol: see benzene
Benzex: see benzene hexachloride
Benzide: see sodium azide
Benzilan: see chlorobenzilate
Benz-O-Chlor: see chlorobenzilate
benzoepin: see edosulfan
Benzofurolin: see resmethrin
benzol: see benzene
benzole: see benzene
benzolene: see benzene
2-benzothiazolethiol: see MBT
1-benzothiazol-2-yl-3-methylurea: see benzthiazuron
N-(2-benzothiazolyl)-*N'*-methylurea: see
    benzthiazuron
BENZTHIAZURON
Benzyfuroline: see resmethrin
6-BENZYLADENINE
6-benzylamino-purine: see 6-benzyladenine
5-benzyl-3-furylmethyl[1*R,Cis*]-chrysanthemate: see
    cismethrin
5-benzyl-3-furylmethyl[1*R,trans*]-chrysanthemate:
    see bioresmethrin
5-benzyl-3-furylmethyl[1*RS,cis,trans*]-2,2-dimethyl-3-
    (2,2-dimethylvinyl)cyclopropanecarboxylate: see
    resmethrin
5-benzyl-3-furylmethyl(*E*-(1*R*)-*cis*-2,2-dimethyl-3-(2-
    oxothiolan-3-
    ylidenemethyl)cyclopropanecarboxylate: see
    kadethrin
Beosit: see endosulfan
Bercema: see zineb
Bercema NMC50: see carbaryl
Bermat: see chlordimeform
Betafume: see 2-butanamine
Betanal 475: see desmedipham
Betanal AM: see desmedipham
Betanal Perfekt: see ethofumesate, phenmedipham
Betanal Progress: see desmedipham, ethofumesate,
    phenmedipham
Betanal Tandem: see ethofumesate, phenmedipham
Betanex: see desmedipham
Betaron: see ethofumesate, phenmedipham
Betasan: see bensulide
Bethanol-475: see desmedipham
bethrodine: see benefin
Bettaquat-B: see chlormequat chloride
Bexol: see lindane
Bexton: see propachlor
BFV: see formaldehyde
BH Dalapon: see dalapon
BH MCPA: see MCPA
BHC: see benzene hexachloride
Bhopal Chemical: see methyl isocyanate
Bical: see mercuric chloride, mercurous chloride
bicarburet of hydrogen: see benzene
Bicep: see atrazine, metolachlor
Bidrin: see dicrotophos
BIFENOX

BIFENTHRIN
Big Dipper: see diphenylamine
Bilobran: see monocrotophos
BINAPACRYL
Binnell: see benefin
bioallethrin: see allethrin
S-bioallethrin: see allethrin
Biobenzyfuroline: see bioresmethrin
Bio Flydown: see permethrin
Bioguard: see thiabendazole
Bio Lawn Weedkiller: see 2,4-D, dicamba,
    ioxynil
Bio Long Last: see dimethoate, permethrin
bioMET: see tributyltin oxide
Bionex: see azinphosethyl
Bioquin: see oxine-copper
bioresmethrin: see resmethrin
Biosolomycin: see oxytetracycline hydrochloride
Bio Systemic Insecticide: see dimethoate
Biothion: see temephos
Biothrin: see permethrin
Biotrol: see *Bacillus thuringiensis* var. *kurstaki*
Biotrol VHZ: see see Nuclear polyhedrosis virus
biphenate: see bifenthrin
biphenyl: see diphenyl
3-(3-(1-1'-biphenyl)4-yl-1,2,3,4-tetrahydro-1-
    napthalenyl)-4-hydroxy-2H-1-benzopyran-2-one:
    see difenacoum
Birlane: see chlorfenvinphos
2,3,4,5-bis(2-butylene)tetrohydro-2-furaldehyde: see
    MGK R11
1,3-bis(carbamoylthio)-2-(N,N-dimethylamino
    propane: see cartap
bis(2-chloroethyl) ether: see dichloroethyl ether
2,2-bis(p-chlorophenyl)-1,1-dichloroethane: see
    DDD
1,1-bis(4-chlorophenly)ethanol: see chlorfenethol
3,6-bis(2-chlorophenly)-1,2,4,5-tetrazine: see
    clofentezine
1,1-bis(p-chlorophenyl)-2,2,2-trichloroethanol: see
    dicofol
bis(dimethylamino)fluorophosphine oxide: see
    dimefox
bis(dimethyldicarbamato)zinc: see ziram
bis(dimethylthiocarbamoyl)disulfide: see thiram
1-[[bis(4-flurophenyl)methylsilyl]methyl]-1H-1,2,4-
    triazole: see flusilazole
bis(4-flurophenyl)(methyl)(1H-1,2,4-triazole-]-
    ylmethyl]silane: see flusilazole
bis(hydrazine)bis(Hydrogensulfato)copper: see
    cupric hydrazinium sulfate
bis(2-hydroxyl[1,1:pri-biphenyl]-3-carboxylate-
    O,O)copper: see copper bis(3-phenylsalicylate)
bishydroxycoumarin: see Dicumarol
2,4-bis(isopropylamino)-6-methoxy-s-triazine: see
    prometon
bis[3-(methoxycarbonyl)-2-thioureido]benzene: see
    thiophanate methyl

2,2-bis(p-methoxyphenyl)-1,1,1-trichloroethane: see
    methoxychlor
bis(pentachloro-2,4-cyclopentadiene-1-yl): see
    dienochlor
bis(3-phenylsalicylato)copper: see copper bis(3-
    phenylsalicylate)
bis(tri-n-butyltin)oxide: see tributyltin oxide
bis(tributyltin) oxide: see tributyltin oxide
bis(tris-β,β-dimethylphenethyl)tin)oxide: see
    fenbutatin-oxide
Bithin: see bithionol
BITHIONOL
Black Flag Ant Trap: see chlordecone
Black Fungicide: see ferbam
Black Leaf 40 (nicotine sulfate): see nicotine
Bladafum: see sulfoTEPP
Bladen Extra: see methyl parathion
Bladex: see cyanazine
Bladotyl: see cyanazine, mecoprop
Blagal: see cyanazine, MCPA
BLA-S: see blasticidin S
BLASTICIDIN S
Blastmycin: see antimycin A3
Blatex: see hydramethylnon
Blattanex Residual Spray: see dichlorvos, propoxur
Blazer 2L: see acifluorfen
Blazer 2S: see acifluorfen
Blazine: see atrazine, cyanazine
Blazon: see chlorothalonil
Blecar MN: see mancozeb
Blex: see pirimiphos-methyl
Blightox: see zineb
Blitex: see zineb
Blitox: see copper oxychloride
Blizene: see zineb
Bloc: see fenarimol
Blue Copperas: see copper sulfate
Blue Shield: see copper hydroxide
Blue Vitriol: see copper sulfate
Bluestone: see copper sulfate
BMC: see *Bacillus thuringiensis* var. *israelensis*
BMC: see carbendazim
B-nine: see daminozide
BoAna: see famphur
Bolda: see carbendazim, maneb, sulfur
Bolero: see benthiocarb
BOMYL
Bonide: see red squill
Bonide Ryatox: see ryania
Bonide Topzol Rat Baits & Killing Syrup: see red squill
boracic acid: see boric acid
Borascu: see borax
Borate: see borax
BORAX
Bor-Dax: see Bordeaux mixture
BORDEAUX MIXTURE
Bordermaster: see MCPA
Borea: see bromacil

Borer Kill: see lindane
Borer-sol: see ethylene dichloride
BORIC ACID
Borocil: see borax
Borofax: see boric acid
Borolin: see picloram
Borospray: see borax
borsaure: see boric acid
Botran: see dichloran
Botrilex: see PCNB
Bovinox: see trichlorfon
Boygon: see propoxur
Brake: see fluridone
Brascop: see copper oxychloride
Brasoran: see aziprotryne
Brassicol: see PCNB
Brassoron: see aziprotryne
Bravo: see chlorothalonil
Bravocarb: see carbendazim, chlorothalonil
Bravo C/M: see chlorothalonil
Bremen blue: see basic copper carbonate
Bremen gree: see basic copper carbonate
Brestan: see triphenyltin acetate
Brevinyl: see dichlorvos
Brevinyl E50: see dichlorvos
brick oil: see creosote (coal tar)
Brifur: see carbofuran
Brigade: see bifenthrin
brimstone: see sulfur
Bripoxur: see carbofuran
Briten: see trichlorfon
Brittox: see bromoxynil, ioxynil, mecoprop
Broadshot: see 2,4-D, dicamba, triclopyr
Brocide: see ethylene dichloride
BRODIFACOUM
Brofene: see bromophos
Broma: see bromacil
BROMACIL
Bromacil-lithium
BROMADIOLONE
Bromaflor: see ethephon
Bromakil: see bromadiolone
Bromard: see bromadiolone
Bromatrol: see bromadiolone
bromchlophos: see naled
BROMETHALIN
Bromex: see dichlofenthion
Bromex: see chlorbromuron, naled
Brominal: see bromoxynil
Brominil: see bromoxynil
3-[3-(4'-bromo[1,1'-biphenyl]-4-yl)-3-hydroxy-1-
    phenylpropyl]-4-hydroxy-2H-1-benzopyran-2-
    one: see bromadiolone
3-(4-'bromo(1,1'-bipheny)-4-yl)1,2,3,4-tetrahydro-1-
    naphthalenyl-4-hydroxy-2H-1-benzopyran-2-one:
    see brodifacoum
3-[3-(4'-bromobiphenyl-4-yl)-1,2,3,4-tetrahydro-1-
    naphthyl]-4-hydroxycoumarin: see brodifacoum

5-bromo-3-sec-butyl-6-methyluracil: see bromacil
3-(4-bromo-3-chlorophenyl)-1-methoxy-1-methylurea:
    see chlorbromuron
N'-(4-bromo-3-chlorophenyl)-N-methoxy-N-
    methylurea: see chlorbromuron
O-(4-bromo-2,5-dichlorophenyl)-O,O-
    diethylphosphorothioate: see bromophos-ethyl
O-(4-bromo-2,5-dichlorophenyl)-O,O-
    dimethylphosphorothioate: see bromophos
O-(4-bromo-2,5-dichlorophenyl) O-
    methylphenylphosphonothioate: see leptophos
bromoethane: see methyl bromide
bromofos: see bromophos
bromofos-ethyl: see bromophos-ethyl
Bromofume: see ethylene dibromide
Brom-O-Gas: see methyl bromide
5-bromo-6-methyl-3-(1-methylpropyl)uracil: see
    bromacil
Bromone: see bromadiolone
β-bromo-β-nitromethyleneglycol: see bronopol
2-bromo-2-nitropropan-1,3-diol: see bronopol
3-[-[p-(p-bromophenyl)-hydroxyphenethyl]-benzyl-4-
    bydroxycoumarin]: see bromadiolone
BROMOPHOS
BROMOPHOS-ETHYL
BROMOPROPYLATE
Bromorat: see bromadiolone
Bromox: see bromacil
BROMOXYNIL
Bronate: see bromoxynil, MCPA
Bronco: see alachlor, glyphosate
Bronocot: see bronopol
BRONOPOL
Bronosol: see bronopol
Bronotak: see bronopol
Brophene: see bromophos
brown copper oxide: see copper oxide
Brozone: see methyl bromide
Brulan: see tebuthiuron
Brush Bullet: see tebuthiuron
Brush Buster: see dicamba
Brush-Off: see bromacil
Brushtox: see 2,4,5-T
Brygou: see propoxur
BSZ: see zinc coposil
B.t.: see Bacillus thuringiensis (Berliner)
BTC: see benzalkonium chloride
BTS 27419: see amitraz
BTV: see Bacillus thuringiensis var. kurstaki
Buckle: see triallate, trifluralin
Buctril: see bromoxynil
Buctril 20: see bromoxynil
Buctril 21: see bromoxynil
Buctril D: see bromoxynil, dicamba, MCPA
Buctril M: see bromoxynil, MCPA
Buctril MA: see bromoxynil, MCPA
Buctril ME4: see bromoxynil
BUFENCARB

Bug-geta: see metaldehyde
Bug Time: see *Bacillus thuringiensis* var. *kurstaki*
Bullet: see alachlor
Bullseye: see amitrole, atrazine, diuron
Burts & Harvey Mecoprop: see mecoprop
Bushwacker: see tebuthiuron
BUTACHLOR
Butacide: see piperonyl butoxide
BUTAM
Butamin: see tetramethrin
2-BUTANAMINE
butanedioic acid mono (2,2-dimethyl hydrazide):
    see daminozide
Butanex: see butachlor
Butanox: see butachlor
2-butanoxyethyl ester of triclopyr: see triclopyr
2-butenedioc acid, diethylester: see diethyl
    fumarate
butichlorfos: see bromoxynil
Butinox: see tributyltin oxide
Butoss: see deltamethrin
Butoxone: see 2,4-D,2,4-DB,MCPA
Butoxone SB: see 2,4-DB
BUTOXYCARBOXIM
α-[2-(2-butoxyethyoxy)ethoxy]-4,5-(methylenedioxy)-
    2-propyltoluene: see piperonyl butoxide
N-(butoxymethyl)-2-chloro-2′,6-diethylacetanilide:
    see butachlor
N-(butoxymethyl)-2-chloro-N-(2,6-
    diethylphenyl)acetamide: see butachlor
BUTOXY POLYPROPYLENE GLYCOL
BUTRALIN
sec-butylamine: see 2-butanamine
2-(tert-butylamino)-4-(ethylamino)-6-(methylthio)-s-
    triazine: see terbutryn
BUTYLATE
N-sec-butyl-4-tert-butyl-2,6-dinitrobenzamine: see
    butralin
(butylcarbityl)(6-propylipiperonyl) ether: see
    piperonyl butoxide
3-tert-butyl-5-chloro-6-methyl uracil: see terbacil
4-tert-butyl-2-chlorophenyl
    methylmethylphosphoramidate: see crufomate
2-tert-butyl-4-(2,4-dichloro-5-isopropoxyphenyl)-
    $\Delta^{21,3,4}$-oxadiazoline-5-one: see oxadiazon
(10*E*, 14*E*, 16*E*, 22*Z*)-(1*R*,4*S*,5′*S*,6*S*,6′*R*, 8*R*,
    12*S*,13*S*,20*R*,21*R*,24*S*)-6′[(*S*)-sec-butyl]-21,24-
    dihydroxy-5′,11,13,22-tetramethyl-2-oxo-3,7,19-
    trioxatetracyclo[15.6.1.1^{4,8}.0^{20,24}]pentacosa-
    10,14,16,22-tetraene-6-spiro-2′-(5′,6′-dihydro-
    2′*H*-pyran)-12-yl 2,6-dideoxy-4-*O*-
    (2,6-dideoxy-3-*O*-methyl-α-L-*arabino*-
    hexopyranosyl)-3-*O*-α-L-*arabino*-hexo-
    pyranoside (i) mixture with (10*E*,14*E*,16*E*,22*Z*)-
    (1*R*,4*S*,5′*S*,6*S*,6′*R*,8*R*,12*S*,13*S*,20*R*,21*R*,24*S*)-21,22-
    dihydroxy-6′-isopropyl-5′,11,13,22-tetramethyl-2-
    oxo-3,7,19-trioxatetracyclo[15.6.1^{4,8}.0^{20,24}]pentacosa-
    10,14,16,22-tetraene-6-spiro-2′-(5′,6′-dihydro-

2′*H*-pyran)-12-yl2,6-dideoxy-4-*O*-(2,6-dideoxy-3-
    O-methyl-α-L-*arabino*-hexopyranosyl)-3-*O*-
    methyl-α-L-*arabino*hexopyranoside (ii) (4:1): see
    abamectin
5-*n*-butyl-2-dimethylamino-4-hydroxy-6-
    methylpyrimidine: see dimethirimol
N-*sec*-butyl-2,6-dinitroanaline: see butralin
2-*sec*-butyl-4,6-dinitrophenol: see dinoseb
2-*tert*-butyl-4,6-dinitrophenol: see dinoterb
2-*tert*-butyl-4,6-dinitrophenyl acetate: see dinoterb
    acetate
2-*sec*-butyl-4,6-dinitrophenylisopropyl carbonate:
    see dinobuton
2-*sec*-butyl-4,6-dinitrophenyl-3-methyl-2-butenoate:
    see binapacryl
5-butyl-2-ethylamino-6-methylpyrimidin-4-ol: see
    ethirimol
5-butyl-2-(ethylamino)-6-methyl-4-pyrimidinol: see
    ethirimol
5-butyl-2-(ethylamino)-6-methyl-4(1*H*)-pyrimidone:
    see ethirimol
N-butyl-N-ethyl-α,α,α-trifluoro-2,6-dinitro-*p*-
    toluidine: see benefin
[2*R*-(2*R**,3*S** 6*S**,7*R**,8*R**)]-8-butyl-3-
    [[30(formylamino)-2-hydroxybenzoyl]amino]-2,6-
    dimethyl-4,9-dioxo-1,5-dioxonan-7-yl 3-
    metylbutanoate: see antimycin A3
n-butyl-hydroxyfluorene-9-carboxylate: see flurecol-
    butyl
2-(*p*-*tert*-butylphenoxy)cyclohexyl 2-propynyl sulfite:
    see propargite
2(*p*-*tert*-butylphenoxy)-1-methylethyl-2′-chloroethyl
    sulfite: see Aramite
1-(5-*tert*-butyl-1,3,4-thiadiazol-2-yl)-1,3-dimethylurea:
    see tebuthiuron
S-[*tert*-butylthiomethyl]*O*,*O*-
    diethylphosphorodithioate: see terbufos
butyl 2-[4-[[5-(trifluoromethyl)-2-
    pyridinyl]oxy]phenoxy]propananoate: see
    fluazifop-butyl
butyphos: see DEF
Butyrac 118: see 2,4-DB
Butyrac ester: see 2,4-DB
Bux: see bufencarb
Bux-Ten Granular: see bufencarb
C-106: see chlorfenson
C 1414: see monocrotophos
C-1983: see chloroxuron
C-2059: see fluometuron
C-6313: see chlorbromuron
C-6989: see fluorodifen
C709: see dicrotophos
C 8949: see chlorfenvinphos
C-Sn-9: see tributyltin oxide
CO7019: see aziprotryne
Cab-O-Sil: see silica aerogel
cacodylic acid: see arsenic
Cadan: see cartap

Caddy: see cadmium chloride
Cadminate: see cadmium succinate
CADMIUM
cadmium carbonate: see cadmium
cadmium chloride: see cadmium
cadmium decanedioate: see cadmium sebacate
cadmium oxide: see cadmium
cadmium sebacate: see cadmium
cadmium succinate: see cadmium
cadmium sulfate: see cadmium
Cagro: see ethephon
Caid: see chlorophacinone
cajeputene: see d-limonene
Cakucap: see dinocap
calcitriol: see cholecalciferol
CALCIUM
calcium acid methanearsenate: see arsenic
calcium acid methyl arsenate: see calcium acid
    methanearsenate
calcium arsenate: see arsenic
calcium o-arsenate: see calcium arsenate
calcium arsenite: see arsenic
calcium chlorate: see cadmium
calcium cyanide: see calcium
calcium hypochlorite: see calcium
calcium methanearsenate: see calcium acid
    methanearsenate
calcium phosphide: see calcium
calcium polysulfide: see calcium
calcium propanersenate: see arsenic
calcium propionate: see calcium
calcium propyl arsonate: see calcium
    propanersenate
calcium sulfate: see calcium
Cal Cop 1-: see copper ammonium carbonate
Caldan: see cartap
Calgo-gran: see mercurous chloride, mercuric
    chloride
Calo-clor: see mercurous chloride, mercuric
    chloride
Calogreen: see mercurous chloride
Calomel: see mercurous chloride
CAMA: see calcium acid methanearsenate
2-camphanone: see camphor
camphechlor: see toxaphene
Camphoclor: see toxaphene
Camphofene Huileux: see toxaphene
CAMPHOR
camphor tar: see naphthalene
Canadien 2000: see bromadiolone
Candex: see asulam, atrazine
Cannon: see alachlor, trifluralin
canoethylene: see acrylonitrile
Canogard: see dichlorvos
Canopy: see chlorimuron, metribuzin
Caocobre: see copper oxide
Capfos: see fonofos
Capro 57: see copper oxychloride sulfate

Caprolin: see carbaryl
Capsine: see dinitrocresol
Capsolane: see dichlormid, EPTC
Captaf: see captan
CAPTAFOL
CAPTAN
captane: see captan
Captanex: see captan
captaphol
Captax: see MBT
Captec: see dicapthon
Carbacide: see carbaryl
Carbacryl: see acrylonitrile
Carbadine: see zineb
Carbam: see metam-sodium
Carbamate: see ferbam
Carbamine: see carbaryl
CARBANOLATE
Carbaril: see carbaryl
CARBARYL
Carbate: see carbendazim
Carbatene: see metiram
Carbatox: see carbaryl
CARBENDAZIM
carbendazol: see carbendazim
Carbicron: see dicrotophos
Carbofos: see malathion
carbofos: see malathion
CARBOFURAN
carbolic acid: see phenol
carbon bichloride: see tetrachloroethylene
carbon bisulfide: see carbon disulfide
carbon dichloride: see tetrachloroethylene
CARBON DISULFIDE
4,4-carbonimidoylbis[N,N-
    dimethylbenzenamine]monohydrochloride: see
    auramine
carbon monoxide: see chlordane, heptachlor,
    methylene chloride
carbon oil: see benzene
CARBON TETRACHLORIDE
CARBOPHENOTHION
Carbophos: see malathion
Carboxide: see ethylene dibromide
Carboxide: see ethylene oxide
CARBOXIN
Carbyen: see barban
Carfene: see azinphosmethyl
Carma: see carbofuran, isofenphos
Carpene: see dodine
Carpidor: see trifluralin
Carpolin: see carbaryl
CARTAP
Cartox: see ethylene dibromide
Cartox: see ethylene oxide
Caryne: see barban
Carzol: see formetanate hydrochloride
Cascade: see flufenoxuron

Casoron G: see dichlobenil
caustic barley: see sebadilla
Cavdilla: see sabadilla
CBN: see barban
CCA: see chromated copper arsenate
CCC: see chlormequat chloride
CCN52: see cypermethrin
CDAA
CDB 90: see trichloroisocyanuric acid
CDB Clearon: see sodium dichloroisocyanurate
    dihydrate
CDEC
CDT: see simazine
CeCeCe: see chlormequat chloride
Cekiuron: see diuron
Cekubacillina: see *Bacillus thuringiensis* var.
    kurstaki
Cekubaryl: see carbaryl
Cekudazim: see carbendazim
Cekudifol: see dicofol
Cekufon: see trichlorfon
Cekumeta: see metaldehyde
Cekumethion: see methyl parathion
Cekusan: see dichlorvos, simazine
CELA-S-1942: see bromophos
CELA S-2225: see bromophos-ethyl
Celatom: see diatomaceous earth
Celfume: see methyl bromide
Celite: see diatomaceous earth
Celmide: see ethylene dibromide
Celphos: see aluminum phosphide
Cenol Flea Powder: see rotenone
Cenol Garden Dust: see rotenone
Cent-7: see isoxaben
Cepha: see ethephon
Ceravax: see carboxin
Cercobin M: see thiophanate methyl
Cerone: see ethephon
Certan: see *Bacillus thuringiensis* var. *aizawei*
Certosan: see dinitrocresol, metoxuron
Certrol: see ioxynil
Certrol PA: see dichlorprop, ioxynil, MCPA
CES: see Aramite
Cestacide: see niclosamide
CET: see simazine
Ceylon citronella oil: see citronella
CF 125: see chlorflurecol
CFV: see chlorfenvinphos
CGA 112913: see chlorfluazuron
CGA-12223: see isazophos
CGA 26351: see chlorfenvinphos
CGA 913: see chlorfluazuron
CHA-KEM-CO: see DBCP
Champion: see copper hydroxide
Chardol: see 2,4-D
Check Mate: see MSMA
Chemagro 9010: see propoxur
Chemagro B-1776: see DEF

Chem Bam: see nabam
Chemical 109: see antu
Chem-Mite: see rotenone
Chem-Neb: see maneb
Chem-O-Bam: see amobam
Chemosect DNOC: see dinitrocresol
Chemox PE: see dinitrophenol
Chempar: see copper oxychloride
Chempar Amitrole: see amitrole
Chem Pels: see sodium arsenite
Chem-Penta: see pentachlorophenol
Chem-phene: see toxaphene
Chem-Sen 56: see sodium aresenite
Chem-Tol: see pentachlorophenol
Chemtrol: see pentachlorophenol
Chem Zineb: see zineb
Chiefmate: see flucythrinate, phenthoate
Chinufur: see carbofuran
Chip-Cal Granular: see calcium arsenate
Chipco-26109: see iprodione
Chipco Buctril: see bromoxynil
Chipco Crab-kleen: see bromoxynil
Chipco Turf Fungicide MCPP: see mecoprop
Chipco Turf Herbicide D: see 2,4-D
Chiptox: see MCPA
chloor: see chlorine
Chlor: see chlorine
CHLORAMBEN
chlorambene: see chloramben
CHLORANIL
chloranocryl: see dicryl
Chlorasol: see ethylene dichloride
Chlorax: see sodium chlorate
CHLORBENSIDE
CHLORBROMURON
Chlordan: see chlordane
CHLORDANE
CHLORDECONE
CHLORDIMEFORM
chlore: see chlorine
Chlorex: see dichloroethyl ether
CHLORFENETHOL
Chlorfenidim: see monuron
CHLORFENSON
CHLORFENSULPHIDE
CHLORFENVINPHOS
CHLORFLUAZURON
CHLORFLURECOL
chlorflurecol-methyl: see chlorflurecol
chlorflurenol: see chlorflurecol
chlorflurenol-methyl: see chlorflurecol
chloride of lime: see calcium hypochlorite
CHLORIMURON
chlorinat: see barban
CHLORINATED ISOCYANURATES
chlorinated naphthalenes: see PCB's
CHLORINE
Chlor Kil: see chlordane

CHLORMEPHOS
CHLORMEQUAT
chlormequat chloride: see chlormequat
Chlormite: see chloropropylate
chlornitrofen: see CNP
N-chloroacetyl-N-(2,6-diethylphenyl)glycine: see
    diethatyl
2-chloroacrolein: see CDEC, diallate
2-chloroallyldiethyldithiocarbamate: see CDEC
4-chloroaniline: see diflubenzuron
CHLOROBENZILATE
p-chlorobenzyl p-chlorophenyl sulfide: see
    chlorbenside
2-chloro-4,6-bis(ethylamino)-s-triazine: see simazine
chlorobromuron: see chlorbromuron
4-chloro-2-butyl N-(3-chlorophenyl)carbamate: see
    barban
4-chloro-2-butynyl m-chlorocarbanilate: see barban
5-chloro-N-(2-chloro-4-nitrophenyl)-2-
    hydroxybenazmide, 2-aminoethanol salt: see
    niclosamide
4-chloro-α-(4-chlorophenyl)-α-
    methylbenzenemethanol: see chlorfenethol
chlorocholine chloride: see chlormequat chloride
Chlorocide: see chlorbenside
2-chloro-4-(cyclopropylamino)-6-(isopropylamino)-s-
    triazine: see cypraxine
6-chloro-N-cyclopropyl-N'-(1-methylethyl)-1,3,5-
    triazine-2,4-diamine: see cyprazine
α-chloro-N,N-diallylacetimide: see CDAA
2-chloro-1-(2,4-dichlorophenyl) vinyl
    diethylphosphate: see chlorfenvinphos
O-[2-chloro-1-(2,5-dichlorophenyl)vinyl] O,O-diethyl
    phosphorothioate: see Akton
2-chloro-α-
    [(diethoxyphosphinothioyloxyimino)phenyl-
    acetonitrile: see chlorphoxim
2-chloro-2-diethylcarbamoyl-1-methylvinyl
    dimethylphosphate: see phosphamidon
2-chloro-2',6'-diethyl-N-(methyoxymethyl)-
    acetanilide: see alachlor
S-[2-chloro-1-(1,3-dihydro-1,3-dioxo-2H-isoindol-2-
    yl)ethyl]O,O-diethylphosphorodithioate: see
    dialifor
S-6-chloro-2,3-dihydro-2-oxo-1,3-benzoxazol-3-
    ylmethyl: see phosalone
6-chloro-N²,N⁴-di-isopropyl-1,3,5,triazine-2,4-
    diamine: see propazine
2-chloro-3-dimethoxyphosphinoyloxy-N,N-
    diethylbut-2-enamide: see phosphamidon
5-chloro-3-(1,1-dimethylethyl)-6-methyl-2,4-(1H,3H)-
    pyrimidinedione: see terbacil
2-chloro-4,5-dimethylphenylmethylcarbamate: see
    carbanolate
2-chloro-N,N-di-2-propenyl-acetamide: see CDAA
1-chloro-2,3-epoxypropane: see epichlorohydrin
2-chloro-N-ethoxymethyl-6'-ethylacet-o-toluidide:
    see acetochlor

chloro-N-(ethoxymethyl)-N-(2-ethyl-6-methylphenyl)
    acetamide: see acetochlor
2-chloro-4-ethylamino-6-isopropylamino-s-triazine:
    see atrazine
2-(4-chloro-6-ethylamino-s-triazin-2-ylamino)-2-
    methylpropionitrile: see cyanazine
N-(2-chloroethyl)-2,6-dinitro-N-propyl-4-
    (trifluoromethyl)aniline: see fluchloralin
N-(2-chloroethyl)-2,6-dinitro-N-propyl-4-
    (trifluoromethyl)benzenamine: see fluchloralin
2-chloro-6'-ethyl-N-(2-methoxy-1-methylethyl)acet-o-
    toluidide: see metolachlor
2-(chloroethyl)phosphonic acid: see ethephon
N-(2-chloroethyl)-α,α,α-trifluoro-6-dinitro,N-propyl-p-
    toluidine: see fluchloralin
2-chloroethyl-N,N,N-trimethylammonium chloride:
    see chlormequat chloride
2-chloroethyl-N,N,N-trimethylammonium ion: see
    chlormequat
chlorofenizon: see chlorfenson
chlorofos: see trichlorfon
5-chloro-2-hydroxydiphenylmethane: see O-benzyl-
    p-chlorophenol
Chloro-IPC: see chlorpropham
2-chloro-N-isopropylacetanilide: see propachlor
CHLOROFLUOROCARBONS
CHLOROFORM
2-[[[[(4-chloro-6-methoxyprimidin-2-
    yl)amino]carbonyl]amino]sulfonyl]benzoate: see
    chlorimuron
2-chloro-N-[(4-methoxy-1,3,5-triazin-2-
    yl)aminocarbonyl]benzenesulfonamide: see
    chlorsulfuron
4-chloro-5-methylamino-2-(α,α,α-trifluro-m-
    tolyl)pyridazin-3(2H)-one: see norflurazon
S-(chloromethyl)O,O-diethylphosphorodithioate:
    see chlormephos
O-(5-chloro-1-methylethyl)-1H-1,2,4-triazol-3-yl O,O-
    diethylphosphorothioate: see isazophos
(4-chloro-2-methylphenoxy)acetic acid: see MCPA
2-(4-chloro-2-methylphenoxy)propionic acid: see
    mecoprop
N-(4-chloro-2-methylphenyl)-N,N-
    dimethylformamidine: see chlordimeform
CHLORONEB
chloronebe: see chloroneb
O-2-chloro-4-nitrophenyl,O,O-
    dimethylphosphorothioate: see dicapthon
4-chloro-2-oxo-3-benzothiazoline acetic acid: see
    benazolin
4-chloro-2-oxo-3-(2H)-benzothiazol acetic acid: see
    benazolin
4-chloro-2-oxobenzothiazolin-3-yl-acetic acid: see
    benazolin
CHLOROPHACINONE
chlorophen: see pentachlorophenol
Chlorophenothane: see DDT
(4-chlorophenoxy)acetic acid: see 4-CPA

cismethrin: see resmethrin
Citcop: see copper linoleate
CITRONELLA
Citrusperse: see zinc coposil
Clandelite: see aresenic trioxide
Clanex: see pronamide
Clarosan: see terbutryn
Classic: see chlorimuron
Cleansweep: see diquat, paraquat
Clenecorn Plus: see dichlorprop, mecoprop
Clinicide: see carbaryl
CLOFENTEZINE
CLOMAZONE
Clonitralid: see niclosamide
Clonitralide: see niclosamide
CLOPYRALID
Clor Chem T-590: see toxaphene
cloro: see chlorine
Clorophene: see O-benzyl-p-chlorophenol
Cloroxone: see 2,4-D
Clortocar Ramato: see chlorothalonil
Clortosip: see chlorothalonil, copper oxychloride,
    maneb
CMPP: see mecoprop
CMU: see monuron
CNA: see dichloran
CNP
coal naphtha: see benzene
coal tar creosote: see creosote (coal tar)
coal tar oil: see creosote (coal tar)
Cobex: see dinitramine
Cobexo: see dinitramine
Cobox Blue: see copper oxychloride
Cobredon: see basic copper carbonate
cocoa fatty acids: see soap
Codal: see metholachlor, prometryn
College Brand Weed Killer: see 2,4-D
colloidal arsenic: see arsenic
Colloidox: see copper oxychloride
Collunosol: see 2,4,5-trichlorophenol
Colsul: see sulfur
Comac Parasol: see copper hydroxide
Combat: see hydramethylnon
Combat: see isoxaben
Combine: see teburthiuron
Combinex: see permethrin, thiram
Command: see clomazone
Commando: see flamprop-isopropyl
Commence: see clomazone, trifluralin
Comodor: see butam
Compel: see fluridone
Compitox: see mecoprop
Compo: see difenacoum
Composan: see ethephon
Compound 1080: see sodium fluoroacetate
Compound 1081: see fluoroacetamide
Compound 118: see aldrin
Compound 269: see endrin

Compound 338: see chlorobenzilate
Compound 4072: see chlorfenvinphos
Compound 497: see dieldrin
Compound 604: see dichlone
Compound 7744: see carbaryl
Compounds 53-CS-17: see heptachlor epoxide
Conquer Liquid Vegetation Killer: see prometon
Conquest: see atrazne, cyanazine
Contain: see imazapyr
Contrac: see bromadiolone
Contraven: see terbufos
Coopex: see permethrin
Copace-E: see copper sulfate
Cophamate: see copper oxychloride
Copharten: see niclosamide
Cop-O-Zinc: see zinc coposil
Cop-O-cide: see copper linoleate
Copoloid: see copper linoleate
Copophos: see basic copper carbonate
Copox: see copper oxide
COPPER
Copper 8: see oxine-copper
copper acetoarsenite: see arsenic
Copper A Compound: see copper oxychloride
copper ammonium carbonate: see copper
copper ammonium sulfate: see sulfate
copperas: see ferrous sulfate
copper bis(3-phenylsalicylate): see copper
copper carbonate: see copper
copper carbonate hydroxide: see basic copper
    carbonate
copper chelate: see copper
copper chloride dihydrate: see copper
copper (II) chloride oxidehydrate: see copper
    oxychloride
copper chromated arsenate: see chromated copper
    arsenate
copper citrate: see copper
Copper-Count-N: see copper ammonium carbonate
Copper-Cure: see copper napthenate
copper dihydrazine disulfate: see cupric
    hydrazinium sulfate
copper (II) dihydraziniumdisulfate: see cupric
    hydrazinium sulfate
copper-ethylenediamine complex
Copperfine-Zinc: see copper sulfate
Copper Hydro Bordo: see Bordeaux mixture
copper hydroxide: see copper
copper (II) hydroxide: see copper hydroxide
copper linoleate: see copper
copper napthenate: see copper
Coppernate: see copper napthenate
copper nitrate: see copper
copper oxalate: see copper
copper oxide: see copper
Copper Oxinate: see oxine-copper
copper oxychloride: see copper
copper oxychloride sulfate: see copper

copper-3-phenyl salicylate: see copper bis(3-phenylsalicylate)
copper salts of fatty androsin acid, 20–25% copperabietate, 8–12% copper linoleate and copper oleate: see copper linoleate
copper salt of napthenic acid: see copper napthenate
Coppersan: see copper oxychloride
Copper-Sandoz: see copper oxide
Copper Sardez: see copper oxicde
copper sulfate: see copper
copper sulfate monohydrate: see copper
copper sulfate monohydrate: see copper sulfate
copper sulfate pentahydrate: see copper sulfate
copper tea complex: see copper
copper triethanolamine complex: see copper tea complex
copper zinc chromate: see copper
Cop-R-Nap: see copper napthenate
Coprantol: see copper oxychloride
Copravit: see copper oxychloride
Copro 53: see copper oxychloride sulfate
Cop Tox: see copper oxychloride
Coptox: see copper oxychloride
Co-Ral: see coumaphos
Coraza: see diphenylamine
Corbit: see anthraquinone
Corliss: see phorate terbufos
Cornox Plus: see 2,4,-D, dicamba, 2,4,5-T
Cornox-Plus: see mecoprop
Cornox RD & RK: see dichlorprop
Cornoxynil: see bromoxynil, dichlorprop
Corodane: see chlordane
Coromate: see ferbam
Corosul D and S: see sulfur
Corothion: see parathion
Corozate: see ziram
corrosive sublimate: see mercuric chloride
Corry's Slug Death: see metaldehyde
Corsaid: see permethrin
Corsair: see permethrin
Cos: see copper oxychloride
Cosan: see sulfur
Cotnion-Ethyl: see azinphosethyl
Cotnion-methyl: see azinphosmethyl
Cotodon: see dipropetryn, metolachlor
Cotofor: see dipropetryn
Cotogard: see fluometuron, prometryn
Cotoran multi: see fluometuron, metolachlor
Cotoran Multi 50WP: see fluometuron
Cotton Aide HC: see cacodylic acid
Cottonex: see fluometuron
coumafene: see warfarin
COUMAFURYL
COUMAPHOS
Counter: see terbufos
Counter 15G Soil Insesticide: see terbufos
Counter Plus: see terbufos

Coxysan: see copper oxychloride
Coxysul: see copper oxychloride sulfate
CP 15336: see diallate
CP 23426: see triallate
CP 31393: see propachlor
CP 50144: see alachlor
CP 53619: see butachlor
CP-6343: see CDAA
CP67573: see glyphosate
4-CPA
CPCBS: see chlorfenson
CR-3029: see maneb
Crab Grass Killer: see arsenic acid
Crackdown: see deltamethrin
Crag Fly Repellent: see butoxy polypropylene glycol
Crag Fungicide 658: see copper zinc chromate
Crag Fungicide 974: see dazomet
Crag Nemacide: see dazomet
Crag Sevin: see carbaryl
Creasote: see creosote (wood tar)
CREDAZINE
CREOSOTE (COAL TAR)
CREOSOTE (WOOD TAR)
creosote oil: see creosote (coal tar)
creosotum: see creosote (coal tar)
cresols: see cresylic acid
Cresopur: see benazolin
CRESYLIC ACID
cresylic creosote: see creosote (coal tar)
Criscobre: see copper hydroxide
Crisfolatan: see captafol
Crisfolatan: see captaphol
Crisodrin: see monocrotophos
Cristoxo: see toxaphene
Crittox: see zineb
Crittox MZ: see mancozeb
Croak: see fluometuron, MSMA
Croneton: see ethiofencarb
Crop Rider: see 2,4-D
Crop Saver: see malathion permethrin
Croptex Chrome: see fenuron
Croptex Onyx: see bromacil
Croptex Ruby: see fenuron
Crossbow: see triclopyr
Crotilin: see 2,4-D
Crotothane: see dinocap
CROTOXYPHOS
CRUFOMATE
CRYOLITE
Cryptocidal Soap: see soap
Cryptogil ol: see pentachlorophenol
Crystal Zineb: see zineb
Crysthion: see azinphosethyl
CS-56: see copper oxychloride sulfate
CTR 6669: see carbendazim
Cu 56: see copper oxychloride
Cucumber Dust: see calcium arsenate

Cudgel: see fonofos
Cudrox: see copper hydroxide
Cudrox: see copper oxychloride
Cuidrox: see copper hydroxide
Cuman: see ziram
Cunilate 2472: see oxine-copper
Cupramar: see copper oxychloride
Cupravit: see copper oxychloride
Cupravit Blue: see copper hydroxide
Cupravit Green: see copper oxychloride
cupric arsenite: see copper
cupric carbonate: see basic copper carbonate
cupric citrate: see copper citrate
cupric hydrazinium sulfate: see copper
cupric hydroxide: see copper hydroxide
cupric subacetate: see basic cupric acetate
cupric subcarbonate: see basic copper carbonate
Cuprin: see copper oxychloride
Cuprinol Brown: see copper napthenate
Cuprinol Green: see copper napthenate
Cuprocaffaro: see copper oxychloride
Cuprocide: see copper oxide
Cuprocitrol: see copper citrate
Cupro-Euparene: see copper oxychloride,
    dichlofluanid
Cupro-Phynebe (Super Mixy): see copper
    oxychloride, zineb
Cuproquin: see oxine-copper
Cuprosana: see copper oxychloride
Cuprosan Blue: see copper oxychloride
cuprous oxide: see copper oxide
Cuprovinol: see copper oxychloride
Cuproxol: see copper oxychloride
Curaterr: see carbofuran
Curbiset: see chlorflurecol
Curex Flea Duster: see rotenone
Curitan: see dodine
Curpinol: see copper napthenate
Curtail: see clopyralid, 2,4-D
Curzate M: see cymoxanil, maneb
Custos: see carbendazim
Cutlass: see dikegulac sodium
Cyaforce: see hydramethylnon
Cyanamid: see calcium cyanamide
CYANAZINE
1-(2-cyano-2 methoxyimino-aetyl)-3-ethylurea: see
    cymoxanil
canoethylene: see acrylonitrile
(RS)-α-cyano-4-fluoro-3-[phenoxybenzyl(1RS)-
    cis,trans-3-(2,2-dichlorovinyl)-2,2-
    dimethylcyclopropanecarboxylate: see
    cyfluthrin
Cyanogas: see calcium cyanide
(RS-α-cyano-3-phenoxybenzyl(RS)-2-(4-chlorophenyl)-
    3-methylbutyrate: see fenvalerate
α-cyano-3-phenoxybenzyl 3-(2-chloro-3,3,3-
    trifluoropropenyl)-2,2-
    dimethylcyclopropanecarboxylate (1:1 mixture

of (Z)-(1S,3S)-R-ester and (Z)-(1R,3R) S-ester: see
    λ-cyhalothrin
(RS)-α-cyano-3-phenoxybenzyl(R)-2-(2-chloro-α,α,α-
    trifluoro-p-toluidino)-3-methylbutyrate: see
    fluvalinate
(RS)-α-cyano-3-phenoxybenzyl N-(2-chloro-α,α,α-
    trifluoro-p-tolyl)-(D)-valinate: see τ-fluvalinate
(RS)-α-cyano-3-phenoxybenzyl1(RS)-cis,trans-3-(2,2-
    dichlorovinyl)-2,2-
    dimethylcyclopropanecarboxylate: see
    cypermethrin
(RS)-α-cyano-3-phenoxybenzyl(S)-2-(4-
    difluoromethoxyphenyl)-3-methylbutyrate: see
    flucythrinate
(S)-α-cyano-3-phenoxybenzyl(R,3S)-2,2,-dimethyl-3-
    [(RS)-1,2,2,2-
    tetrabromoethyl]cyclopropanecarboxylate: see
    tralomethrin
(RS)-α-cyano-3-phenoxybenzyl 2,2,3,3,-
    tetramethylcyclopropanecarboxylate: see
    fenpropathrin
[1R-[1α(S*-cyano(3-phenoxyphenyl)methyl3-(2,2-
    dibromoethenyl)-2,2-
    dimethylcyclopropanecarboxylate: see
    deltamethrin
CYANOPHENPHOS
O-4-cyanophenyl O,O-dimethylphosphorothioate:
    see cyanophos
O-(4-cyanophenyl) O,O-dimethylphosphorothioate:
    see cyanophos
O-p-cyanophenyl O-ethyl phenylphosphonothioate:
    see cyanophenphos
O-(4-cyanophenyl) O-ethyl
    phenylphosphonothioate: see cyanophenphos
CYANOPHOS
Cyanotril: see dimethoate flucythrinate
Cyanox: see cyanophos
CYAP: see cyanophos
Cybolt: see flucythrinate
CYCLOATE
Cyclodan: see endosulfan
CYCLOHEXANE
CYCLOHEXANONE
cyclohexatriene: see benzene
CYCLOHEXIMIDE
3-cyclohexyl-6-(dimethylamino)-1-methyl-1,3,5-
    triazine-2,4-(1H,3H)-dione: see hexazinone
3-cyclohexyl-6-(dimethylamino)-1-methyl-s-triazine-
    2,4(1H,3H)-dione: see hexazinone
2-cyclohexyl-4,6-dinitrophenol: see dinex
Cyclon: see hydramethylnon
Cyclon: see hydrogen cyanide
Cyclone: see diquat, paraquat
3-cyclo-octyl-1,1-dimethylurea: see cycluron
N-cyclopropyl-1,3,5-triazine-2,4,6-triamine: see
    cyromizine
Cyclosan: see mercurous chloride
CYCLURON

Cycocel: see chlormequat chloride
Cycocel-Extra: see chlormequat chloride
Cycogan Extra: see chlormequat chloride
Cycogen: see chlormequat chloride
Cyfen: see fenitrothion
Cyflee: see cythioate
CYFLUTHIRN
Cygard: see phorate terbufos
Cygon: see dimethoate
λ-CYHALOTHRIN
cyhexan: see cyhexatin
cyhexatin: see tin
Cykuthoate: see dimethoate
Cymag: see sodium cyanide
Cymbus: see cypermethrin
CYMOXANIL
Cymperator: see cypermethrin
Cynkotox: see zineb
Cynock: see cyanophos
Cynogan: see bromacil
CYP: see cyanophenphos
Cypercopal: see cypermethrin
Cyperkill: see cypermethirn
CYPERMETHRIN
cypermethrine: see cypermethrin
Cypon EC: see crotoxyphos
Cypona: see dichlorvos
CYPRAZINE
Cyprex: see dodine
Cyprex 65W: see dodine
Cyprokylt: see copper oxychloride
Cyprozine: see cyprazine
CYROMAZINE
Cyrux: see cypermethrin
Cytel: see fenitrothion
CYTHIOATE
Cythion: see malathion
Cythrin: see flucythrinate
Cytrol: see amitrole
Cytrol Amitrole-T: see amitrole, ammonium thiocyanate
Cyuram DS: see thirma
Cyuthion: see malathion
D 1221: see carbofuran
D 50: see 2,4-D
D 735: see carboxin
2,4-D
DAC 893: see DCPA
Dacamine: see 2,4-D
Dacamine: see 2,4,5-T
Dacobre: see chlorothalonil
Daconil 2787: see chlorothalonil
Dacthal: see DCPA
Dagadip: see carbophenothion
Dagger: see imazamathabenz
Dailon: see diuron
Daisen: see zineb
DALAPON

Dalapon-Na: see dalapon
DAMINOZIDE
Danathion: see fenitrothion
Danex: see trichlorfon
Danitol: see fenpropathrin
Dantril: see bromoxynil, ioxynil, dichlorprop, MCPA
DAPA: see fenaminosulf
Dapacril: see binapacryl
Daphene: see dimethoate
Dardo: see fomesafen
Dasanit: see fensulfothion
DATC: see diallate
DATC-BW: see triallate
DAZOMET
Dazzel: see diazinon
2,4-DB
2,4-DB sodium: see 2,4-DB
DBCP
DBP: see dibutyl phthalate
2,3-DCDT: see diallate
DCMO: see carboxin
DCMU: see diuron
DCNA: see dichloran
DCP: see dichloropropene
DCPA
DCPC: see chlorfenethol
DCPE: see chlorfenethol
D-D
D-D: see 1,2-dichloropropane, dichloropropene
D-D92: see dichloropropene
DDD
DDDM: see dichlorophen
DDE
DDM: see dichlorophen
DDT
DDVF: see dichlorvos
DDVP: see dichlorvos
Deadline: see bromadiolone
dead oil: see creosote (coal tar)
Debroussaillant 4323 DP: see picloram, dichlorprop
Debucol: see fenitrothion
Decabane: see dichlobenil
decachlorobis (2,4-cyclopentadiene-1-yl): see dienochlor
1,1a,3,3a,4,5,5,5a,5b,6-decachloro-octachloro-1,3,4-metheno-2H-cyclobutano(cd)pentalen-2-one: see chlordecone
decamethrin: see deltamethrin
Deccoscald 282: see diphenylamine
Deccotane: see 2-butanamine
Dechlorane: see mirex
Decimate: see DCPA, propachlor
Decis: see deltamethrin
Decis D: see dimethoate, deltamethrin
Decrotox: see crotoxyphos
Dedelo: see DDT
Dedevap: see dichlorvos
Dedisol C: see dichloropropene

Ded-Weed: see silvex
Ded-Weed: see 2,4,5-T
Ded-Weed 40: see 2,4-D
Ded-Weed Aero Ester: see 2,4-D
DEET
DEF
DEF defoliant: see DEF
Defithion: see methyl parathion
deflubenzon: see diflubenzuron
De-Fol-Ate: see sodium chlorate
Degesch Calcium Cyanide A-Dust: see calcium
    cyanide
DeGreen: see DEF
Dekrysil: see dinitrocresol
Deksan: see thriam
Delicia: see aluminum phosphide
Delnav: see dioxathion
Delsekte: see deltamethrin
Delsene: see carbendazim
Delsene M: see carbendazim
Delsene M: see carbendazim, maneb
Delsene MX 200: see carbendazim, mancozeb
Delta: see chlorophacinone
DELTAMETHRIN
deltamethrine: see deltamethrin
Deltaphos: see deltamethrin, triazophos
delta^{4}tetrahydrophthalimide
Deltic: see dioxathion
DEMETON
DEMETON-METHYL
Demise: see 2,4-D
Demon: see cypermethrin
Demosan: see chloroneb, thiram
De-Moss: see soap
Demox: see demeton
Denapon: see carbaryl
Denkaphon: see trichlorfon
O-2-deoxy-2-methylamino-α-L-glucopyranosyl-(1→2)-
    O-5-deoxy-3-C-formyl-α-L-lyxofuranosyl-
    (1→N^3,N^3-diamidino-D-strepamine: see
    streptomycin
depallethrin: see bioallethrin
Derbac: see carbaryl
Dermacid: see MBT
Derriban: see dichlorvos
Derribante: see dichlorvos
Derrin: see rotenone
Derringer: see resmethrin, piperonyl butoxide
desbromoleptophos: see leptophos
Des-i-cate: see endothall
DESMDIPHAM
DESMETRYN
desmetryne: see desmetryn
Desormone: see dichlorprop
Dessin: see dinobuton
Destral: see 2,4-D, dalapon, diuron
Destruxol: see ethylene dichloride
Destruxol Orchid Spray: see nicotine

Destun: see perfluidone
Detal: see dinitrocresol
Dethdiet: see red squill
Dethmor: see warfarin
Detmol: see permethrin
Detox 25: see lindane
Detrans: see deltamethrin, esbiothrin, piperonyl
    butoxide
Deturi Ratones: see bromadiolone
Devicarb: see carbaryl
Devigon: see dimethoate
Devikol: see dichlorvos
DEX: see EXD
Dexon: see fenaminosulf
Dextrone: see diquat, paraquat dichloride
Dexuron: see diuron, paraquat
DHB: see benzene hexachloride
Diacon: see methoprene
Dialam: see asulam, diuron
DIALIFOR
dialifos: see dialifor
dialiphor: see dialifor
DIALLATE
N,N-diallyl-2,2-dichloroacetamide: see dichlormid
N,N-diallyl-2-chloroacetamide: see CDAA
Diamekta 50%: see DDT
Diametan B: see cymoxanil, propineb, traisdimefon
DIAMIDFOS
diammonium ethylenebisdithiocarbamate: see
    amobam
Dianat: see dicamba
Dianex: see methoprene
dianisyltrichloroethane: see methoxychlor
Diapadrin: see dicrotophos
DIATOMACEOUS EARTH
20,25-diazacholesteroldihydrochloride: see
    azacosterol
Diazajet: see diazinon
diazasterol: see azacosterol
Diazatol: see diazinon
Diazide: see diazinon
DIAZINON
Diazinon: see diazinon
Diazitol: see diazinon
diazoben: see fenaminosulf
Diazol: see diazinon
dibasic, heptahydrate: see sodium arsenate [III]
dibasic lead arsonate: see lead arsenate
dibasic sodium arsenate: see sodium arsenate [III]
dibasis sodium arsenate: see sodium arsenate [II]
Dibrom: see naled
Dibrome: see ethylene dibromide
dibromochloropropane: see DBCP
1,2-dibromo-3-chloropropane: see DBCP
1,2-dibromo-2,2-dichloroethyldimethylphosphate:
    see naled
1,2-dibromoethane: see ethylene dibromide
3,5-dibromo-4-hydroxybenzonitrile: see bromoxynil

1,3-dichloropropene: see dichloroene

1,3-dichloro-1-propene: see dichloropropene

2,2-dichloropropionic acid: see dalapon

2,2-dichloropropionic acid, sodium salt: see sodium salt of dalapon

3,6-dichloro-2-pyridinecarboxylic acid: see clopyralid

2,4'-dichloro-α-(pyrimidin-5-yl) benzhydril alcohol: see fenarimol

1,2-dichloro-1,1,2,2-tetraluorethane: see chlorofluorocarbons

dichloro-s-triazinetrione: see dichloroisocyanuric acid

1,3-dichloro-s-triazine-2,4,6(1H,3H,5H)trione: see dichloroisocyanuric acid

1,3-dichloro-s-triazine-2,4,6(1H,3H,5H)trione sodium salt: see sodium dichloroisocyanurate

1,3-dichloro-s-triazine-2,4,6(1H,3H,5H)trione potassium salt: see potassium dichloroisocyanurate

1,3-dichloro-s-triazine-2,4,6(1H,3H,5H)trione sodium salt dihydrate: see sodium dichloroisocyanurate

4,4'-dichloro-α-(trichloromethyl)benzhydorl: see dicofol

2,2-dichlorovinyl dimethylphosphate: see dichlorvos dihydrate

DICHLORPROP

DICHLORVOS

diclofention: see dichlofenthion

DICLOFOP

DICLOFOP-METHYL

diclofop: see diclofop-methyl

dicloran: see dichloran

DICOFOL

Dicontal Neu: see fenitrothion, trichlorfon

dicopper chloride trihydroxide: see copper oxychloride

dicopper oxide: see copper oxide

Dicotox: see 2,4-D

Dicoumarin: see Dicumarol

Dicoumarol: see Dicumarol

DICROTOPHOS

Dicumarin: see Dicumarol

DICUMAROL

Dicumarol: see Dicumarol

dicyclohexylamine salt: see dinex

Didakene: see tetrachloroethylene

Didimac: see DDT

di(N,N-dimethylcocoamine): see endothall

di(N,N-dimethyltridecyl amine): see endothall

DIELDRIN

dieldrine: see dieldrin

Dielmoth: see dieldrin

DIENOCHLOR

diesel oil: see fuel oil

diethamine: see dinitramine

DIETHATYL

diethion: see ethion

2-diethoxyphosphinothioloxyimino)-2-phenylacetonitrile: see phoxim

O-2-diethylamino-6-methylpyrimidin-4-yl O,O-diethylphosphorothiote: see pirimiphos-ethyl

O,O-diethyl O-3-chloro-4-methyl-2-oxo-2H-1-benzopyran-7-ylphosphorothioate: see coumaphos

O,O-diethyl S-[(p-chlorophenylthio)methyl]phosphorodithioate: see carbophenothion

diethyl diphenyldichloroethane: see ethylan

diethyldithiobis(thioformate): see EXD

O-[2-diethylamino)-6-methyl-4-pyrimidinyl] O,O-dimethylphosphorothioate: see pirimiphos-methyl

N³N³-DIETHYL-2,4-dinitro-6-trifluoro-methyl-m-phenylenediamine: see dinitramine

O,O-diethyl S-ethylthiomethylphosphorodithiote: see phorate

diethyl fumarate: see malathion

O,O-diethyl O-(2-isopropyl-6-methyl-4-pyrimidinyl)phosphorothioate: see diazinon

N,N-diethyl-3-methylbenzamine: see deet

O,O-diethyl O-[p-methylsulfinyl)phenyl]phosphorothioate: see fensulfothion

diethyl P-nitorphenolester of phosphoric acid: see paraoxon

O,O-diethyl O-p-nitrophenylphosphorothioate: see parathion

O,O-diethyl 3-(4-oxo-1,2,3-benzotriazin-3(4H)-yl)methylphosphorodithioate: see azinphosethyl

diethyl 4,4'-(o-phenylene)bis(3-thioallophanate) (I): see thiophanate ethyl

O,O-diethyl O-1-phenyl-1H-1,2,4-triazol-3-yl phosphorothiote: see triazophos

O,O-diethylphosphorodithioate: see phosalone O-diethylphosphorothioate: see chlorphoxim

O,O-diethyl phthalimidephosphonothioate: see ditalimfos

N,N-diethyl-m-toluamide: see deet

O,O-diethyl O-3,5,6-trichloro-2-pyridyl phosphorothioate: see chlorpyrifos

N⁴,N⁴-diethyl-α,α,α-trifluoro-3,5-dinitrotoluene-2,4-diamine: see dinitramine

Di-Farmon M: see dicamba, mecoprop

DIFENACOUM

difenson: see chlorfenson

Difenthos: see temephos

DIFENZOQUAT

diflubenuron: see diflubenzuron

DIFLUBENZURON

difluron: see diflubenzuron

Difolatan: see captafol

Difosan: see captafol

2,3-dihydro-5-carboxanilido-6-methyl-1,4-oxathiin: see carboxin

O,O-dimethyl S-[2-(methylamine)-2-
oxoethyl]phosphorothioate: see oomethoate
N,N-dimethyl-N'-[3-
[[(methylamino)carbonyl]oxy]phenyl]meth-
animidamide: see formetanate hydrochloride
O,O-dimethyl S-2-(1-methylcarbamoylethylthio)ethyl
phosphorothioate: see vamidothion
O,O-dimethyl S-(N-
methylcarbamoylmethyl)phosphorodithioate:
see dimethoate
dimethyl S-(N-
methylcarbamoylmethyl)phosphorothioate: see
omethoate
N,N-dimethyl-2-methylcarbamoyloxyimino-2-
(methylthio)acetamide: see oxamyl
N,N-dimethyl-N'-(2-methyl-4-chlorophenyl)-
formamidine: see chlordimeform
2,2-dimethyl-N-(methylethyl)-N-
(phenylmethyl)propanamide: see butam
O,O-dimethyl O-(4-
methylmercaptophenyl)phosphate: see GC 6506
3,5-dimethyl-4-(methylthio)phenol methylcarbamate:
see methiocarb
O,O-dimethyl O-3-methyl-4-(methylthio)phenyl: see
fenthion
dimethyl (E)-1-methyl-2-(1-phenylethoxycarbonyl)
vinylphosphate: see crotoxyphos
dimethyl p-(methylthio)phenylphosphate: see GC
6506
O,O-dimethyl-O-[4-methylthio]-m-
tolylphosphorothioate: see fenthion
O,O-dimethyl O-p-nitrophenylphosphorothioate:
see methyl parathion
dimethylnitrosamine: see dicamba
O,O-dimethyl O-(4-nitro-m-tolyl phosphorothioate:
see fenitrothion
dimethylolpropionic acid: see DMPA
O,O-dimethyl S[4-oxo-1,2,3-benzotriazin-
3(4H)ylmethyl]phosphorodithioate: see
azinphosmethyl
3[2-(3,5-dimethyl-2-oxocyclohexyl)-2-hydroxyethyl]-
glutarimide: see cyclohexmide
5,5-dimethylperhydropyrimidin-2-one 4-
trifluoromethyl-α-(4-
trifluoromethylstyryl)cinnamylidenehydrazone:
see hydramethylnon
N'-(2,4-dimethylphenyl)-N-((2,4-
dimethylphenyl(imino)methyl-N-
methylmethamimidamid: see amitraz
dimethyl[1,2-
phenylene)bis(iminocarbonothioyl)]bis[carbamate]:
see thiophanate methyl
dimethyl 4,4'-o-phenylenebis(3-thioallophanate):
see thiophanate methyl
N-(2,6-dimethylphenyl)-N-(methoxyacetyl)-
alaninemethyl ester: see metalaxyl
1,1-dimethyl-3-phenylurea: see fenuron

dimethyl phosphate ester of 3-hydroxy-N,N-
dimethyl-cis-crotonamide: see dicrotophos
dimethyl phosphate of 3-hydroxy-N-methyl-cis-
crotonamide: see monocrotophos
O,S-dimethylphosphoramidothioate: see acephate-
met
O,O-dimethylphosphorodithioate, S-ester with 4-
(mercaptomethyl)-2-methoxy-Δ²-1,3,4-
thiodiazolin-5-one: see methidathion
S-(O,O-dimethylphosphorodithioate): see phosmet
O,O-dimethylphosphorothioate O-ester with p-
hydroxybenzonitrile: see cyanophos
DIMETHYL PHTHALATE
N-(1,1-dimethyl-2-propynyl) 3,5-dichlorobenzamide:
see pronamide
O,O-dimethyl O-p-sulfamoylphenyl
phosphorothioate: see cythiate
DIMETHYL SULFOXIDE
dimethyl 2,3,5,6-tetrachloro-
1,4benzenedicarbosylate: see DCPA
dimethyltetrachloroterepthalate: see DCPA
3,5-dimethyl-2H-1,3,5-thiadiazine-2-thione: see
dazomet
dimethyl (2,2,2-trichloro-1-
hydroxyethyl)phosphonate: see trichlorfon
O,O-dimethyl O-(2,4,5-trichlorophenyl)
phosphorothioate: see ronnel
1,1-dimethyl-3-(α,α,α-trifluoro-m-tolyl)urea: see
fluometuron
Dimilin: see diflubenzuron
Dimilin IG: see diflubenzuron
Dimilin W-25: see diflubenzuron
Dimite: see chlorfenethol
dimpylate: see diazinon
DINEX
Dinitrall: see dinoseb
DINITRAMINE
dinitro: see dinoseb
dinitroamine: see dinitramine
dinitrobutyl phenol: see dinoseb
2,4-dinitor-6-sec-butylphenol: see dinoseb
4,6-dinitor-o-sec-butylphenol: see dinoseb
DINITROCRESOL
4,6-dinitro-o-cresol: see dinitrocresol
dinitrocyclohexylphenol: see dinex
4,6-dinitro-o-cyclohexylphenol: see dinex 3,5-
dinitro-N⁴N⁴-dipropylsulfaniliamide: see oryzalin
2,6-dinitro-N-ethyl-N-(2-methyl-2-propenyl)-α,α,α-
trifluoro-p-toluidine: see ethalfluralin
DINITROPHENOL
2,4-dinitrophenol: see dinitrophenol
α-dinitrophenol: see dinitrophenol
N³N³-2,4-dinitrol-6-(trifluoromethyl)-1,3-
benzendiamine: see dinitramine
Dinitro Weed Killer: see dinoseb
2,6-dinitro-3,4-xylidine: see pendimethalin
DINOBUTON

DMU: see diuron
DN Dust No. 12: see dinex
DN-111: see dinex
DN-289: see dinoseb
DN-75: see dinex
DNBP: see dinoseb
DNOC: see dinitrocresol
DNOCHP: see dinex
DNOSBP: see dinseb
DNTBP: see dinoterb
Docklene: see dicamba, MCPA-sodium
Docofen: see fenitrothion
dodecachlorooctahydro-1,3,4-methene-2*H*-
    cyclobuta[c,d]pentalene: see mirex
dodecylguanidine acetate: see dodine
dodin: see dodine
DODINE
dodine acetate: see dodine
doguadine: see dodine
Dojyopicrin: see chloropicrin
Dokirin: see oxine-copper
Dol: see benzene hexachloride
Dolmix: see benzene hexachloride
Dolochlor: see chloropicrin
Domatol: see amitrole
Doom: see tetramethrin
Doom: see milky spore disease
Dorlone II: see dichloropropene
Dormone: see 2,4-D
Dormycin: see oxine-copper
Dorytox: see dieldrin
Dotan: see chlormephos
Double Strength: see silvex
Doubledown: see disulfoton, fonofos
Dovip: see famphur
Dow 1329: see DMPA
Dow Crabgrass Killer: see DMPA
Dow General Weedkiller: see dinoseb
Dow Pentachloropheno DP-2 Antimicrobial: see
    pentachlorophenol
Dow Pentachlorophenol DP-2 Antimicrobial: see
    pentachlorophenol
Dow Selective Weedkiller: see dinoseb
Dowco 118: see DMPA
Dowco 132: see crufomate
Dowco 139: see mexacarbate
Dowco 179: see chlorpyrifos
Dowco 186: see fentin hydroxide
Dowco 199: see ditalimfos
D213: see cyhexatin
D233: see triclopyr
D290: see clopyralid
Dowco-169: see diamidfos
Dowfume: see methyl bromide
Dowfume EDB: see ethylene dibromide
Dowfume MC 33: see chloropicrin
Dowicide 2: see 2,4,5-trichlorophenol
Dowicide 2S: see 2,4,6-trichlorophenol

Dowicide 7: see pentachlorophenol
Dowicide EC-7: see pentachlorophenol
Dowicide G: see pentachlorophenol
Dowpon: see dalapon
Dowspray 17: see dinex
Dozer: see fenuron
DP-2: see pentachlorophenol
2-(2,4-DP): see dichlorprop
2,4-DP: see dichlorprop
D&P 77 Dust: see formaldehyde
DPA: see diphenylamine
3,6-DPA: see clopyralid
D-264 Plus Captan: see captan
DPX 1108: see fosamine ammonium
DPX-F6025: see chlorimuron
DPX-H 6573: see flusilazole
DPX-W4189: see chlorsulfuron
Dragnet: see permethrin
Dragon: see permethrin
Drat: see chlorophacinone
Drawinol: see dinobuton
Draza: see methiocarb
DRAZOXOLON
Drexel Captan: see captan
Drexel Captan Plus Molybdenum: see captan
Dri-Die: see silica aerogel
Drianone: see silica aerogel
Drinafog: see endrin
Drinox: see aldrin
Drinox H-34: see heptachlor
Drione: see silica aerogel
Drop-Leaf: see sodium chlorate
DS: see penta-*s*-triazinetrione
DSE: see nabam
DSMA: see arsenic
DSMA: see daminozide
Du-Dusit: see bromacil, dichlobenil
Du-Ter: see fentin hydroxide
DuPont 1991: see benomyl
DuPont Herbicide 732: see terbacil
Dual: see metolachlor
Dudubitoke: see dieldrin
Duo-Kill: see crotoxyphos
Duo-Kill: see dichlorvos
Duplosan DP-D: see 2,4-D, dichlorprop
Duplosan DP-M: see dichlorprop, MCPA
Duplosan Super: see 2,4-D, dichlorprop, MCPA
Duracide 15: see piperonyl butoxide tetramethrin
Duraphos: see mevinphos
Duratox: see demeton-methyl
Duravos: see dichlorvos
Durotox: see pentachlorophenol
Dursban: see chlorpyrifos
Dutch Liquid: see ethylene dichloride
Duter: see fentin hydroxide
DW 3418: see cyanazine
D-Weed-O: see bromacil
3-D Weedone: see 2,4-D

Dwell: see etridiazol
Dybar: see fenuron
Dycarb: see bendiocarb
Dyclomec 4G: see dichlobenil
Dyclomec G2: see dichlobenil
Dyfonate: see fonofos
Dylox: see trichlorfon
Dymec: see 2,4-D
Dymid: see diphenamid
Dyna-Form: see formaldehyde
Dyna-carbyl: see carbaryl
Dynex: see diuron
Dynone-II: see dinex
Dyrene: see anilazine
E-103: see tebuthiuron
E-1059: see demeton
E1-171: see fluridone
E-3314: see heptachlor
Earthcide: see PCNB
Eastern States Duocide: see warfarin
Ecothrin: see tetramethrin
Ectiban: see permethrin
Ectoral: see ronnel
Ectrin: see fenvalerate
Edabrom EC: see ethylene dibromide
EDB: see ethylene dibromide
E-D-Bee: see ethylene dibromide
EDC: see ethylene dichloride
Edesol: see ethylene dibromide
Edge: see ethalfluralin
EDIFENPHOS
Effix: see flamprop-isopropyl
Effusan: see dinitrocresol
Efuzin: see dodine
Ekagom TV: see thiram
Ekalux: see fenthion
Ekamet: see etrimfos
Ekamet ULV: see etrimfos
Ekanon: see disulfoton
Ekatin TD: see disulfoton
Eksmin: see permethrin
Ektafos: see dicrotophos
El 110: see benefin
EL-119: see oryzalin
EL-161: see ethalfluralin
EL-222: see fenarimol
EL-614: see bromethalin
Elastrel: see dichlorvos
Elcar: see Nuclear polyhedrosis virus
Eldrinol: see dieldrin
Elgetol 30: see dinitrocresol
Elgetol 318: see dinoseb
Elmpro: see thiabendazole
Elocron: see dioxacarb
Elset: see isoxaben
Elvaren: see dichlofluanid
Elvaron: see dichlofluanid
Embafume: see methyl bromide

Embathion: see ethion
Embutox: see 2,4-DB
Emerald green: see copper acetoarsenite
emetic: see paraquat
EmmatosAC 4049: see malathion
Emo-Nik: see nicotine
emulsamine-E3: see 2,4-D
Endocel: see endosulfan
ENDOD
Endosan: see binapacryl
ENDOSULFAN
Endothal: see endothall
endothal: see endothall
ENDOTHALL
Endox: see dicamba, mecoprop
3,6-endoxohexahydrophthalicacid: see endothall
Endrix: see endrin
Endricol: see endrin
ENDRIN
Endrotox: see endrin
Endyl: see carbophenothion
Enide: see diphenamid
Enpar: see endrin
Entex: see fenthion
Envel: see endrin
Envert 171: see 2,4-D, dichlorprop
EP 30: see pentachlorophenol
EP 332: see formetanate hydrochloride
EP-333: see chlordimeform
EP-475: see desmedipham
Eparen: see dichlofluanid
ephirsulphonate: see chlorfenson
EPICHLOROHYDRIN
Epigon: see permethrin
EPN
1,2-epoxyethane: see ethylene oxide
Eptam: see EPTC
EPTC
Equigard: see dichlorvos
Equigel: see dichlorvos
Equino-Acid: see trichlorfon
Equitdazin: see carbendazim
Eradicane E: see dichlormid, EPTC
Eradicane G: see dichlormid, EPTC
Erbitox Gratto: see dicamba, MCPA-sodium
Erbitox LV: see 2,4-D
ERBON
Erbon R: see erbon
Esbiol: see S-bioallethrin
esbiothrin: see allethrin
esdepallethrin: see S-bioallethrin
Estanox: see toxaphene
Esteron: see silvex
Esteron 44: see 2,4-D
Esteron 6E: see 2,4-D
Esteron 76BE: see 2,4-D
Esteron 99: see 2,4-D
Estonmite: see chlorfenson

Estrella: see amitraz
Estrosel: see dichlorvos
Estrosol: see dichlorvos
ETACELASIL
etephon
Ethanaminium: see chlormequat chloride
ETHALFLURALIN
Ethanox: see ethion
ethazole: see etridiazol
ETHEPHON
Etheverse: see ethephon
Ethimeton: see disulfoton
ETHIOFENCARB
ethiofencarp: see ethiofencarb
Ethiol: see ethion
ETHIOLATE
ETHION
ethiophencarp: see ethiofencarb
ETHIOZIN
ETHIRIMOL
Ethisul: see metiram
Ethodon: see ethion
ETHOFUMESATE
ethohexadiol: see ethyl hexanediol
ETHOPROP
ethoprophos: see ethoprop
2-ethoxy-2,3-dihydro-3,3-dimethyl-5-
    benzofuranylmethanosulphonate: see
    ethofumesate
6-ethoxy-1,2-dihydro-2,2,4-trimethylquinoline: see
    ethoxyquin
4-[1-[2-(ethoxyethoxy)ethoxy]ethoxy]-1,2-
    methyldioxybenzene(8Cl): see acetaldehyde
O-6-ethoxy-2-ethylpyrimidin-4-yl O,O-
    dimethylphosphorothioate: see etrimfos
O-(6-ethoxy-2-ethyl-4-pyrimidinyl) O,O-
    dimethylphosphorothioate: see etrimfos
ETHOXYQUIN
ethoxyquine: see ethoxyquin
5-ethoxy-3-trichloromethyl-1,2,4-thiadiazole: see
    etridiazole
Ethrel: see ethephon
2-(ethylamino)-4-(isopropylamino)-6-(methylthio)-s-
    triazine: see ametryn
ETHYLAN
ethyl analogue of thiram: see disulfiram
ethylazinphos: see azinphosethyl
ethylbenzene: see rotenone
S-ethylbis(2-methylpropyl)carbamothiote: see
    butylate
(±)-ethyl 2-[4-[(6-chloro-2-benoxazolyl)oxy]phenoxy]
    propanoate: see fenoxaprop-ethyl
S-ethylcyclohexylethylcarbamothioate: see cycloate
S-ethylcyclohexylethylthiocarbamate: see cycloate
ethyl 4,4'-dichlorobenzilate: see chlorobenzilate
S-ethyldiethylcarbamothioate: see ethiolate
S-ethyldiethylthiocarbamate: see ethiolate

S-ethyldiisobutylthiocarbamate: see butylate
O-ethyl-S,S-diphenyl phosphorodithioate: see
    edifenphos
O-ethyl S,S-dipropylphosphorodithioate: see
    ethoprop
S-ethyldipropylthiocarbamate: see EPTC
S-ethyldipropylthiolcarbamate: see EPTC
ethylene aldehyde: see acrolein
ethylenebisdithiocarbamate: see mancozeb
ethylene chloride: see ethylene dichloride
ETHYLENE DIBROMIDE
ETHYLENE DICHLORIDE
1,1'-ethylene-2,2'-dipyridyllium dibromide: see
    diquat
ETHYLENE OXIDE
ethylene tetrachloride: see tetrachloroethylene
ethylene thiourea: see amobam, mancozeb, maneb,
    metiram, nabam, zineb
S-ethyl-N-ethylthiocyclohexanecarbamate: see
    cycloate
ETHYL FORMATE
ethylformic acid: see propionic acid
Ethyl Guthion: see azinphosethyl
ETHYL HEXANEDIOL
2-ethyl-1,3-hexanediol: see ethyl hexanediol
ethyl-m-hydroxycarbanilate carbanilate (ester): see
    desmedipham
ethyl mercaptan: see EPTC
2-ethyl-mercaptomethyl-phenyl-N-methylcarbamate:
    see ethiofencarb
ethyl methanoate: see ethyl formate
N-ethyl-N'-(1-methylethyl)-6-(methylthio)-1,3,5-
    triazine-2,4-diamine: see ametryn
ethyl 3-methyl-4-(methylthio)phenyl(1-
    methylethyl)phosphoramidate: see fenamiphos
N-ethyl-N-(2-methyl-2-propenyl)-2,6-dinitro-4-
    (trifluoromethyl)benzeneamine: see
    ethalfluralin
N-[3-(1-ethyl-1-methylpropyl)-5-isoxazolyl]-2,6-
    dimethoxybenzamide: see isoxaben
O-ethyl)-4-(methylthio)phenyl S-
    propylphosphorodithioate: see sulprofos
ethyl 4-methylthio-m-
    tolylisopropylphosphoramidate: see
    fenamiphos
O-ethyl O-p-nitrophenyl phenylphosphonothioate:
    see EPN
ethyl p-nitrophenylthionobenzenephosphonate: see
    EPN
ethyl[3[[(phenylamino)carbonyl]oxy]phenyl]carbamate:
    see desmedipham
O-ethyl S-phenylethylphosphonodithioate: see
    fonofos
O-ethyl S-phenyl(RS)-ethylphosphonodithioate: see
    fonofos
N-(1-ethylpropyl)-N-nitroso-3,4-dimethyl-2,6-
    dinitrobenzamine: see N-nitrosopendimethalin

S-2-ethylsulfinylethyl O,O-dimethylphosphorothiote: see oxydemeton-methyl

3-ethylthio-4-amino-6-*tert*-butyl-1,2,4-triazine-5-one: see ethiozin

2-ethylthio-4,6-bis(isopropylamino)-s-triazine: see dipropetryn

6-(ethylthio)-N,N'-bis(1-methylethyl)-1,3,5-triazine-2,4-diamine: see dipropetryn

S-2-(ethylthio)ethylphosphorodithioate: see disulfoton

S-2-(ethylthio)ethylphosphorothioate: see demeton

2-[(ethylthio)methyl]phenylmethylcarbamate: see ethiofencarb

α-ethylthio-o-tolylmethylcarbamate: see ethiofencarb

ethyl 3-trichloromethyl-1,2,4-thiadiazol-5-yl ether: see etridiazole

ETO: see ethylene oxide

Etox: see ethylene oxide

ETRIDIAZOL

ETRIMFOS

Etrolene: see ronnel

ETU: see ethylene thiourea

Euparen: see dichlofluanid

Euparene: see dichlofluanid

Euparin: see dichlofluanid

Eurex: see cycloate

Evercide: see permethrin

Evik: see ametryn

Exbiol: see allethrin

Excel: see fenoxaprop-ethyl

EXD

Exotherm: see chlorothalonil

Exotherm Termil: see chlorothalonil

Expar: see permethrin

Experimental Fungicide 658: see copper zinc chromate

Exporsan: see bensulide

Extar A: see dinitrocresol

Extra: see MCPA

Extrazine: see atrazine, cyanazine

Extrin: see fenvalerate

E-Z-off D: see DEF

Falisilvan: see fenuron

Fall: see sodium chlorate

Fallow Master: see dicamba, glyphosate

Famfos: see famphur

Famid: see dioxacarb

Famophos: see famphur

FAMPHUR

Far-Go: see triallate

Farmacel: see chlormequat chloride

Farmon: see dichlorprop, MCPA

Farmon Condox M: see dicamba, mecoprop

Farmon PDQ: see diquat, paraquat

Fasco Fascrat Powder: see warfarin

Fasco Gransil-X: see bromacil

Fasco-Terpene: see toxophene

Faso Wy-Hoe: see chlorpropham

Fatal: see warfarin

Fatsco Ant Poison: see sodium arsenate [I]

FBHC: see benzene hexachloride

FDA 1446: see allethrin

FENAC

FENAMINOSULF

FENAMIPHOS

FENARIMOL

Fenasal: see niclosamide

Fenatrol: see fenac

fenbutatin-oxide: see tin

Fence Rider: see 2,4,5-T

fenchlorphos: see ronnel

Fenidim: see fenuron

fenidin: see fenuron

Fenimine: see 2,4-D

Fenitox: see fenitrothion

FENITROTHION

Fenkill: see fenvalerate

Fenom: see cypermethrin

Fenoprop: see silvex

fenothrin: see phenothrin

Fenoxan: see clomazone

FENOXAPROP-ETHYL

FENPROPATHRIN

fenpropathrine: see fenpropathrin

Fenstan: see fenitrothion

FENSULFOTHION

FENTHION

fentin acetate: see triphenyltin acetate

fentin hydroxide: see tin

Fenulon: see fenuron

fenulon: see fenuron

FENURON

FENVALERATE

fenvalethrin: see fenvalerate

Fenzol: see fenarimol

FERBAM

ferbame: see ferbam

Ferbert: see ferbam

Fermate: see ferbam

Fermide 850: see thiram

Fermocide: see ferbam

Fernasan: see thiram

Ferncot: see copper oxychloride

Fernesta: see 2,4-D

Fernimine: see 2,4-D

Fernoxone: see 2,4-D

Ferradow: see ferbam

ferriamide: see mirex

ferric dimethyldithiocarbamate: see ferbam

FERROUS SULFATE

Fersone: see 2,4-D

Fettel: see dicamba, mecoprop, triclopyr

Ficam: see bendiocarb

Filariol: see bromophos-ethyl

Finesse: see chlorsulfuron
Fintrol: see antimycin A
Fisons B25: see barban
FLAC: see calcium arsenate
FLAMPROP-ISOPROPYL
flamprop-M-isopropyl: see flamprop-isopropyl
Flectron: see cypermethrin
Flex: see fomesafen
Flexidor: see isoxaben
Flocron: see dioxacarb
Flordimex: see ethephon
Florel: see ethephon
Florencol: see flurecol-butyl
flour sulfur: see sulfur
flowers of sulfur: see sulfur
Flu: see fluenethyl
FLUAZIFOP-BUTYL
FLUCHLORALIN
FLUCYTHRINATE
FLUENETHYL
Fluenyl: see fluenethyl
FLUFENOXURON
FLUOMETURON
Fluorakil 100: see fluoroacetamide
FLUOROACETAMIDE
2-fluroacetamide: see fluoroacetamide
fluorobutylstannane: see tributyltin fluoride
fluorocitrate: see fluoroacetamide, sodium
      fluoroacetate
FLUORODIFEN
fluorodiphen: see fluorodifen
2-fluoroethyl 4-biphenylacetate: see fluenethyl
2-fluoroethyl(4-biphenyl)acetate: see fluenethyl
2-fluoroethyl[1,1'-biphenyl]-4-acetate: see fluenethyl
FLURECOL-BUTYL
flurecol-n-butylester: see flurecol-butyl
flurenol: see flurecol-butyl
flurenol-n-butylester: see flurecol-butyl
FLURIDONE
FLUSILAZOLE
Flutrin: see dimethoate, flucythrinate
FLUVALINATE
τ-FLUVALINATE
Fly Bait Grits: see Bomyl
Fly Fighter: see dichlorvos
Fly-Die: see dichlorvos
FM 10242: see carbofuran
FMC 1240: see ethion
FMC 17370: see resmethrin
FMC 249: see allethrin
FMC 33297: see permethrin
FMC 5273: see piperonyl butoxide
FMC 5462: see endosulfan
FMC 54800: see bifenthrin
FMC-57020: see clomazone
FMC 9044: see binapacryl
FMC 9102: see metiram
FMC 9260: see tetramethrin

Focal: see carbendazim
Fogard: see atrazine
Fogard L: see atrazine
Fogard S: see atrazine, simazine
Folbex: see chlorobenzilate
Folcid: see captafol
Folcid: see captaphol
Folcord: see cypermethrin
Folidol E-65: see parathion
Folimat Combi: see methyl, parathion omethoate
Folimat T: see omethoate, tetradifon
Folithion: see fenitrothion
Folnit: see folpet
Folosan: see PCNB
Folpan: see folpet
FOLPET
Folplan: see folpet
Folprame: see copper, oxychloride folpet
Folsystem: see folpet
Foltan: see folpet
Foltapet: see captafol, folpet
Foltazip: see folpet
Foltene: see folpet
Foltimil: see copper, sulfate folpet
FOMESAFEN
FONOFOS
For-Cop-80NC: see copper ammonium carbonate
For-Mal: see malathion
FORMALDEHYDE
FORMETANATE HYDROCHLORIDE
formic aldehyde: see formaldehyde
formic acid: see formaldehyde
3-[[3-(formylamino)-2-hydroxybenzoyl]amino]-8-
      hexyl-2,6-dimethyl-4,9-dioxo-1,5-dioxonan-7-yl 3-
      methylbutanoate: see antimycin A
N-formyl-chloro-o-toluidine
formyl trichloride: see chloroform
For-Synm: see resmethrin
Foray: see Bacillus thuringiensis var. kurstaki
Forca: see tefluthrin
Force: see tefluthrin
Fore: see mancozeb
Foremost Weed-Away: see bromacil
Foremost Weed-Buster: see bromacil
Foremost Weed-Zapper: see bromacil
Forlin: see lindane
Formagene: see paraformaldehyde
Formalin: see formaldehyde
Formalina: see formaldehyde
Format: see clopyralid
Formula 40: see 2,4-D
Formula 40: see sulfoTEPP
Formula GH-200: see sulfoTEPP
N-formyl-chloro-o-toluidine: see chlordimeform
Forron: see 2,4,5-T
Forte: see dichlobenil, simazine
Fortrol: see cyanazine
Forturf: see chlorothalonil

Forza: see tefluthrin
FOSAMINE AMMONIUM
FOSETYL-AL
Fos-Fall "A": see DEF
fosfamid: see dimethoate
Fostion MM: see dimethoate
Fostox Metil: see methyl parathion
Fosuex: see TEPP
Foumarin: see coumafuryl
Framed: see simazine
Fratol: see sodium fluoroacetate
French green: see copper acetoarsenite
Freon Genetron: see chlorofluorocarbons
Frucote: see 2-butanamine
Fruitdo: see oxine-copper
Fruitone: see 4-CPA
Fruitone A: see 2,4,5-T
Fruitone T: see silvex
Frumin G: see disulfoton
Frumin-Al: see disulfoton
FT 2M: see Bordeaux mixture, mancozeb
Fubol: see mancozeb
Fubol: see mancozeb, metalaxyl
Fuclasin Ultra: see ziram
fuel oil: see petroleum oils
Fuklasin: see ziram
Fuller's Earth: see diatomaceous earth
fumarin
Fumasol: see coumafuryl
Fumazone: see DBCP
Fumetobac: see nicotine
Fumigant-1: see methyl bromide
Fumo-Gas: see ethylene dibromide
Fundal 500: see chlordimeform
Fundal Forte: see chlordimeform, formetanate
    hydrochloride
Fundex: see chlordimeform
Fungchex: see mercuric chloride, mercurous
    chloride
Fungi-Bordo: see Bordeaux mixture
Fungi-Rhap: see copper oxide
Fungicide-531: see cadmium-calcium-copper-zinc-
    chromate complex
Fungifen: see pentachlorophenol
Fungitrol: see folpet
Fungo: see thiophanate methyl
Fungol: see pentachlorophenol
Furacarb: see carbofuran
Furadan: see carbofuran
Furadan 15G: see carbofuran
Furado: see mancozeb
3-[1-(2-furanyl)-3-oxobutyl]-4-hydroxy-2*H*-1-
    benzopyran-2-one: see coumafuryl
Furloe: see chlorpropham
Furmarin: see coumafuryl
furmarin: see coumafuryl
Furore: see fenoxaprop-ethyl

3-[1-(2-furyl)-3-oxobutyl]-4-hydroxycoumarin: see
    fumarin
Fusilade: see fluazifop-butyl
Fusilade 2000: see fluazifop-butyl
Fusilade Five: see fluazifop-butyl
Fusilade Super: see fluazifop-butyl
Fussol: see fluoroacetamide
FW-293: see dicofol
Fydulan G: see dichlobenil, sodium salt of dalapon
Fydulex G: see dichlobenil, sodium salt of dalapon
Fydusit: see bromacil dichlobenil
Fyfanon: see malathion
Fytolan: see copper oxychloride
Fytospore: see cymoxanil, mancozeb
G 23992: see chlorobenzilate
G 24163: see chloropropylate
G 24480: see diazinon
G 34360: see desmetryn
G-25: see chloropicrin
G-25804: see chloranil
G-31435: see prometon
G-34162: see ametryn
G-4: see dichlorophen
G-444-E: see chloranil
G-52 and 56: see TEPP
gamma isomer of benzenehexachloride: see
    lindane
gamma isomer of BHC: see lindane
gamma-BHC: see lindane
gamma-HCH: see lindane
Galar: see bromacil, diuron
Galaxy: see acifluorfen, bentazon
Galben M: see benalaxyl, mancozeb
Galecron: see chlordimeform
Galgo-quat: see paraquat dichloride
Galipan: see benazolin
Gallery: see isoxaben
Galtak: see benazolin
Gamid: see dioxacarb
Gamit: see clomazone
Gammaspra: see lindane
Gammcide: see lindane
Gamonil: see carbaryl
Ganocide: see drazoxolon
Gard-Star: see permethrin
Garden Fume: see DBCP
Gardentox: see diazinon
Garlon: see silvex
Garlon: see silvex, triclopyr
Garlon 3A: see triclopyr
Garlon 4: see triclopyr
Garrathion: see carbophenothion
Garvox: see bendiocarb
Gatinon: see benzthiazuron
Gatnon: see benzthiazuron
GC 1189: see chlordecone
GC 1293: see mirex

GC 3707: see Bomyl
GC 6506
Gearphos: see methyl parathion
Gebutox: see dinoseb
Geigy 30,027: see atrazine
Geigy 338: see chlorobenzilate
Geigy LO-V Brush Killer No. 300: see 2,4-D
Gemini: see chlorimuron, linuron
Geniphene: see toxaphene
Genite 883: see chlorfenson
Genitox: see DDT
Genthion: see parathion
Geofos: see parathion
Geonter: see terbacil
Germain's: see carbaryl
Gerstley: see borax
Gesafram: see prometon
Gesafram 50: see prometon
Gesapax: see ametryn
Gesapax H: see ametryn, 2,4-D
Gesapax combi: see ametryn, atrazine
Gesapon: see DDT
Gesaprim: see atrazine
Gesaprim D: see atrazine, 2,4-D
Gesaprim S: see atrazine, simazine
Gesaprim combi: see atrazine terbutryn
Gesaran 2079: see methoproptryne, simazine
Gesarex: see DDT
Gesarol: see DDT
Gesatop: see simazine
Gexane: see lindane
Gexane: see benzene hexachloride
Gioallolio: see dinitrocresol
Giror: see amitrole, ammonium thiocyanate,
    paraquat
Glazd Penta: see pentachlorophenol
Glean: see chlorsulfuron
Glean 20DF: see chlorsulfuron
Glean T: see chlorsulfuron
Glean TP: see chlorsulfuron
glycophene: see iprodione
GLYPHOSATE
glyphosate trimesium: see glyphosate
Go-Go-San: see pendimethalin
Goal: see oxyfluorfen
Gold: see fenuron
Gold Crest: see diphacinone
Gopha + Rid: see zinc phosphide
Gramazine: see paraquat simazine
Gramevin: see dalapon
Gramixel: see diuron, paraquat
Gramocel: see diuron, paraquat
Gamonol: see monolinuron, paraquat dichloride
Gramoxone: see MCPA, paraquat dichloride
Gramoxone Methyl Sulfate: see paraquat
    bis(methylsulfate)
Gramoxone Special: see paraquat dichloride
Gramuron: see diuron, paraquat

Grandal: see niclosamide
Grandslam: see methiocarb
Granosan: see ethylene dibromide
Granosan: see carbon tetrachloride, ethylene
    dichloride
Granovil 75: see ethylene dibromide
Graslam: see asulam, MCPA, mecoprop
Grass-B-Gone: see fluazifop-butyl
Grassland Weedkiller: see benazolin
Grazon: see clopyralid
Green Cross Amine 80: see 2,4-D
Green Cross Warble Powder: see rotenone
Green Cross kerbam: see ferbam
Green Light: see DBCP
green vitriol: see ferrous sulfate
Grindor: see atrazine ethalfluralin
Gro-Tone Liquid: see copper linoleate
Groundhog: see amitrole, diquat, paraquat,
    simazine
Grug Attack: see milky spore disease
Grundier Arbezol: see pentachlorophenol
GS 13005: see methidathion
GS 14260: see terbutryn
GS 19851: see bromopropylate
GS-16068: see dipropetryn
Gusathion: see azinphosmethyl
Gusathion A: see azinphosethyl
Guthion: see azinphosmethyl
Gy-Phene: see toxaphene
Gypsine: see lead arsenate
Gyron: see DDT
Gy-TET40: see TEPP
H-170: see bentranil
H-22234: see diethatyl
H-722: see credazine
Habco Bromex: see bromacil
Habco Hychlor: see bromacil
Hache Uno Super: see fluazifop-butyl
HAG 107: see tralomethrin
Haipen: see captafol
Haipen: see captaphol
Halizam: see metaldehyde
Hanane: see dimefox
Hard-Hitter: see permethrin
Harness: see acetochlor
Harness: see bromoxynil
HCB: see hexachlorobenzene
HCCH: see benzene hexachloride
HCDD: see heptachlorodibenzo-p-dioxin,
    hexachlorodibenzo-p-dioxin
HCH: see benzene hexachloride
HCN: see hydrogen cyanide
Heapoudre: see benzene hexachloride
Hedonal: see 2,4-D
Hedonal DP: see dichlorprop
Hedonal-MCPP: see mecoprop
Helarion: see metaldehyde
Helene "Clean Up": see bromacil, sodium chlorate

Helmiantin: see niclosamide
Hemoxone: see dicamba, dichlorprop, MCPA
HEOD: see dieldrin
Heolite: see anthraquinone
HEPTACHLOR
heptachlor epoxide: see heptachlor
heptachlore: see heptachlor
heptachlorodibenzo-p-dioxin: see
    pentachlorophenol
heptachlorodibenzofuran
2,3,4,5,6,7,7-heptachloro-1a,1b,5,5a,6,6a-hexahydro-
    2,5-methano-2H-indeno[1,2-b]oxirene: see
    heptachlor epoxide
heptachlorotetrahydro-4,7-methanoindene: see
    heptachlor
1,4,5,6,7,8,8-heptachloro-3a,4,7,7a-tetrahydro-4,7-
    methanoindene: see heptachlor
Heptagran: see heptachlor
Heptamul: see heptachlor
Herald: see fenpropathrin
Herb-All: see MSMA
Herbadox: see pendimethalin
Herbaron B: see dicamba, mecoprop, triclopyr
Herbazin: see simazine
Herbazolin: see benazolin
Herbex: see simazine
Herbicide 273: see endothall
Herbicide 282: see endothall
Herbicide 976: see bromacil
Herbizan #5: see EXD
Herbizole: see amitrole
Herbogil: see dinoterb
Herbogil Liquid D: see dinoterb
Herbon Yellow: see chlorpropham, fenuron
Herbox: see simazine
Herboxy: see simazine
Hercon Disrupt Gypsy Moth: see disparlure
Hercon Luretape Gypsy Moth: see disparlure
Hercules 14503: see dialifor
Hercules 22234: see diethatyl
Hercules 9573: see terbucarb
Hercules AC 528: see dioxathion
Herkol: see dichlorvos
Herrifex DS: see mecoprop
Herrisol: see dicamba, MCPA
Hexablanc: see benzene hexachloride
hexabutyldistannoxane: see tributyltin oxide
hexachlor: see benzene hexachloride
hexachloran: see benzene hexachloride
HEXACHLOROBENZENE
1,2,3,4,5,6-hexachlorobenzene: see
    hexachlorobenzene
1,2,3,4,5,6-hexachlorocyclohexane: see benzene
    hexachloride
1,2,3,4,5,6-hexachlorocyclohexane: see lindane
hexachlorocyclopentadiene: see chlordane
2a,2,2,4,5,5a-hexachlorodecahydro-2,4,6-metheno-

2H-cychlopenta[4,5]pentaleno[1,2-]oxirene: see
    photodieldrin
hexachlorodibenzo-p-dioxin: see
    pentachlorophenol
hexachlorodibenzofuran: see pentachlorophenol,
    2,4,6-trichlorophenol
HEXACHLOROPHENE
hexachloroepoxyoctahydro-endo-endo-
    dimethanonapthalene: see endrin
hexachloroepoxyoctahydro-endo,exo-
    dimethanonaphthalene: see dieldrin
hexachlorohexahydro-endo-exo-
    dimethanonaphthalene: see aldrin
1,2,3,4,10,10,-hexachloro-1,4,4a,5,8a-hexahydro-1,4-
    endo-exo-5,8-dimethanonaphthalene: see aldrin
6,7,8,9,10,10-hexachloro-1,5,5α,6,9,9α-hexahydro-6,9-
    methano-2,4,3-benzodioxathiepin 3-oxide: see
    endosulfan
hexachlorohexhydromethano-2,4,3-
    benzodioxathiepin oxide: see endosulfan
1,2,3,4,10,10-hexachloro-6,7-epoxy-1,4,4a,5,6,7,8,8a-
    octahydro-1,4-endo,endo-5,8-
    dimethanonapthalene: see endrin
1,2,3,4,10,10-hexachloro-6,7-epoxy-1,4,4a,5,6,7,8,8a-
    octahydro-1,4-endo-exo-5,8-
    dimethanonaphthalene: see dieldrin
Hexadrin: see endrin
hexaethyl tetraphosphate: see TEPP
Hexaferb: see ferbam
Hexafor: see benzene hexachloride
hexahydrobenzene: see cyclohexane
hexakis(β,β-dimethylphenethyl)tin)oxide: see
    fenbutatin-oxide
hexakis(2-methy-2-phenylpropyl)distannoxane: see
    fenbutatin-oxide
hexamethylene: see cyclohexane
Hexamul: see benzene hexachloride
hexanapthene: see cyclohexane
Hexasul: see sulfur
Hexathane: see zineb
Hexathion: see carbophenothion
Hexavin: see carbaryl
HEXAZINONE
Hexazir: see ziram
Hexide: see hexachlorophene
Hex-Nema: see dichlofenthion
Hexthir: see thiram
Hexyclan: see benzene hexachloride
hexylthiocarbam: see cycloate
HHDN: see aldrin
Hi-Dep: see 2,4-D
Hi-Yield Dessicant H-10: see arsenic acid
Hibor: see bromacil
Hibrom: see naled
Hico CCC: see chlormequat chloride
Higalcoton: see fluometuron
Hilco X: see bromacil

Hinosan: see edifenphos
Hitrun: see thiophanate-methyl, vinclozolin
Hizarocin: see cycloheximide
HK-80 Weed Killer: see bromacil
Hododrex: see dieldrin
Hoe 17411: see carbendazim
Hoe 2671: see endosulfan
Hoe 2727: see monolinuron
HOE 2784: see binapacryl
HOE 2810: see linuron
Hoegrass: see diclofop-methyl
Hoelon: see diclofop-methyl
Holtox: see atrazine, cyanazine
Hopkins Allyl Alcohol: see allyl alcohol
Hopkins Urox-'B': see bromacil
Horbadox: see pendimethalin
Hormatox: see dichlorprop
Hormocel: see chlormequate chloride
Hormosalt: see 2,4-D
Hormotox: see 2,4-D
Hormotuho: see MCPA
Hox 1901: see ethiofencarb
HpCDD: see heptachlorodibenzo-p-dioxin
HRS-16: see dienochlor
Hungazin DT: see simazine
HxCDD: see hexachlorodibenzo-p-dioxin
Hyamine 3500: see benzalkonium chloride
Hyban: see dicamba, mecoprop
Hydon: see bromacil, picloram
Hydout: see endothall
HYDRAMETHYLNON
hydrated cupric oxide: see copper hydroxide
hydrazoic acid: see potassium azide, sodium azide
hydrochloric acid: see hydrogen chloride
hydrocyanic acid: see hydrogen cyanide
hydrocyanic acid, sodium salt: see sodium cyanide
Hydrocyclin: see oxytetracycline hydrochloride
hydrogen bromide
hydrogen chloride: see chlordane, dienochlor,
    heptachlor, lindane
hydrogen cyanide: see calcium cyanide, sodium
    cyanide
hydrogen sulfide: see calcium polysulfide, dazomet
Hydrothal 47: see endothall
Hydrothol 191: see endothall
hydroxybenzene: see phenol
4-hydroxy-3,5-diiodobenzonitrile: see ioxynil
9-hydroxyfluorene-9-carboxylic acid butyl ester: see
    flurecol-butyl
4-hydroxy-3-[3-oxo-1-(2-furyl)butyl] coumarin: see
    coumafuryl
N-hydroxymethyl methiocarb sulfoxide: see
    methiocarb
1-hydroxy-2(1H)-pyridinethione, sodium salt (CA):
    see sodium omadine
2-hydroxyl-2,3-propanetricarboxylic acid copper
    salt: see copper citrate

4-hydroxy-2,5,6-trichloroisophtalonitrile: see
    chlorothalonil
hydroxytriphenylstannane: see fentin hydroxide
Hygrass: see dicamba, mecoprop
Hylemox: see ethion
Hymec: see mecoprop
hypochlorous acid: see trichloroiosocyanuric acid
Hyprone: see dicamba, MCPA
Hysan "600" Weed Killer: see bromacil
Hysward: see dicamba, MCPA
Hytrol: see amitrole, 2,4-D, diuron, simazine
Hytrol O: see cyclohexanone
Hyvar L: see bromacil
Hyvar X: see bromacil
Hyvar X-L (with bromacil-lithium): see bromacil
Ibertox Pasa: see dinitrocresol
Icone: see λ-cyhalothrin
idall-zinc: Rumetan
Igran: see terbutryn
IKI-7899: see chlorfluazuron
Ikurin: see AMS
Illoxan: see diclofop-methyl
Illoxol: see dieldrin
IMAZABETHABENZ
IMAZAPYR
2-imidazolidinethione: see ethylene thiourea
Impact: see hydramethylnon
Imperator: see permethrin
Imperator: see cypermethrin
Inemacury: see fenamiphos
infusorial earth: see diatomaceous earth
Insect & Mite Houseplant Mist: see methoprene
Insectiban: see permethrin
Insectipen: see cyfluthrin
Insectophene: see endosulfan
"Insect powder": see pyrethrum
Insectrin: see endrin
Insyst-D: see disulfoton
Inverton 245: see 2,4,5-T
Iometan: see niclosamide
IOXYNIL
Ipersan: see trifluralin
IPRODIONE
iron protosulfate: see ferrous sulfate
iron(II) sulfate: see ferrous sulfate
iron (2+)sulfate: see ferrous sulfate
iron tris(dimethyldithiocarbamate): see ferbam
iron vitriol: see ferrous sulfate
Isathrine: see resmethrin
ISAZOPHOS
Iscobrome: see methyl bromide
Iscobrome D: see ethylene dibromide
Iscothane: see dinocap
Iso-Cornox: see mecoprop
Isocarb: see propoxur
isodiazinon: see diazinon
ISOFENPHOS

Kilumal: see fenpropathrin
Kilval: see vamidothion
Kilvar: see vamidothion
Kisparmone: see disparlure
Kiwi Luster: see dichloran
Klartan: see τ-fluvalinate
Klartan: see τ-fluvalinate
Klorex: see sodium chlorate
Knave: see disulfoton
Knew: see copper linoleate
Knock Out: see isoxaben
Knockmate: see ferbam
Knox Out 2FM: see diazinon
Knoxweed: see dinoseb
K-O: see deltamethrin, esbiothrin, piperonyl
    butoxide
Koban: see etridiazol
K-Obiol: see deltamethrin piperonyl butoxide
Kobu: see PCNB
Kobutol: see PCNB
Kocide: see copper hydroxide
Kocide 101: see copper hydroxide
Kocide 404S: see copper hydroxide
Kolo-100: see dichlone sulfur
Kolofog: see sulfur
Kolospray: see sulfur
Koltar: see oxyfluorfen
Komeen: see copper-ethylenediamine complex
Konker: see carbendazim, vinclozolin
Kontal: see niclosamide
Kop Karb: see basic copper carbonate
Kop-Mite: see chlorobenzilate
Kop-Thiodan: see endosulfan
Kop-Thion: see malathion
KopFume: see ethylene dibromide
Kopsol: see DDT
Korlan: see ronnel
K-Otek: see deltamethrin
Kothrin: see deltamethrin
K-Othrin: see deltamethrin
Kotol: see benzene hexachloride
Koyoside: see cryolite
K Pin: see picloram
K-Pool: see copper tea complex
Krater: see asulam, diuron
Krenite: see dinitrocresol
Krenite Brush Control Agent: see fosamine
    ammonium
Krenite S: see fosamine ammonium
Kripid: see antu
Kroma-Clor: see cadmium succinate
Krovar I: see bromacil, diuron
Krovar II: see bromacil, diuron
Krumkil: see coumafuryl
Kryocide: see cryolite
krysid: see antu
ksylene: see xylene
K-Tea: see copper tea complex

KUE 13032c: see dichlofluanid
Kumulus S: see sulfur
Kupratsin: see zineb
Kuprite: see copper oxide
Kuron: see silvex
Kurosal: see silvex
Kusakira: see credazine
Kusatol: see sodium chlorate
Kwell: see lindane
Kwit: see ethion
Kyadrin: see dieldrin
Kylar: see daminozide
Kylar-85: see daminozide
Kypchlor: see chloradane
Kypfarin: see warafarin
Kypfos: see malathion
Kypman 80: see maneb
Kypzin: see zineb
Labilite: see thiophanate methyl
Laddok: see atrazine, bentazon
Lambast: see butachlor
Lambrol: see fluenethyl
Lancer: see flamprop-isopropyl
Landmaster: see 2,4-D, glyphosate
Lanex: see fluometuron
Lanirat: see bromadiolone
Lannate: see methomyl
Laptran: see ditalimfos
Lariat: see alachlor
Larvacide: see chloropicrin
Larvadex: see cyromazine
Larvatrol: see *Bacillus thuringiensis* var. *kurstaki*
Lasso: see alachlor, atrazine
Laurel: see trifluralin
laurylguanidine acetate: see dodine
Lauxtol: see pentachlorophenol
Lauxtol A: see pentachlorophenol
Lawn Groom: see 2,4-D
Lawn-Keep: see 2,4-D
Lawnsman: see dicamba, dichlorprop, MCPA
Lazo: see alachlor
LE 79-519: see permethrin
LE 79600: see cypermethrin
lead acid arsonate: see lead arsenate
lead arsenate: see arsenic
Lebaycid: see fenthion
Legumex: see MCPA
Legumex Estra: see benazolin
Leivasom: see trichlorfon
Lektran: see ethiozin
Lemonene: see diphenyl
Lenetemul: see 2,4-D, dichlorprop, MCPA, MCPP
Lepister: see chlorpyrifos, flucythrinate
Lepit: see chlorophacinone
Lepton: see leptophos
LEPTOPHOS
Leptox: see *Bacillus thuringiensis* var. *kurstaki*
Lesan: see fenaminosulf

Lethalaire: see parathion
Lethalaire G-57 Aerosol Insecticide: see sulfoTEPP
Lethalaire G-58: see chlorfenson
Lethox: see carbophenothion
Lexone Sencor: see metribuzin
Lextra: see linuron, trifluralin
Ley-Cornox: see benazolin, 2,4-DB, MCPA
Leymin: see benazolin
Lico-40: see TEPP
Lignasan: see carbendazim
Lignum: see atrazine, dalapon
Lihocin: see chlormequat chloride
lime sulfur: see calcium polysulfide
Limit: see CDAA
d-LIMONENE
Lindacol: see benzene hexachloride
Lindafor: see lindane
Lindagam: see lindane
LINDANE
Lindaterra: see lindane
Line Rider: see 2,4,5-T
Lintex: see niclosamide
Lintox: see lindane
Linruex: see linuron
LINURON
Lipan: see dinitrocresol
Liphadione: see chlorophacinone
Liquamycin: see oxytetracycline hydrochloride
Liranox: see mecoprop
Liro-paraquat: see paraquat dichloride
Liroprem: see pentachlorophenol
Lirotan: see zineb
Lithane: see 2,4-D
LM 91: see chlorophacinone
LM-637: see bromadiolone
Lonacol: see zineb
Longlife Plus: see paraquat dichloride
Lonocol M: see maneb
Lontrel: see clopyralid, 2,4-D
Lontrel Plus: see clopyralid, dichlorprop, MCPA, mecoprop
Lorothiodol: see bithionol
Lorox: see linuron
Lorsban: see chlorpyrifos
Luprosil: see propionic acid
Lurat: see coumafuryl
M 2060: see fluenethyl
M-44: devices: sodium cyanide
M-74: see disulfoton
MAA: see methyl arsonic acid
Mach-Nic: see nicotine
Machete: see butachlor
Macondray: see 2,4-D
Mafu Strip: see dichlorvos
MAGNESIUM PHOSPHIDE
Magnetic 70, 90 and 95: see sulfur
Magnicide H Herbicide: see acrolein
Magnum: see ethofumesate

Maintain: see chlorflurecol
Maintain CF 125: see chlorflurecol
Maizor: see atrazine, ethalfluralin
Maki: see bromadiolone
Malachite: see basic copper carbonate
Malacide: see malathion
Malagram: see malathion
Malakill: see malathion
Malamar: see malathion
malaoxon: see malathion
Malaphos: see malathion
Malaspray: see malathion
Malatal: see malathion
MALATHION
Malathiozoo: see malathion
Malathon: see malathion
Malatox: see malathion, parathion
Malaude: see malathion
maldison: see malathion
Malic: see endosulfan
Malix: see endosulfan
Malmed: see malathion
Maloran: see chlorbromuron
MAMA: see ammonium methanearsonate
Mancobleu: see copper oxychloride, mancozeb
MANCOZEB
Maneba: see maneb
MANEB
Manebe: see maneb
Manebgan: see maneb
manebza: see maneb
Manesan: see maneb
Manex: see maneb
magnanese ethylenebisdithiocarbamate: see maneb
Manosil: see niclosamide
Manzate: see maneb
Manzate 200: see mancozeb
Manzati: see maneb
manzeb: see mancozeb
Manzin: see mancozeb
Manzin: see maneb
Maposol: see metam-sodium
Maralate: see methoxychlor
Marks 4-CPA: see 4-CPA
Marksman: see atrazine, dicamba
Marlate: see methoxychlor
Marmer: see diuron
Martin's Mar-Frin: see warfarin
Marvex: see dichlorvos
Matacil: see aminocarb
Mataven: see difenzoquat
Mataven: see flamprop-isopropyl
Mathieson 466: see cupric hydrazinium sulfate
Mato: see niclosamide
Matox: see hydramethylnon
Matrak: see difenacoum
Mavrik: see fluvalinate
Mavrik 2E: see τ-fluvalinate

Mavrik 2F: see τ-fluvalinate
Mavrik B: see τ-fluvalinate, thiometon
Mavrik HR: see τ-fluvalinate
Maxforce: see hydramethylnon
Mayclene: see clopyralid, dichlorprop, MCPA
Maytril: see bromoxynil, ioxynil, mecoprop
MB 10064: see bromoxynil
MB 2878: see 2,4-DB
MB 9057: see asulam
MB-8873: see ioxynil
MBC: see sodium chlorate
MBC: see carbendazim
MBR-825: see perfluidone
MBT
MC 1053: see dinobuton
MC 1478: see CNP
MC 2188: see chlormephos
MC1108: see dinoterb acetate
MCB: see carbendazim
MCPA
MCPP: see mecoprop
MDBA: see dicamba
M-Diphar: see maneb
MEB: see maneb
MeBr: see methyl bromide
Mecomec: see mecoprop
Mecoper: see mecoprop
Mecopex: see mecoprop
MECOPROP
Mediben: see dicamba
Megatox: see fluoroacetamide
melamine: see cyromazine
Meidon 15 Dust: see carbaryl, EPN
Meldane: see coumaphos
Melitoxin: see Dicumarol
Melprex: see dodine
Menaphtham: see carbaryl
Mendrin: see endrin
Menite: see mevinphos
Meothrin: see fenpropathrin
MEP: see fenitrothion
Mephanac: see MCPA
Mephetol Extra: see dicamba, dichlorprop, MCPA
mepiquat chloride
Mepro: see mecoprop
2-mercaptobenzothiazole: see MBT
mercaptodimethur: see methiocarb
mercaptofos: see demeton
N-(mercaptomethyl)phthalimide S-(O,O-
    dimethylphosphorodithioate): see phosmet
mercaptophos: see fenthion
mercaptothion: see malathion
Mercuram: see thiram
MERCURIC CHLORIDE
MERCUROUS CHLORIDE
MERCURY
mercury chloride: see mercuric chloride
mercury (II) chloride: see mercuric chloride

mercury monochloride: see mercurous chloride
Merfusan: see mercuric chloride
Merge 823: see MSMA
Merit: see clomazone
Merpafol: see captafol
Merpafol: see captaphol
Merpan: see captan
Mersil: see mercuric chloride, mercurous chloride
Mertax: see MBT
Mertect: see thiabendazole
Mesoranil: see aziprotryne
Mesurol: see methiocarb
metacetaldehyde: see metaldehyde
metafos: see methyl parathion
METALAXYL
METALDEHYDE
metalkamate: see bufencarb
metallic arsenic: see arsenic
METAM-SODIUM
metason: see metaldehyde
metaxon: see MCPA
methamidophos: see acephate-met
methanal: see formaldehyde
methanearsonic acid: see methyl arsonic acid
methanearsonic acid, calcium salt: see calcium acid
    methanearsenate
methanearsonic acid, disodium salt: see DSMA
methanearsonic acid, monoammonium salt: see
    ammonium methanearsonate
methanearsonic acid, monosodium salt: see MSMA
METHIDATHION
METHIOCARB
methiocarb sulfoxide: see methiocarb
METHOMYL
METHOPRENE
N-(2-methoxyacetyl)-N-(2,6-xylyl)-DL-alaninate: see
    metalaxyl N-(mercaptomethyl): see phosmet
2-methoxy-4H-1,2,3-benzodioxaphosphorin-2-sulfide:
    see dioxabenzofos
6-methoxy-N,N-bis(1-methylethyl)-1,3,5-triazine-2,4-
    diamine: see prometon
2-(methoxycarbonylamino)-benzimidazole: see
    carbendazim
3-methoxycarbonylaminophenyl 3'-
    methylcarbanilate: see phenmedipham
2-methoxycarbonyl-1-methylvinyl dimethyl
    phosphate: see mevinphos
METHOXYCHLOR
2-methoxy-3,6-dichlorobenzoic acid: see dicamba
(E,E)-11-methoxy-3,7,11-trimethyl-2,4-
    dodecadienoate: see methoprene
methylacetic acid: see propionic acid
methyl aldehyde: see formaldehyde
methyl [(4-aminophenyl)sulfonyl]carbamate: see
    asulam
methylarsenic acid: see methyl arsonic acid
methyl arsonic acid: see arsenic
methylarsonic acid, disodium salt: see DSMA

methylarsonic acid, monoammonium salt: see ammonium methanearsonate

methylarsonic acid, monosodium salt: see MSMA

methylazinphos: see azinphosmethyl

methylbenzene: see toluene

methyl-2-benzimidazolecarbamate phosphate: see carbendazim

methyl 1*H*-benzimidazol-2-yl-carbamate: see carbendazim

α-methylbenzyl(α)-3-hydroxycrotonate ester with dimethyl phosphate: see crotoxyphos

2-methylbiphenyl-3-ylmethyl(Z)-(1*RS*,3*RS*)-3-)2-chloro-3,3,3-trifluoroprop-1-enyl)-2,2-dimethylcyclopropanecarboxylate: see bifenthrin

methyl 1-(butylcarbamoyl)-2-benzimidazole carbamate: see benomyl

METHYL BROMIDE

methylcarbamate 1-naphthalenol: see carbaryl

methyl 2-chloro-9-hydroxyfluorene-9-carboxylate: see chlorflurecol

methyl 3-chloro-9-hydroxyfluorene-9*H*-carboxylic acid: see chlorflurecol

2-(2-methyl-4-chlorophenoxy)propionic acid: see mecoprop

1-(2-methylcyclohexyl)-3-phenylurea: see siduron

methyl demeton: see demeton-methyl

methyl 5-(2,4-dichlorophenoxy)-2-nitro benzoate: see bifenox

methyl 2-(4-(2,4-dichlorophenoxy)phenoxy)proprionate: see diclofop-methyl

methyl 3-(dimethoxyphosphinoloxy)but-2-enoate: see mevinphos

S-methyl N',N'-dimethyl-*N*-(methylcarbamoyloxy)-1-thiooxamimidate (I): see oxamyl

2-methyl-4,6-dinitrophenol: see dinitrocresol

N-methyl-2,4-dinitro-*N*-(2,4,6-tribromophenyl)-6-(trifluoromethyl)-benzenamine: see bromethalin

2,2'-methylenebis(4-chlorophenol): see dichlorophen

S',S'-methylenebis(O,O-diethylphosphorodithioate): see ethion

3,3'-methylenebis(4-hydroxycoumarin): see Dicumarol

2,2'-methylenebis(3,4,6-trichlorphenol): see hexachlorophene

METHYLENE CHLORIDE

1,2-methylenedioxy-4-propylbenzene: see dihydrosafrole

methylene oxide: see formaldehyde

S',S'-methylene O,O,O',O-tetraethyldi(phosphorodithioate): see ethion

2-(1-methylethoxy)phenolmethyl carbamate: see propoxur

1-methylethyl 4-bromo-α-(4-bromophenyl)-α-hydroxybenzeneacetate: see bromopropylate

4-(1-methylethyl)-2,6-dinitro-*N*,*N*-dipropylbenzanamine: see isopropalin

1-methylethyl 2-[ethoxy[(1-methylethyl)amino]phosphinothioyl]oxy]benzoate: see isofenphos

(*E*,*E*-1-methylethyl-11-methoxy-3,7,11-trimethyl-2,4-dodecadienoate: see methoprene

1-methylethyl 2-(1-methylpropyl)-4,6-dinitrophenyl carbonate: see dinobuton

(1-methylethyl)phosphoramidothoic acid O-(2,4-dichlorphenyl) O-2,4-(dichlorphenyl) O-methyl ester: see DMPA

2-(1-methylheptyl)-4,6-dinitrophenyl crotonate: see dinocap

N',N'-[(methylimino)dimethylidyne] bis[2,4-xylidine]: see amitraz

methyl isocyanate: see bendiocarb, metam-sodium

O'-methylisopropylphosphoroamidothioate: see DMPA

methyl isothiocyanate: see dazomet

S-methyl N-[(methylcarbamoyl)oxythioacetimidate: see methomyl

methyl 3-(3-methylcarbaniloyloxy)carbanilate: see phenmedipham

1-methyl-4-(1-methylethenyl)cyclohexene: see *d*-limonene

exo-1-methyl-4-(1-methylethyl)-2-[(2-methylphenyl)methoxy]-7-oxabicyclo[2.2.1]heptane: see cinmethylin

2-methyl-2-(methylsulfinyl)propionaldehyde O-(methylcarbamoyl) oxime: see aldicarb sulfoxide

2-methyl-2-(methylsulfonyl)propionaldehyde O-(methylcarbamoyl) oxime: see aldicarb sulfone

2-methyl-2-(methylthio)propionaldehyde O-(methylcarbamoyl)oxime: see aldicarb

methyl mustard oil: see methyl isothiocyanate

cis-7,8-*epoxy*-2-methyloctadecane: see disparlure

METHYL PARATHION

(3-(2-methylphenoxy)pyridazine: see credazine

methyl N-phenylacetyl-*N*-2,6-xylyl-DL-alaninate: see benalaxyl

1-methyl-3-phenyl-5-(trifluoro-*m*-tolyl)-4-pyridone: see fluridone

1,3-(1-methyl-2-pyrrolidyl)pyridine: see nicotine

methyl sulfanilylcarbamate: see asulam

3-(methylsulphonyl)-2-butanone O-[(methylammino)carbaryl]oxime: see butoxycarboxim

methylthioacetate: see acephate

4-(methylthio)-3,5-xylylmethyl-carbamate: see methiocarb

methyl viologen: see paraquat

N-methyl-*N'*-2,4-xylyl-*N*-(N-2,4-xylyl-formimidoyl)formamidine: see amitraz

Met-Systox: see demeton-methyl

Meta: see metaldehyde

Metaphos: see methyl parathion
Meth-O-Gas: see methyl bromide
Methacide: see toluene
Methar 30: see DSMA
Methasen: see ziram
Methoxane: see MCPA
Methoxide: see methoxychlor
Methoxo: see methoxychlor
Methoxy-DDT: see methoxychlor
Methyl Guthion: see azinphosmethyl
Methyl-bladen: see methyl parathion
Methylsystox: see demeton-methyl
metil-triazotion: see azinphosmethyl
METIRAM
metmercapturon: see methiocarb
METOLACHLOR
METRIBUZIN
metrifonate: see trichlorfon
Metro: see fonofos
Meuturon 4L: see fluometuron
Mevinox: see mevinphos
MEVINPHOS
MEXACARBATE
Mexprex: see dodine
Mezene: see ziram
Mezuron: see aziprotryne
MF-344: see etridiazol
MGK R11
Micasin: see chlorfensulphide
Micide: see zineb
Micro-Fume: see dazomet
Micromite: see diflubenzuron
Micropcop: see copper oxychloride
Microzul: see chlorophacinone
Midox: see chlorbenside
Midstream: see diquat
Mifaslug: see metaldehyde
Mikal: see folpet, fosetyl-al
Mikantop: see dimethoate, fenvalerate
Mil-Col: see drazoxolon
Milbam: see ziram
Milbex: see chlorfenethol
Milcurb: see dimethirimol
Milcurb Super: see ethirimol
Mild mercury chloride: see mercurous chloride
Mildane: see dinocap
Mildex: see dinocap
Mildothane: see thiophanate methyl
Milgo E: see ethirimol
MILKY SPORE DISEASE
Milky Spore Powder: see milky spore disease
milky white disease: see milky spore disease
Miller Blue Mold Dust: see ferbam
Miller Liquid Weedaway: see 2,4-D
Miller's Fumigrain: see acrylonitrile
Milmer 1: see oxine-copper
Milocep: see metolachlor, propazine
Milstem: see ethirimol

Miltox: see zineb
Minc: see chlormequat chloride
minderal naptha: see benzene
mineral oil: see petroleum oils
mineral spirits: see petroleum oils
Minerva: see bromoxynil, dichlorprop, ioxynil, MCPA
Miracle: see 2,4-D
Miral: see isazophos
MIREX
Mitaban: see amitraz
Mitigan: see dicofol
mitis green: see copper acetoarsenite
Mitox: see chlorbenside
Mitran: see chlorfenson
Mixte: see dicofol, methyl parathion
mixture of calcium hydroxide and copper sulfate: see Bordeaux mixture
mixture of cresols (methylphenols) obtained from coal tar: see cresylic acid
mixture: m-(ethylpropyl)phenylmethylcarbamate and m-(1-methylbutyl)phenylmethylcarbamate: see bufencarb
mixture of (ethylenebi(dithiocarbamate))zinc with ethylenebis(dithiocarbamicacid), bimolecular and trimolecular cyclic anhydrosulfides and disulfides: see metiram
mixture of pyrethrin I & II, cinerin I & II, jasmolin I & II: see pyrethrum
mixture of various chlorinated camphenes: see toxaphene
MK-936: see abamectin
MLT: see malathion
MMA: see ammonium methanearsonate
MnEBD: see maneb
MO: see CNP
Mo-338: see CNP
Mobilawn: see dichlofenthion
Mocap: see ethoprop
Modown DG: see bifenox, 2,4-D
Mole and Gopher Bait: zinc phosphide
Molluscicide Bayer 73: see niclosamide
Molutox: see niclosamide
Mon 0573: see glyphosate
Moncide: see cacodylic acid
Mondak: see dicamba, MCPA
Monitor: see acephate-met
monoammonium methanearsonate: see ammonium methanearsonate
monoammonium methylarsonate: see ammonium methanearsonate
monobromomethane: see methyl bromide
Monocron: see monocrotophos
MONOCROTOPHOS
mono(N,N-dimethylcocoamine): see endothall
mono(N,N-dimethyltridecylamine): see endothall
monofluoroacetic acid: see fluenethyl
monohydroxybenzene: see phenol

MONOLINURON
monomethylarsinic acid: see methyl arsonic acid
monomethylformamide: see fluridone
n-monomethylformamide: see
    monomethylformamide
monomethyl pyrazole: see difenzoquat
M-One: see *Bacillus thuringiensis* var. *san diego*
monosodium acid methanearsonate: see MSMA
monosodium methanearsonate: see MSMA
monosodium methylarsonate: see MSMA
Monsur: see carbaryl
Montar: see cacodylic acid
Montrel: see crufomate
Monurex: see monuron
Monurox: see monuron
MONURON
Monuruon: see monuron
MONURON TCA: see monuron
Moosuran: see pentachlorophenol
Mopari: see dichlorvos
Morfaron: see bromadiolone
Morkit: see anthraquinone
Morlex: see chlorpropham, ethofumesate, fenuron,
    propham
Morocide: see binapacryl
Moscade: see fenvalerate
moth balls: see naphthalene
moth flakes: see naphthalene
motor benzol: see benzene
Motox: see toxaphene
Mouse Blues: see coumafuryl
Mouse-con: see zinc phosphide
Mowdown: see bifenox, 2,4-D
Moxie: see methoxychlor
Mr. Triple Zero: see bromacil
MSMA: see arsenic
Mudekan: see trifluralin
Multamat: see bendiocarb
Multicide: see tetramethrin
Multicide Concentrate F-2271: see phenothrin
Multitox: see endrin
Muriol: see chlorophacinone
Murvin: see carbaryl
Musal: see bromadiolone
Muscatox: see coumaphos
Muscatox: see cyflurthrin, phoxim
Mycodifol: see captafol
Mycodifol: see captaphol
Mycodifol MZ: see folpet, mancozeb
Mylone: see dazomet
N-2790: see fonofos
N-521: see dazomet
Nabac: see hexachlorophene
NABAM
nabame: see nabam
Nac: see carbaryl
Nadone: see cyclohexanone
Nainit: see chlormequat chloride

NALED
Namekil: see metaldehyde
Nankor: see ronnel
naphtha: see petroleum oils
2-Naphthacene carboxamide 4(dimethylamino)-
    1,4,4a,5,5a,6,11,12a-octahydroxy-6-methyl-1, 11-
    dioxo-monohydrochloride: see oxytetracycline
    hydrochloride
NAPHTHALENE
NAPHTHALENEACETIC ACID
naphthalene oil: see creosote (coal tar)
1-naphthalenylthiourea: see antu
naphthalin: see naphthalene
1-naphthol: see carbaryl
α-naphthol: see 1-naphthol
2-(1-naphthyl)acetic acid: see naphthaleneacetic
    acid
1-naphthyl methylcarbamate: see carbaryl
N-1-naphthylphthalamic acid (I): see naptalam
α-naphthylthiocarbamide: see antu
1-(1-napthyl)-2-thiourea: see antu
α-napthyl thiourea: see antu
Naphtox: see antu
N-1-naphtylphthalamic acid (I): see naptalam
NAPTALAM
naramycin: see cycloheximide
Navadel: see dioxathion
NC 8438: see ethofumesate
NC-21314: see clofentezine
NCI-C566417: see boric acid
NCN: see copper napthenate
Necarboxylic acid: see allethrin
Necatorina: see carbon tetrachloride
Neguvon: see trichlorfon
Nellite: see diamidfos
Nema: see tetrachloroethylene
Nemacide: see dichlofenthion
Nemacur: see fenamiphos
Nemacur O: see isofenphos, fenaminophos
Nemacur P: see fenamiphos
Nemafume: see DBCP
Nemagon: see DBCP
Nematox: see 1,2-dichloropropane,
    dichloropropene
Nematox II: see dichloropropene
Nentosol: see ethylene dibromide
nendrin: see endrin
Neo-Pynamin 5/1/30: see allethrin, phenothrin,
    piperonyl butoxide
Neo-Pynamin Forte Aerosol: see phenothrin,
    tetramethrin
Neoban: see barban
Neobor: see borax
Neocid: see DDT
Neocidol: see DDT
Neocidol: see diazinon
Neoron: see bromopropylate
Neosorexa: see difenacoum

Neostanox: see fenbutatin-oxide
Nephis: see ethylene dibromide
Nephocarp: see carbophenothion
Nerkol: see dichlorvos
Nespor, Polyram M: see maneb
Netal: see bromophos
Nettle-Ban: see dicamba, 2,4-D, 2,4,5-T
Neutrion: see methyl parathion, tetradifon
Neutrocrop: see copper sulfate
New Leaf Black Fungicide: see ferbam
Nexagan: see bromophos-ethyl
Nexion: see bromophos
NF-44: see thiophanate methyl
NIA 10242: see carbofuran
NIA 1240: see ethion
NIA 17370: see resmethrin
NIA 249: see allethrin
NIA 33297: see permethrin
NIA 9044: see binapacryl
NIA 9102: see metiram
Niagar Carbamate: see ferbam
Niagara 1240: see ethion
Niagara Am Sol: see 2,4-D
Niagara P.A. Dust: see nicotine
Niagara Z-C Spray: see ziram
Niagaramite: see Aramite
Niagrathal: see endothall
Nialate: see ethion
Nic-Dust: see nicotine
NICLOSAMIDE
Nic-Sal: see nicotine
Nico-Fume: see nicotine
Nicocide: see nicotine
NICOTINE
Nicouline: see rotenone
Nifos T: see TEPP
Niklor: see chloropicrin
Nimitex: see temephos
Nimitox: see temephos
Niocides: see MBT
Niomil: see bendiocarb
Nipsan: see diazinon
Niran: see chlordane
Niran: see parathion
Nitrador: see dinitrocresol
NITRAPYRIN
Nitrochloroform: see chloropicrin
p-nitro-o-chlorophenyl dimethyl thionophosphate:
    see dicapthon
p-nitro-m-cresol: see fenitrothion
Nitro Kleenup: see dinitrophenol
2-nitro-1-(4-nitrophenoxy)-4-(trifluoromethyl)
    benzene: see flourodifen
p-nitrophenol: see para-nitrophenol
p-nitrophenyl α,α,α-trifluoro-2-nitro-p-tolyl ether:
    see fluorodifen
Nitropone C: see dinoseb
4-nitropyridine N-oxide: see Avitrol 100

N-nitrosamide: see pronamide
nitrosamines: see dinoseb, metolachlor, picloram
N-nitrosoamide: see pronamide
nitrosocarbaryl: see carbaryl
N-nitroso-di-n-propylamine: see trifluralin
N-nitrosodimethylamine: see thiram, ziram
N-nitrosodipropylamine: see EPTC
N-nitrosoglyphosate: see glyphosate
N-nitrosonornicotine: see nicotine
N-nitrosopendimethalin: see pendimethalin
n-nitroso propoxur: see propoxur
Nix: see permethrin
Nix-Scald: see ethoxyquin
Nixone: see dinoterb
NIZ 5462: see endosulfan
Nobencutan: see thiram
No Bunt: see hexachlorobenzene
Nogos: see dichlorvos
Nomersan: see thiram
No-Pest Insecticide Strip: see dichlorvos
Nopocide: see chlorothalonil
5-norbornene-2,3-dimethanol-1,4,5,6,7,7-
    hexachlorocyclicsulfite: see endosulfan
Nordox: see copper oxide
Norex: see chloroxuron
NORFLURAZON
Norosac 4G: see dichlorbenil
Norosac 10 G: see dichlorbenil
Nortron: see ethofumesate
No Scald: see diphenylamine
Novabac: see Bacillus thuringiensis var. kurstaki
Novathion: see fenitrothion
Novege: see erbon
Novigam: see lindane
Novon: see erbon
Novozin N 50: see zineb
Novozir: see zineb
Noxfish: see rotenone
Noxide: see sodium azide
NRDC 104: see resmethrin
NRDC 107: see bioresmethrin
NRDC 143: see permethrin
NRDC 149: see cypermethrin
NRDC 161: see deltamethrin
NTM: see dimethyl phthalate
NU 831: see tralomethrin
Nu-Bait II: see methomyl
NUCLEAR POLYHEDROSIS VIRUS
Nu-Cop: see copper sulfate
Nu-Lawn Weeder: see bromoxynil
Nudrin: see methomyl
Nuodex 84: see MBT
Nurelle: see cypermethrin
Nustar: see flusilazole
Nutra-Spray Basic Copper Carbonate: see basic
    copper carbonate
Novacron: see monocrotophos
Nuvan: see dichlorvos

oxygen analogue of azinphosmethyl: see
    azinphosmethyl
oxymethylene: see formaldehyde
OXYTETRACYCLINE HYDROCHLORIDE
Oxytril 4: see bromoxynil, dichlorprop, ioxynil,
    MCPA
Oxytril CM: see bromoxynil, ioxynil
Oxytril P: see bromoxynil, dichlorprop, ioxynil
P-1053: see dinobuton
Paarlan: see isopropalin
Padan: see cartap
Paddox: see dicamba, MCPA, mecoprop
Padquat: see chlormequat chloride
Pakhtaran: see fluometuron
Palado: see sodium salt of glyphosate
pallethrine: see allethrin
Palmarol: see endrin
Pamosol 2 Forte: see zineb
Panam: see carbaryl
Panatac: see clofentezine
Panoram: see thiram
Pansoil: see etridiazol
Paraban: see bromophos-ethyl
Parable: see diquat, paraquat
parachlorophenoxyacetic acid: see 4-CPA
Paracide: see paradichlorobenzene
Paracol: see diuron, paraquat
Para Crystals: see paradichlorobenzene
PARADICHLOROBENZENE
Paradow: see paradichlorobenzene
Paradusto: see parathion
Paraform: see paraformaldehyde
PARAFORMALDEHYDE
Paralindex: see lindane, methyl parathion
Paramoth: see paradichlorobenzene
para-nitrophenol: see methyl parathion, parathion
Para Nuggets: see paradichlorobenzene
paraoxon: see parathion
PARAQUAT
PARAQUAT DICHLORIDE
Parasol: see copper hydroxide
Parasoufre Acaricide: see dicofol, methyl parathion,
    sulfur
Paraspra: see parathion
PARATHION
parathion methyl: see methyl parathion
Parch: see pentachlorophenol, prometon
Pardi Weedol: see diquat, paraquat
Paris green: see copper acetoarsenite
Partner: see bromoxynil
Partner-Mon-9848: see alachlor
Partox: see chlorophacinone
Partron M: see methyl parathion
Parzate: see nabam
Parzate: see zineb
Parzate C: see zineb
Pasta Caffaro: see copper oxychloride
Pasturol: see dicamba, MCPA-sodium

Patap: see cartap
Pathclear: see amitrole, diquat, simazine, paraquat
Paturyl: see 6-benzyladenine
Paf-Off M: see flucythrinate, methomyl
PCA: see pentachloroaniline
PCBs
PCMS: see pentachlorophenylmethylsulfide
PCNB
PCP: see pentachlorophenol
PCPA: see 4-CPA
PCPCBS: see chlorfenson
P.D.I.C.: see potassium dichloroisocyanurate
PDU: see fenuron
PDV: see fenuron
pedinex: see dinex
Pencal: see calcium arsenate
penchlorol: see pentachlorophenol
PENDIMETHALIN
Penite: see sodium arsenite
Pennamine D7: see 2,4-D
Pennant: see metolachlor
Penncap-E: see parathion
Penphene: see toxaphene
penta: see pentachlorophenol
Pentac: see dienochlor
Pentachlorin: see DDT
pentachloroaniline: see PCNB
pentachlorobenzene: see hexachlorobenzene,
    lindane
PENTACHLOROPHENOL
pentachlorodibenzo-p-dioxin: see 2,4,5-
    trichlorophenol
pentachlorodibenzofuran: see PCBs, 2,4,5-
    trichlorophenol
pentachlorofenol: see pentachlorophenol
pentachlorofenolo: see pentachlorophenol
pentachloronitrobenzene
pentachloronitrobenzene: see PCNB
2,3,4,5,6-pentachlorophenol: see pentachlorophenol
pentachlorophenylmethylsulfide: see PCNB
pentachlorphenate: see pentachlorophenol
Pentacon: see pentachlorophenol
Pentagen: see PCNB
Penta General Weed Killer: see pentachlorophenol
Penta-Kil: see pentachlorophenol
pentanol: see pentachlorophenol
penta-s-triazinetrione: see chlorinated isocyanurates
Penta Plus 40: see pentachlorophenol
Penta Pres 1-10: see pentachlorophenol
Penta Preservative Ready-To-Use P: see
    pentachlorophenol
Penta Ready: see pentachlorophenol
Pentasol: see pentachlorophenol
Pentech: see DDT
α-pentyl-β-phenylacrylaldehyde: see acrolein
Penwar: see pentachlorophenol
per: see tetrachloroethylene
Peratox: see pentachlorophenol

perc: see tetrachloroethylene
perchloethylene: see oxyfbuorfen,
   tetrachloroethylene
perchlor: see tetrachloroethylene
perchlorobenzene: see hexachlorobenzene
perchloroethylene: see tetrachloroethylene
perchloromethane: see carbon tetrachloride
Perclene: see tetrachloroethylene
Perecol: see copper oxychloride
Perecot: see copper oxide
Perenox: see copper oxide
Perfekthion: see dimethoate
PERFLUIDONE
perfluoridone: see perfluidone
Perfmid: see tebuthiuron
Perigen: see permethrin
Perizin: see coumaphos
perk: see tetrachloroethylene
Perma Guard*: see diatomaceous earth
Permacide: see pentachlorophenol
Permagard: see pentachlorophenol
Permandine: see permethrin
Permasect: see permethrin
Permectrin: see permethrin
PERMETHRIN
Permit: see permethrin
Perosin: see zineb
Perthane: see ethylan
Perthrine: see permethrin
Pesguard: see allethrin, phenothrin, piperonyl
   butoxide
Pesguard ANS: see fenitrothion, tetramethrin
Pesguard FS: see phenothrin, tetramethrin
Pesguard Insect Killer: see phenothrin, tetramethrin
Pesguard NS: see fenitrothion, tetramethrin
Pesguard NSB: see fenitrothion, piperonyl
   butoxide, tetramethrin
Pesguard NX: see phenothrin, tetramethrin
Pesguard Plant Spray: see phenothrin, tetramethrin
Pestex: see dieldrin
Pestmaster: see methyl bromide
petroleum naphtha: see naphtha
PETROLEUM OILS
P.F. Harris Famous Tablets: see boric acid
PH 60-40: see diflubenzuron
Phaltan: see folpet
Pharoid: see methoprene
PHC: see propoxur
Phenacide: see toxaphene
Phenadox-X: see diphenyl
phenamiphos: see fenamiphos
Phenatox: see toxaphene
phene: see benzene
phenic acid: see phenol
phenisobromolate: see bromopropylate
PHENMEDIPHAM
PHENOL
PHENOTHRIN

d-phenothrin: see phenothrin
phenothrine: see phenothrin
Phenox: see 2,4-D
3-phenoxybenzyl (±)-cis,trans-chrysanthemate: see
   phenothrin
3-phenoxybenzyl (1RS)-cis,trans-3-(2,2-dichlorovinyl)-
   2,2-dimethylcyclopropanecarboxylate: see
   permethrin
3-phenoxybenzyl (1RS)-cis-trans-2,2-dimethyl-3-(-2-
   methylprop-1-enyl)cyclopropanecarboxylate: see
   sumithrin
phentinoacetate: see triphenyltin acetate
phenyl alcohol: see phenol
phenylbenzene: see diphenyl
N-phenylbenzenamine: see diphenylamine
N-phenylcarbamate: see desmedipham
phenyl N,N'-dimethylphosphorodiamidate: see
   diamidfos
3-phenyl-1,1-dimethylurea: see fenuron
1-phenylethyl 3-
   (dimethoxyphosphinoyloxy)isocrotonate: see
   crotoxyphos
(E)-1-phenylethyl 3-[(dimethoxyphosphinyl)oxy]-2-
   butenoate: see crotoxyphos
phenyl hydride: see benzene
phenyl hydroxide: see phenol
phenylmethane: see toluene
[1R-[1α,3α(E)]]-[5-(phenylmethyl)-3-furanyl]methyl3-
   [(dihydro-2-oxo-3(2H)-thienylidene)methyl-2,2-
   dimethcyclopropanecarboxylate: see kadethrin
Phenylphenol, O-phenylphenol: see phenol
Pherocon GM: see disparlure
PHORATE
PHOSALONE
Phosdrin: see mevinphos
Phosfene: see mevinphos
phosgene gas: see carbon tetrachloride, chlordane,
   chlorofluorocarbons, chloroform, dienochlor,
   lindane, methylene chloride, trichloroethene
PHOSMET
PHOSPHAMIDON
phosphine gas: see aluminum phosphide, calcium
   phosphide, magnesium phosphide, zinc
   phosphide
N-(phosphomethyl)glycine, isopropyl amine salt:
   see isopropylamine salt of glyphosate
N-(phosphonomethyl)glycine, sodium salt: see
   sodium salt of glyphosate
N-(phosphonomethyl)glycine: see glyphosate
Phostoxin: see aluminum phosphide
Phosvel: see leptophos
Phosvin: see zinc phosphide
Phosvit: see dichlorvos
photodieldrin: see dieldrin
photomirex: see mirex
PHOXIM
PHPH: see diphenyl
phthalthrin: see tetramethrin
Phygon: see dichlone

Phygon Paste: see dichlone
Phygon Seed Protectant: see dichlone
Phygon XL: see dichlone
Phytar 138: see cacodylic acid
Phyton 27: see copper sulfate
Pic-Clor: see chloropicrin
Picfume: see chloropicrin
Picket: see permethrin
PICLORAM
piclorame: see picloram
Picride: see chloropicrin
Pillarstin: see carbendazim
Pillarzo: see alachlor
pimelic ketone: see cyclohexanone
PIPERONYL BUTOXIDE
PIRIMIPHOS-ETHYL
PIRIMIPHOS-METHYL
PKhNB: see PCNB
Planotox: see 2,4-D
Plant Pin: see butoxycarboxim
Plantdrin: see monocrotophos
Planters Blue Mold Dust: see ferbam
Plantgard: see 2,4-D
Plantonit: see terbutryn
Plictran: see cyhexatin
Plondrel: see ditalimfos
Plus: see trichloroisocyanuric acid
Polybor 3: see borax
Polybor chlorate: see sodium chlorate
polychlorinated biphenyls: see PCBs
polychlorocamphene: see toxaphene
polymer of acetaldehyde: see metaldehyde
polymerized formaldehyde: see paraformaldehyde
Polymone: see dichlorprop
Polymone 60: see 2,4-D mecoprop
Polymone X: see 2,4-D, dichlorprop
poly(oxy-1,2-ethanediyl),α-isooctadecyl-ω-hydroxy:
    see Arosurf
polyoxyethyleneamine: see glyphosate
polyoxymethylene: see paraformaldehyde
polypropylene glycolmonobutylether: see butoxy
    polypropylene glycol potassium azide: see
    potassium azide
Polyram: see metiram
Polyram Ultra: see thiram
Polyram Z: see zineb
Polyram-Combi: see metiram
Polytox: see dichlorprop
Polytrin: see cypermethrin
Pomarsol Forte: see thiram
Pomarsol S Forte: see zineb
Pomarsol Z Fote: see ziram
Pomex: see carbaryl
Pondmaster: see isopropylamine salt of glyphosate
POTASSIUM AZIDE
POTASSIUM BROMIDE
potassium dichloroisocyanurate: see chlorinated
    isocyanurates

POTASSIUM PERMANGANATE
potassium salts of selected fatty acids: see soap
Pounce: see permethrin
Powder & Root: see rotenone
PP 021: see formesafen
PP 321: see λ-cyhalothrin
PP 383: see cypermethrin
PP-511: see pirimiphos-methyl
PP 557: see permethrin
PP-910: see paraquat bis(methylsulfate)
PP 993: see tefluthrin
PP009: see fluazifop-butyl
Pramex: see permethrin
Pramitol: see prometon
Pramitol 25E: see prometon
Preban: see terbutryn
Prebane: see terbutryn
precipitated sulfur: see sulfur
Precor: see methoprene
Precor Residual Fogger with Adulticide: see
    methoprene
Prefar: see bensulide
Prefix D: see dichlobenil
Preflan: see tebuthiuron
Prefmid: see tebuthiuron
Preforan: see fluorodifen
Prefox: see cyprazine, ethiolate
Preglone: see diquat, paraquat
Prelude: see paraquat dichloride
Premalin: see monolinuron
Premolox: see fenuron
Premazine: see simazine
Premerge: see dinoseb
Premgard: see resmethrin
Premol B: see 2,4-D, MCPA
Prenetol GDC (alkaline solution): see dichlorophen
prentox: see dichlorvos
Prep: see ethephon
Presan: see bensulide
Preservit: see dazomet
Preventol: see 2,4,5-trichlorophenol
Preventol GD: see dichlorophen
Preweed: see chlorpropham
Pride: see fluridone
Priglone: see diquat, paraquat
Primagram: see atrazine, metolachlor
Primatol: see atrazine
Primatol AD: see amitrol, atrazine, 2,4-D
Primatol S: see simazine
Primatol SE: see amitrole, simazine
Primextra: see atrazine, metolachlor
Primicid: see drazoxolon, pirimiphos-ethyl
Princep: see simazine
Printazol N: see 2,4-D, MCPA, picloram
Printazol Total: see 2,4-D, MCPA, mecoprop,
    picloram
Printop: see simazine
Proban: see cythioate

PROCYMIDONE
Prodalumnol Double: see sodium arsenite
Prodaram: see ziram
Profume: see methyl bromide
Pro-Gro: see carboxin
Prokarbol: see dinitrocresol
Prokil: see cryolite
Promar: see diphacinone
PROMETON
prometone: see prometon
PROMETRYN
PRONAMIDE
PROPACHLOR
propachlore: see propachlor
propanearsonic acid, calcium salt: see calcium
    propanersenate
PROPANIL
propanoic acid: see propionic acid
PROPARGITE
PROPAZINE
2-propenal: see acrolein
propenenitrile: see acrylonitrile
Propenex-Plus: see mecoprop
1-propenol-3: see allyl alcohol
2-propene-1-ol: see allyl alcohol
PROPHAM
prophos: see ethoprop
PROPIONIC ACID
Propionic Acid Grain Preserver: see propionic acid
PROPOXUR
S-propyldipropylthiocarbamate (I): see vernolate
propylene dichloride: see 1,2-dichloropropane
propyzamide: see pronamide
Protex: see paraquat dichloride
Proturf: see metalaxyl
Prowl: see pendimethalin
Proxol: see trichlorfon
prussic acid: see hydrogen cyanide
PS: see chloropicrin
Punch: see carbendazim, flusilazole
Puralin: see thiram
Puritan 3925: see bromacil
Py-Kill: see tetramethrin
Pydrin: see fenvalerate
Pyfos: see TEPP
Pynamin: see allethrin
Pynamin Forte: see d-cis/trans-allethrin
Pynosect: see resmethrin
Pynosect 6,10,25,PCO,WT: see permethrin
Pyrenone: see piperonyl butoxide
Pyrescel: see allethrin
Pyresin: see allethrin
Pyretherm: see resmethrin
PYRETHRUM
Pyrid: see fenvalerate
pyridinol: see chlorpyrifos
pyrimidinone: see hydramethylnon
Pyrinex: see chlorpyrifos

pyrobenzol: see benzene
pyrobenzole: see benzene
Pyrobor: see borax
Pyrocide: see allethrin
Pyro-Phos: see TEPP
Q-137: see ethylan
Qikron: see chlorfenethol
QINA
Quadmec: see 2,4-D dicamba, mecoprop, MSMA
Quamlin: see permethrin
Quelatox: see fenthion
Queletox: see fenthion
Quick: see chlorphacinone
Quilan: see benefin
Quinolate: see oxine-copper
Quinolate 15: see oxine-copper
Quinolate 20: see oxine-copper
Quinolate AC, Fs, Quinolate AC Kara: see
    anthraquinone, oxine-copper
Quinolate MG SAFI: see endosulfan, oxine-copper,
    lindane
Quinolate Triple Kara: see anthraquinone, lindane,
    oxine-copper
Quinolate V 4 X AC, FS, DS: see anthraquinone,
    carboxin, oxine-copper
Quinolate V 4 X Triple: see lindane, oxine-copper
Quinorexone: see dicamba, mecoprop
Quinoxone: see 2,4-D
Quintex: see fenuron
Quintox: see cholecalciferol
quintozene: see PCNB
88R: see Aramite
R11: see MGK R11
R-1303: see carbophenothion
R1513: see azinphosethyl
R-1910: see butylate
R-2063: see cycloate
R-4461: see bensulide
Rabyon: see carbaryl
Radapon: see dalapon
Rad-E-Cate 25: see cacodylic acid
Radex: see paraquat dichloride
Radocon: see simazine
Radozone TL: see amitrole, ammonium thiocyanate
Raid Ant & Roach Killer: see propoxur
Raid Ant & Roach Killer: see dichlorvos
Raid Mothproofer: see ethylan
Raid Solid Insect Killer: see dichlorvos
Raid Wasp & Hornet Killer: see propoxur
Raid Weed Killer: see 2,4-D
Rambo: see alachlor, atrazine
Ramik: see diphacinone
Ramor: see difenacoum
Ramortal: see bromadiolone
Rampage: see cholecalciferol
Ramrod: see propachlor
Ramucide: see chlorophacinone
Ranac: see chlorophacinone

Ranbeck: see dichlorvos, phosalone
Randox: see CDAA
Rasayanchlor: see butachlor
Rasikal: see sodium chlorate
Rassapron: see amitrole, atrazine, diuron
Rastop: see difenacoum
Rat & Mice Bait: see warfarin
Rat Gard: see warfarin
Rat Nots: see red squill
Rat Snax: see red squill
Rat's End: see red squill
Rat-A-Way: see warfarin
Rat-A-Way: see coumafuryl
Rat-B-Gon: see warfarin
Rat-Death: see warfarin
Rat-Kill: see warfarin
Rat-Mix: see warfarin
Rat-Nix: see warfarin
Rat-O-Cide: see warfarin
Rat-O-Cide Rat Bait: see red squill
Rat-Ola: see warfarin
Rat-Pak: see red squill
Rat-tu: see antu
Ratafin: see coumafuryl
Ratak: see difenacoum
Rataway: see warfarin
Ratimon: see bromadiolone
Ratinus: see bromadiolone
Ratio: see isoxaben
Ratomet: see chlorphacinone
Ratrick: see difenacoum
Rats Squill: see red squill
Ratspax: see red squill
Rattler: see isopropylamine salt of glyphosate
Rattrack: see antu
Rattract: see antu
Raviac: see chlorophacinone
Ravicac: see chlorophacinone
Ravyon: see carbaryl
R-Bix: see paraquat dichloride
RE 4355: see naled
RE-5353: see bufencarb
Reclaim: see clopyralid
Reclaim: see tebuthiuron
Recop: see copper oxychloride
Reddon: see 2,4,5-T
Red Shield: see dieldrin
RED SQUILL
Reflex: see fomesafen
Reflex T: see fomesafen, terbutryn
Regal: see diquat, paraquat
Reglex: see diquat
Reglone: see diquat
Reglox: see diquat, paraquat
Remasan: see maneb
6-12 Repellent: see ethyl hexanediol
Resbuthrin: see bioresmethrin
Residrin: see tetramethrin

Residroid: see permethirn
Resinan: see dichloran
Resistox: see coumaphos
RESMETHRIN
(+)-cis-resmethrin: see cismethrin
(+)-trans-resmethrin: see bioresmethrin
d-trans-resmethrin: see bioresmethrin
resmethrine: see resmethrin
Respond: see resmethrin
Rezifilm: see thiram
RH-2915: see oxyfluorfen
RH-315: see pronamide
Rhodex: see fosetyl-al, mancozeb
Rhodia: see 2,4-D
Rhodiacide: see ethion
Rhodioacuivre: see copper oxychloride
Rhodocide: see ethion
Rhomene: see MCPA
Rhonox: see MCPA
Rhothane: see DDD
Rhyuno oil: see safrole
Ridect Pour-On: see permethrin
Rideon: see diphenamid
Ridomil: see metalaxyl
Ridomil Plus: see copper oxychloride, metalaxyl
Rimidin: see fenarimol
Rimidine Plus: see carbendazim, fenarimol, maneb
Ripcord: see cypermethrin
Ripenthol: see endothall
Riton: see dichlorvos
Riverdale Dibro Granular Weed Killer: see bromacil
Ro-Neet: see cycloate
Roach Prufe: see boric acid
Rodene: see red squill
Rodent Pellets: zinc phosphide
Rodeo: see isopropylamine salt of glyphosate
Rodine: see red squill
Rodine-C: see bromadiolone
Rody: see fenpropathrin
Rogodan 14: see dimethoate, endosulfan
Rogodial: see dimethoate, phenthoate
Rogor: see dimethoate
Rokar X: see bromacil
Rondo: see permethrin
Ronilan: see vinclozolin
Ronilan M: see maneb, vinclozolin
Ronilan S Combi: see sulfur, vinclozolin
Ronilan Spezial: see chlorothalonil, vinclozolin
Ronit: see cycloate
RONNEL
Ronstar: see oxadiazon
Ronstar: see 2,4-D, mecoprop
Rospin: see chloropropylate
Rosuran: see monuron
Rotefive: see rotenone
Rotefour: see rotenone
ROTENONE
Rotessenol: see rotenone

Rotocide: see rotenone
Rotox: see methyl bromide
Rough & Ready Rat Bait & Rat Paste: see red squill
Roundup: see isopropylamine salt of glyphosate
Roundup L&G: see isopropylamine salt of
    glyphosate
Rout: see oryzalin, oxyflurofen
Rovral: see iprodione
Roxion: see dimethoate
Royal TMTD: see thiram
Rozol: see chlorophacinone
RP 17623: see oxadiazon
RPH: see thiabendazole
RP-Thion: see ethion
RS141: see chlordimeform
RU 11484: see bioresmethrin
RU 11705: see bioallethrin
RU 15 525: see kadethrin
RU 16 121: see S-bioallethrin
RU 22974: see deltamethrin
RU 25474: see tralomethrin
RU 27436: see esbiothrin
RU 28173: see allethrin
RU 3054: see S-bioallethrin
RU 48440: see resmethrin
Rubigan: see fenarimol
Ruelene: see crufomate
Rukseam: see DDT
Ruphos: see dioxathion
Rutgers 612: see ethyl hexanediol
ryanodine: see ryania
Ryanex: see ryania
Ryanexcel: see ryania
RYANIA
Ryanicide: see ryania
Rycelan: see oryzalin
Ryzelan: see oryzalin
S-1: see chloropicrin
S-15076: see ethiolate
S-15544: see butam
S 1752: see fenthion
S 22012: see benzthiazuron
S-2539: see phenothrin
S276: see disulfoton
S 3151: see permethrin
S-3206: see fenpropathrin
S-4084: see cyanophos
S-4087: see cyanophenphos
S-5602: see fenvalerate
S 5660: see fenitrothion
S-6115: see cyprazine
S 6176: see ethiolate
S 767: see fensulfothion
S-9115: see cyprazine
Sabacide: see sabadilla
SABADILLA
Sabane Dust: see sabadilla
SAD-85: see daminozide

Sadoplon: see thiram
Safer's Fungicidal Soap: see soap
Safer's Herbicidal Soap: see soap
Safer's Insecticidal Soap: see soap
SAFROLE
SAIsan F: see drazoxolon
Salithion: see dioxabenzofos
Salut: see chlorpyrifos, dimethoate
Salute: see metribuzin, trifluralin
Salvo: see 2,4-D
Samuron: see desmetryn
Sancap: see dipropetryn
Sandofan CM: see copper oxychloride, propineb
Sandolin A: see dinitrocresol
Sanmarton: see fenvalerate
Sanquinon: see dichlone
Sanspor: see captafol
Sanspor: see captaphol
Santobane: see DDT
Santobri: see pentachlorophenol
Santocel C: see silica aerogel
Santophen: see pentachlorophenol
Santophen 1: see O-benzyl-p-chlorophenol
Santoquin: see ethoxyquin
Sanvex: see cartap
Sapecron: see chlorfenvinphos
Sarclex: see linuron
Sariafume: see ethylene dibromide
Sarolex: see diazinon
Satisfar: see etrimfos
Saturn: see benthiocarb
Saturno: see benthiocarb
SBP 1513: see permethrin
SBP-1382: see resmethrin
SC-12937: see azacosterol
Scaldip: see diphenylamine
Scheele's green: see curpic arsenite
Schering 36056: see formetanate hydrochloride
Schering 36268: see chlordimeform
Schwinefurt green: see copper acetoarsenite
Scogal: see cyanazine
Scotlene: see dicamba, MCPA, mecoprop
Scott's O-X-D: see silvex
Scourge: see resmethrin
Scout: see tralomethrin
Scout X-Tra: see tralomethrin
Scrubmaster: see tebuthiuron
Scythe: see paraquat dichloride
SD 14999: see methomyl
SD 15417: see cyanazine
SD 1750: see dichlorvos
SD 3419: see endrin
SD 3562: see dicrotophos
SD 41706: see fenpropathrin
SD 7849: see chlorfenvinphos
SD 9098: see Akton
SD or Shell SD 9129: see monocrotophos
SD-14114: see fenbutatin-oxide

Seccatutto: see diquat, paraquat
Sector: see butralin
Security: see calcium arsenate
Seedox: see bendiocarb
Seedrin: see aldrin
Seffein: see carbaryl
Selinon: see dinitrocresol
Semeron: see desmetryn
Sencoral: see metribuzin
Sencorer: see metribuzin
Sencorex: see metribuzin
Sendran: see propoxur
Sentry: see aldicarb, lindane
Sentry Grain Preserver: see propionic acid
Septene: see carbaryl
Septiphene: see O-benzyl-p-chlorophenol
Serbitox S: see 2,4-D
Seribak: see hexachlorophene
Seritox 50: see dichlorprop, MCPA
sevadilla: see sabadilla
Sevin: see carbaryl
SG-67: see silica aerogel
Shackle: see isopropylamine salt of glyphosate
Shacklet C: see isopropylamine salt of glyphosate
Shamrox: see MCPA
Shed-A-Leaf: see sodium chlorate
Sherpa: see cypermethrin
Shield DPA: see diphenylamine
Shikimol: see safrole
Shikimole: see safrole
Shirlan: see sabadilla
SHL Turf Feed & Weed: see dichlorpropferrous, sulfate, MCPA
Short Stop E: see terbutryn
9,10-Seocholestra-5,7,10(19)-trein-3 betaol: see cholecalciferol
Sialex: see procymidone
Sialite: see dinocap
Siden: see pronamide, simazine
SIDURON
Silbenil: see dichlobenil
Silbos: see thiram, vinclozolin
silica: see diatomaceous earth
SILICA AEROGEL
silica gel: see silica aerogel
siliceous earth: see diatomaceous earth
silicon dioxide: see diatomaceous earth
Silikil: see silica aerogel
Silo: see difenacoum
Silosan: see pirimiphos-methyl
Silox: see silica aerogel
SILVEX
Silvi-Rhap: see silvex
Silvisar: see MSMA
Silvisar: see cacodylic acid
Simadex: see simazine
Simatrol: see amitrole, atrazine, simazine

SIMAZINE
Sinbar: see terbacil
Sinox: see dinitrocresol
Sinox General Subitex: see dinoseb
Sipaxol: see pendimethalin
Siperin: see cypermethrin
Sipquat: see paraquat dichloride
SK-368 Weed Killer: see bromacil
Skeetal: see Bacillus thuringiensis var. israelensis
Slam: see asulam, dalapon
Slaymore: see bromadiolone
Slug Pellets: see metaldehyde
Slug-M: see methiocarb
Smeesana: see antu
Smite 15G: see sodium azide
Smo-Cloud Bug Killer: see methoxychlor
SN 36056: see formetanate hydrochloride
SN 38107: see desmedipham
Snapshot 80: see isoxaben, oryzalin
Snarol Meal: see metaldehyde
SOAP
Sodar: see DSMA
sodium acid arsenate: see sodium arsenate [II]
sodium acid arsenate, heptahydrate: see sodium arsenate [III]
sodium aluminofluoride: see cryolite
sodium arsenate [I]: see arsenic
sodium arsenate [II]: see arsenic
sodium arsenate [III]: see arsenic
sodium arsenate: see sodium arsenate [III]
sodium arsenate: see sodium arsenate [I]
sodium m-arsenate: see sodium arsenate [I]
sodium o-arsenate: see sodium arsenate [I]
sodium arsenate dibasic anhydrous: see sodium arsenate [II]
sodium arseniate: see sodium arsenate [III]
sodium arsenite: see arsenic
sodium m-arsenite: see sodium arsenite
SODIUM AZIDE
sodium biborate: see borax
SODIUM CHLORATE
SODIUM CYANIDE
sodium dichloroisocyanurate: see chlorinated isocyanurates
sodium dichloroisocyanurate dihydrate: see chlorinated isocyanurates
sodium p-(dimethylamino)benzendiazosulfonate: see fenaminosulf
sodium [4-(dimethylamino)phenyl]diazene sulfonate: see fenaminosulf
sodium fluoaluminate: see cryolite
SODIUM FLUOROACETATE
SODIUM HYPOCHLORITE
sodium N-methyldithiocarbamate: see metam-sodium
SODIUM OMADINE
sodium pyroborate: see borax

sodium salt of acifluorfen
sodium salt of asulam
sodium salt of coumafuryl (Fumasol): see coumafuryl
sodium salt of dalapon
sodium salt of 2,3:4,6-di-O-isopropylidene-α-L-xylo-2-hexalofuranosonic acid: see dikegulac sodium
sodium salt of glyphosate: see glyphosate
sodium tetraborate anhydrous: see borax
sodium tetraboratedecahydrate: see borax
Sofril: see sulfur
Soil Fungicide 1823: see chloroneb
Soil-Prep: see metam-sodium
Soilbrom 40: see ethylene dibromide
Soilbrom 85: see ethylene dibromide
Soilbrom-90EC: see ethylene dibromide
Soilfume: see ethylene dibromide
Sok: see carbanolate
Sok: see carbaryl
Soltair: see diquat, paraquat, simazine
Solvigran: see disulfoton
Solvirex: see disulfoton
Somilan: see ethalfluralin
Sonalan: see ethalfluralin
Sonar: see fluridone
Sonar 5P: see fluridone
Sonar A5: see fluridone
Soprabel: see lead arsenate
Sopragram: see lindane, parathion
Sopranebe: see maneb
Soprocide: see benzene hexachloride
Soyex: see fluorodifen
SP 1103: see tetramethrin
Spanon: see chlordimeform
Spanone: see chlordimeform
Spasor: see isopropylamine salt of glyphosate
Spectracide: see diazinon
Speedway: see paraquat dichloride
Spergon: see chloranil
Spersul: see sulfur
Spica 66: see picloram
Spike: see tebuthiuron
Spontox: see 2,4-D, 2,4,5-T
Sporacol: see drazoxolon
Spotrete: see thiram
Spotton: see fenthion
Spra-Cal: see calcium arsenate
Spray-Cop: see copper sulfate
Spray-Tox: see kadethrin
Spraygrow: see diquat, paraquat
Sprayseed: see diquat, paraquat
Spraytop: see diquat, paraquat
Sprigone: see tetramethrin
Spring-Bak: see nabam
Springclene 2: see benazolin
Springcorn Extra: see dicamba, MCPA
Springcorn Plus: see dichlorprop, MCPA
Spritex: see tetramethrin

Sprotive Dust SG-67: see silica aerogel
Sprout Nip: see chlorpropham
Spud-Nic: see chlorpropham
Spur: see τ-fluvalinate
Spyant Ratones: see bromadiolone
Squill: see red squill
SR 406: see captan
SRA 5172: see acephate-met
SRA 7847: see edifenphos
SST: see DEF
ST100: see terbufos
Stabilene Fly Repellent: see butoxy polypropylene glycol
Stannophus: see maneb
Stathion: see parathion
Stauffer: see carbophenothion
Stauffer R-1910: see butylate
Stauufer Captan 80: see captan
Steladone: see chlorfenvinphos
Stempor: see carbendazim
Sting: see isopropylamine salt of glyphosate
Stinger: see clopyralid
Stockade: see cypermethrin
Stockade: see permethrin
stoddard solvents: see petroleum oils
Stomoxin: see permethrin
Stomoxin P: see permethrin
Stomp: see pendimethalin
Stop-Scald: see ethoxyquin
Storm: see acifluorfen, bentazone
STREPTOMYCIN
Streunex: see lindane
strychinidin-10-one: see strychnine
STRYCHNINE
subchloride of mercury: see mercurous chloride
succinic acid 2,2-dimethylhydrazide: see daminozide
sulfinylbis[methane]: see dimethyl sulfoxide
sulfur: see sulfur
Su Seguro Cardidor: see trifluralin
Subdue: see metalaxyl
Suffix BW: see flamprop-isopropyl
Sufonimide: see captafol
Sufonimide: see captaphol
Sulfadene: see MBT
sulfallate: see CDEC
Sulfamate: see AMS
sulfamate: see AMS
Sulfasan: see EXD
Sulfatep: see sulfoTEPP
Sulfemmide: see captafol
Sulfemmide: see captaphol
Sulfex: see sulfur
sulfone analogue of fenthion: see fenthion
Sulforon: see sulfur
sulfotep: see sulfoTEPP
SULFOTEPP: see chlorpyrifos, diazinon
sulfoxide analogue of fenthion: see fenthion

SULFUR
SULFURYL FLUORIDE
Sulkol: see sulfur
SULPROFOS
sultropene: see sulfuryl fluoride
Sumibac: see fenvalerate
Sumiboto: see procymidone
Sumicidin: see fenvalerate
Sumicombi: see fenitrothion, fenvalerate
Sumifleece: see fenvalerate
Sumifly: see fenvalerate
Sumilex: see procymidone
Sumimik: see fenpropathrin
Sumimix: see fenitrothion, fenpropathrin
Sumisclex: see procymidone
Sumithion: see fenitrothion
sumithrin
Sumithrin: see phenothrin
Sumithrin A Plus: see phenothrin, tetramethrin
Sumithrin B Plus: see phenothrin, tetramethrin
Sumithrin Plus: see allethrin, phenothrin
Sumitomo: see fenitrothion, fenvalerate
Sumittick: see fenvalerate
Suncide: see propoxur
Sup'R Flo: see maneb
Sup'R Flo Diuron Flowable: see diuron
Sup'operats: see bromadiolone
Sup'r-Flo Ferbam Flowable: see ferbam
Super Asecho: see bromadiolone
Super Barnon: see flamprop-isopropyl
Super Crab-E-Rad A.M.A.: see ammonium
    methanearsonate
Super Crab-E-Rad-Calar: see calcium acid
    methanearsenate
Super Dal-E-Rad: see calcium acid
    methanearsenate
Super Moxxtox: see dichlorophen
Super Spyant: see bromadiolone
Super Tin: see fentin hydroxide
Super Trimec: see dicamba, dichlorprop, 2,4-D
Super-Caid: see bromadiolone
Super-Cel: see diatomaceous earth
Super-Rozol: see bromadiolone
Superaven: see difenzoquat
Supercarb: see carbendazim
Supersan in Trey Triple Action Lawn Aid: see
    siduron
Supona: see chlorfenvinphos
Supper Suffix: see flamprop-isopropyl
Supracide: see methidathion
Surecide: see cyanophenphos
Surefire: see diuron, paraquat
Surflan: see oryzalin
Susvin: see monocrotophos
Sutan GR: see atrazine, butylate
Sutan Plus: see butylate
Sutan +: see atrazine
Sutazin: see atrazine, butylate

Sutazine + 18-6G: see butylate
Suzu: see triphenyltin acetate
Suzu H: see fentin hydroxide
SW-6701: see credazine
SW-6721: see credazine
Swat: see Bomyl
Swebate: see temephos
Sweeney's Ant-Go: see sodium arsenate [I]
Sweep: see paraquat dichloride
Sylan Methyl: see endosulfan, methyl parathion
Syllit: see dodine
Synklor: see chlordane
Synthin: see resmethrin
Systemox: see demeton
Systol: see dinobuton
Systox: see demeton
Sytasol: see dinobuton
Szklarniak: see dichlorvos
T-1258: see cartap
2,4,5-T
Tackle: see acifluorfen
Tackle 2S: see acefluorfen
Taktic: see amitraz
Talan: see dinobuton
Talbot: see lead arsenate
Talcord: see permethrin
Talent: see asulam, paraquat
Talodex: see fenthion
Talstar: see bifenthrin
Tamaron: see acephate-met
Tamaron: see acephate-met, parathion
Tamazine: see simazine
Tamex: see butralin
Tamogan: see bromadiolone
Tantoo Bomb: see ethyl hexanediol
TAP 94P: see dichlorvos
tar camphor: see naphthalene
Target: see asulam, dalapon
Target MSMA: see MSMA
tar oil: see creosote (coal tar)
Tartan: see asulam, diuron
Task: see dichlorvos
Tat Ant Trap: see chlordecone
Tat Ant Trap: see propoxur
Taterpex: see chlorpropham
Tatoo: see bendiocarb
Taxylone: see methyl parathion, phosalone
Taytox: see copper ammonium carbonate
TBP: see bithionol
TBTO: see tributyltin oxide
TBZ: see thiabendazole
TC-90: see copper linoleate
TCAB: see diuron, linuron
TCCA: see trichloroisocyanuric acid
TCDD: see chloroneb, 2,4-D, DCPA, dichlorprop,
    hexachlorophene, pentachlorophenol, silvex,
    2,4,5-T, 2,4,5-trichlorophenol, 2,4,6-
    trichlorophenol

1,3,6,8-TCDD: see 1,3,6,8-tetrachlorodibenzo-*p*-dioxin
1,3,7,9-TCDD: see 1,3,7,9-tetrachlorodibenzo-*p*-dioxin
2,4,5-TCP: see 2,4,5-trichlorophenol
2,4,6-TCP: see 2,4,6-trichlorophenol
TCPA: see fenac
TD-1881: see thiophanate methyl
TDE: see DDD
Tear Gas: see chloropicrin
Tebulan: see tebuthiuron
Tebutam: see butam
Tebutame: see butam
TEBUTHIURON
Tech DDT: see DDT
Tecto: see thiabendazole
Tedane Extra: see dicofol, dinocap, mancozeb, tetradiforn
TEDP: see sulfoTEPP
TEFLUTHRIN
tefluthrine: see tefluthrin
Teknar: see *Bacillus thuringiensis* var. *israelensis*
Telar: see chlorsulfuron
Telone: see 1,2-dichloropropane, dichloropropene
Telone II: see dichloropropene
Telvar: see monuron TCA
Telvar Monuron Weedkiller: see monuron
TEMEPHOS
temophos: see temephos
Temus: see bromadiolone
Tendust: see nicotine
Tenoram: see chloroxuron
Teep: see TEPP
TEP: see TEPP
TEPP
TERBACIL
TERBUCARB
TERBUFOS
terburyne: see terbutryn
terbutol: see terbucarb
Terbutrex: see terbutryn
TERBUTRYN
Tercyl: see carbaryl
Teremec: see chloroneb
Term-i-Trol: see pentachlorophenol
Termide: see heptachlor
Termide: see chlordane
Termil: see chlorothalonil
Terpal: see ethephon, mepiquat chloride
Terpal C: see chlormequat chloride, ethephon
Terpal M: see chlormequat chloride, ethephon, mepiquat chloride
terraclor: see PCNB
Terr-O-Gas: see chloropicrin, methyl bromide
Terra-Sytam: see dimefox
Terra-Var: see bromacil
Terrachlor: see PCNB
Terrachlor-Super X: see etridiazol

Terracoat: see etridiazol
Terracur: see fensulfothion
Terracur P: see fensulfothion
Terraklene: see paraquat dichloride, simazine
terramicin: see oxytetracycline hydrochloride
terramitsin: see oxytetracycline hydrochloride
Terramycin Hydrochloride: see oxytetracycline hydrochloride
Terraneb: see chloroneb
Terrazole: see etridiazol
Tersan 1991: see benomyl
Tersan 75: see thiram
Tersan LSR: see maneb
Tersan SP Turf Fungicide: see chloroneb
Tersane LSR: see maneb
Tetracap: see tetrachloroethylene
3,4,3',4'-tetrachloroazobenzene: see TCAB
3,4,5,6-tetrachloro-1,2-benzenediol: see tetrachlorocatechol
tetrachloro-1,2-benzenediol: see tetrachlorocatechol
2,3,5,6-tetrachloro-*p*-benzoquinone: see chloranil
1,3,6,8-tetrachlorodibenzo-*p*-dioxin: see 2,4-D, 2,4,5-trichlorphenol
2,3,7,8-tetrachlorodibenzo-*p*-dioxin: see TCDD
tetrachlorocatechol: see pentachlorophenol
2,4,4',5-tetrachlorodiphenylsulfone: see tetradifon
tetrachloroethene: see tetrachloroethylene
TETRACHLOROETHYLENE
1,1,2,2-tetrachloroethylene: see tetrachloroethylene
cis-N-[(1,1,2,2-tetrachloroethyl)thio]-4-cyclohexene-1,2-dicarboximide: see captafol
tetrachlorohydroquinone: see pentachlorophenol
tetrachloroisophthalonitrile: see chlorothalonil
tetrachloromethane: see carbon tetrachloride
tetrachlorophenol: see pentachlorophenol
Tetradusto 100: see TEPP
TETRADIFON
tetramethrine: see tetramethrin
tetrachloropyrocatechol: see tetrachlorocatechol
TETRACHLORVINPHOS
tetraethyl prophosphate: see TEPP
O,O,O,O-tetraethyldithiopyrophosphate: see sulfoTEPP
O,O,O',O-tetraethyl *S,S'*-methylenebisphosphorodithioate: see ethion
tetraethyl thiodiphosphate: see sulfoTEPP
tetraethylthiuram disulfide: see disulfiram
2,3,5,6-tetrafluoro-4-methylbenzyl(Z)-(1RS)-*cis*-3-(2-chloro-3,3,3-trifluroprop-1-enyl]-2,2-dimethylcyclopropanecarboxylate: see tefluthrin
tetrahydro-5,5-dimethyl-2(1H)-pyrimidinone[3-[4-(trifluoromethyl)phenyl]-1-[2-[4-(trifluoromethyl)phenyl]ethenyl]-2-prop enylidene]hydrazone: see hydramethylnon
tetrahydro-3,5-dimethyl-2H-1,3,5-thiadiazine-2-thione: see dazomet

3,4,5,6-tetrahydrophthalimidomethyl (±)-cis,trans-
chrysanthemate: see tetramethrin
Tetralate: see resmethrin, tetramethrin
Tetralex-Plus: see dicamba, MCPA, mecoprop
TETRAMETHRIN
tetramethrin (1R)-isomers: see tetramethrin
r-2,c-4,c-6,c-8-tetramethyl-1,3,5,7-tetroxocane: see
metaldehyde
O,O,O',O'-tetramethyl O,O'-thiodi-p-phenylene
phosphorothioate: see temephos
tetramethylthiperoxydicarbonicdiamide: see thiram
tetramethylthiuram disulfide: see thiram
Tetraspra: see TEPP
Tetron: see TEPP
Tetron 100: see TEPP
Tetropil: see tetrachloroethylene
Tetroxone M: see bromoxynil, dichlorprop, ioxynil,
MCPA
T-Gas: see ethylene oxide
TH 60-40: see diflubenzuron
THIABENDAZOLE
Thibenzole: see thiabendazole
Thifor: see endosulfan
Thimer: see thiram
Thimul: see endosulfan
thiobencarb: see benthiocarb
thiodan: see endosulfan
Thiodow: see zineb
Thioknock: see thiram
Thiolux: see sulfur
Thioneb: see metiram
Thionex: see endosulfan
thiopal: see folpet
Thiophanate M: see thiophanate methyl
THIOPHANATE ETHYL
THIOPHANATE METHYL
thiophos: see parathion
Thiosan: see thiram
Thiotax: see MBT
thiotepp: see sulfoTEPP
Thiotex: see thiram
Thiovit: see sulfur
THIRAM
thirame: see thiram
Thiram Fungicide: see thiram
Thiramad: see thiram
Thirasan: see thiram
thiuram: see thiram
Thiuramin: see thiram
T-H Klean Drop: see chlorpropham
2,2'-thiobis(4,6-dichlorophenol): see bithionol
O,O'-(thio-4,1-phenylene)bis[O,O-
dimethylphosphorothioate]: see temephos
2-(4-thiozolyl)-benzimidazole: see thiabendazole
Thompson's Wood Fix: see pentachlorophenol
Thuramyl: see thiram
Thuricide: see *Bacillus thuringiensis* var. *kurstaki*
T.I.C.A.: see trichloroisocyanuric acid

Tiezene: see zineb
Tiguvon: see fenthion
Tilcarex: see PCNB
TIN
Tin San: see tributyltin chloride complex
Tinamte: see triphenyltin acetate
Tiovel: see endosulfan
Tirade: see fenvalerate
Tirampa: see thiram
Titan: see chlormequat chloride
Tiurolan: see tebuthiuron
TMTD: see thiram
Tobacron: see metobromuron, metolachlor
Tobaz: see thiabendazole
TOLUENE
Toluol: see toluene
3-o-tolyloxypyridazine: see credazine
tomarin: see coumafuryl
Tomato Fix: see 4-CPA
Tomato Hold: see 4-CPA
Tomatotone: see 4-CPA
Tomo-oxiran: see oxine-copper
Topiclor 20: see chlordane
Topidion: see bromadiolone
Topitox: see chlorophacinone
Toppel: see cypermethrin
Topsin M: see thiophanate methyl
Topusyn: see desmetryn
Topzol: see red squill
Torak: see dialifor
Torant: see bifenthrin, clofentezine
Torch 3F: see atrazine, bromoxynil
Tordon: see picloram
Tordon 101 Mixture: see 2,4-D, picloram
Tordon 10K & 22K: see picloram
Tordon RTU: see picloram
Torero: see clofentezine, τ-fluvalinate
Torgal: see picloram
Tormona: see 2,4,5-T
Tornade: see permethrin
Tornado: see fluazifop-butyl, fomesafen
Torocil: see bromacil
Torpedo: see diquat
Torpedo: see permethrin
Torque: see fenbutatin-oxide
Totacol: see diuron, paraquat
Totril: see ioxynil
Touchdown: see glyphosate trimesium
Toxadusto-10: see toxaphene
Toxakil: see toxaphene
TOXAPHENE
Toxaspra-8: see toxaphene
toxynil: see ioxynil
2,4,5-TP: see silvex
TPTA: see triphenyltin acetate
TPTH: see fentin hydroxide
TPTOH: see fentin hydroxide
Tracker: see tralomethrin

Trinol Super: see dicamba, MCPA-potassium,
    mecoprop-potassium
Trinoxol: see 2,4,5-T
Trio: see bromoxynil, 2,4-D, propanil
Triododine: see dodine
Trioneb: see metiram
Triox Vegetation Killer: see pentachlorophenol,
    prometon
Tri-PCNB: see PCNB
triphenyltin acetate: see tin
triphenyltin hydroxide: see fentin hydroxide
Tripomol: see thiram
Triquat: see diquat, paraquat
Triscabol: see ziram
Trithion: see carbophenothion
Tritisan: see PCNB
Tritocol: see carbendazim
Tritofterol: see zineb
Triumph: see isazophos
Tri-VC-13: see dichlofenthion
Trolene: see ronnel
Trona: see borax
Tronabor: see borax
Trooper: see dicamba
Truban: see etridiazol
Trucidor: see vamidothion
Tryosan: see copper napthenate
TS-7236: see fluazifop-butyl
tsitrex: see dodine
Tuads: see thiram
Tubatoxin: see rotenone
Tubothane: see maneb
Tubotin: see fentin hydroxide
Tues: see thiram
Tugon: see trichlorfon
Tulisan: see thiram
Tumbleaf: see sodium chlorate
Tupersan: see siduron
Turbair Dicamate: see mancozeb, zineb
Turbair Systemic Insecticide: see dimethoate
Turbo: see metolachlor, metribuzine
Turcam: see bendiocarb
Turf Fungicide: see benomyl
Turf Fungicide: see maneb
Turflon: see triclopyr
Tuscopper: see copper napthenate
Tutane: see 2-butanamine
Tuver Acaricide: see dicofol, ethion, methyl
    parathion
Twin Light Rat Away: see warfarin
Twinspan: see disulfoton
Tycor: see ethiozin
U-12927: see carbanolate
U-2069: see dichloran
U-36059: see amitraz
U5227: see antu
U 46 DP-M: see dichlorprop, MCPA
U 46 Super: see dichlorprop, MCPA, mecoprop

UC 19786: see dinobuton
UC 21149: see aldicarb
UC-62644: see chlorfluazuron
UC 7744: see carbaryl
UC7744: see carbaryl
Ultima: see bentazon, dichlorprop
Ultra-Clor: see cadmium succinate
Ulvair: see dioxacarb
Ulvair: see monocrotophos
UMDH: see unsymmetrical 1,1-dimethylhydrazine
Umbethion: see coumaphos
Unden: see propoxur
Unicrop: see maneb
Unicrop DNBP: see dinoseb
Unifos: see dichlorvos
Unifume: see ethylene dibromide
Unipon: see dalapon
Uniroyal D 735: see carboxin
Unitox: see chlorfenvinphos
unsymmetrical 1,1-dimethylhydrazine: see
    daminozide
Urab: see fenuron
Uragan: see bromacil
Ureabor: see sodium chlorate
Urox: see monuron TCA
Urox HX: see bromacil
Urox-"B": see bromacil
USB-3584: see dinitramine
Ustaad: see cypermethrin
Ustinex: see amitrole
Utlracide: see methidathion
Vamidoate: see vamidothion
VAMIDOTHION
Vancide: see maneb
Vancide BL or BN: see bithionol
Vancide FE 95: see ferbam
Vancide KS: see fentin hydroxide
Vancide MZ-96: see ziram
Vapona: see dichlorvos
Vapona Flykiller: see bioallethrin, permethrin,
    piperonyl butoxide
Vapona II: see dichlorvos
Vaponite: see dichlorvos
Vapophos: see parathion
Vaptone: see TEPP
V-Bor: see borax
VC-13: see dichlofenthion
V-C-9-104: see ethoprop
VCN: see acrylonitrile
VCS-506: see leptophos
Vectal: see atrazine
Vectobac: see *Bacillus thuringiensis* var. *israelensis*
Vectrin: see resmethrin
Vega: see bentazon, cyanazine, dichlorprop
Vegadex: see CDEC
Vegetox: see cartap
Vegfru Fosmite: see ethion
Vegiven: see chloramben

Velpar: see hexazinone
Velpar Gridmall: see hexazinone
Velpar K: see diuron, hexazinone
Velsicol: see heptachlor epoxide
Velsicol 104: see heptachlor
Velsicol 1068: see chlordane
Vendex: see fenbutatin-oxide
Vengeance: see bromethalin
Ventox: see acrylonitrile
Venturol: see dodine
Veon 245: see 2,4,5-T
Veratrine: see sabadilla
Verfor: see methyl parathion
Vernam: see atrazine vernolate
VERNOLATE
Veromite: see methyl parathion
Vertac Dinitro Weedkiller: see dinoseb
Verthion: see fenitrothion
Vertimec: see abamectin
Verton 2T: see 2,4,5-T
Vi-Cad: see cadmium chloride
Victoria: see formaldehyde
Vigilante: see diflubenzuron
Vikane: see sulfuryl fluoride
Viktor: see clofentezine fenpropathrin
VINCLOZOLIN
vinyl carbinol: see allyl alcohol
vinyl 2-chloroethyl: see propachlor
vinyl cyanide: see acrylonitrile
Vinylphate: see chlorfenvinphos
Violan: see paraquat dichloride
Vioxan: see carbaryl
Viozene: see ronnel
Vipex: see mecoprop
virosin: see antimycin A
Vision: see isopropylamine salt of glyphosate
Visko Rhap: see 2,4,5-T
Vitaflo: see carboxin
Vitamin D3: see cholecalciferol
Vitavax: see carboxin
Vitavax 100: see carboxin
Vitex: see dimethoate
Viticarb: see methyl parathion
Vomiting Gas: see chloropicrin
Voncaptan: see captan
Vondodine: see dodine
Vondozeb: see mancozeb
Vorlan: see vinclozolin
Vorlex 201: see chloropicrin
Vorox: see amitrole
Vorox: see simazine
Vorox Granulat 371: see atrazine
VUAgt-I-4: see thiram
Vulcafor TMTD: see thiram
Vulkacit MTIC: see thiram
Vulklor: see chloranil
Warbex: see famphur
Warbicide: see rotenone

Warfarat: see warfarin
WARFARIN
Warpath: see silica aerogel
Way Up: see pendimethalin
WBA 8107: see difenacoum
Weed Fume: see methyl bromide
Weed-B-Gon: see 2,4-D
Weed-Beads: see pentachlorophenol
Weed-Free G: see bromacil, diuron
Weed-Rhap: see 2,4-D
Weed-Rhap LV-4D: see MCPA
Weedar: see 2,4,5-T
Weedazol: see amitrole
Weedazol T: see amitrole
Weedex A: see atrazine
Weedmaster: see 2,4-D, dicamba
Weedone: see 2,4-D
Weedone: see 2,4,5-T
Weedone 170: see 2,4-D, dichlorprop
Weedone 2,4,5-TP: see silvex
Weedone DCP: see 2,4-D, dichlorprop
Weedone Super BK-32: see 2,4-D, dichlorprop
Weedone TP: see silvex
Weedtrine-D: see diquat
Whip: see fenoxaprop-ethyl
white arsenic: see arsenic trioxide
white tar: see naphthalene
Wipeout: see hydramethylnon
Witox: see EPTC
WL 115110: see flufenoxuron
WL 19805: see cyanazine
WL 28651: see flamprop-isopropyl
WL 41706: see fenpropathrin
WL 43423: see flamprop-isopropyl
WL 43425: see flamprop-isopropyl
WL 43467: see cypermethrin
WL 43775: see fenvalerate
Wofatox: see methyl parathion
Wolmanized: see chromated copper arsenate
wood creosote: see creosote (wood tar)
Wood Ridge Corrosive Sublimate: see mercuric
    chloride
Wood Ridge Mixture 21: see mercuric chloride,
    mercurous chloride
X-All: see amitrole
XE 938: see fenpropathrin
xiloli: see xylene
Xindex: see methyl isothiocyanate
XL-7: see bithionol
X-Pand: see isoxaben
XYLENE
xylenen: see xylene
xylol: see xylene
Y-3: see chlorpropham
Yaltox: see carbofuran
Yanock: see fluoroacetamide
Yasoknock: see sodium fluoroacetate
yellow camphor oil: see safrole

Yellow Cuprocide: see copper oxide
Yomesan: see niclosamide
Z-C Spray: see ziram
Zeapur: see simazine
Zebenide: see zineb
Zebtoc: see zineb
Zebtox: see zineb
Zectran: see mexacarbate
zeidane: see DDT
Zelan: see MCPA
Zep Formula 777: see bromacil
Zephiram: see benzalkonium chloride
Zerdane: see DDT
Zerlate: see ziram
Zestocarp: see niclosamide
Zidan: see zineb
Zilch Liquid Weed Killer: see bromacil
zinc coposil: see copper
zinc dimethyl dithiocarbamate: see ziram
zinc ethylenebisdithiocarbamate: see zineb
zineb-ethylene thiuram disulfide adduct: see
   metiram
Zinc Metiram: see metiram
ZINC PHOSPHIDE
Zincmate: see ziram
ZINEB
Zinosan: see zineb
Zipak: see amitraz, bifenthrin
ZIRAM
Ziram Technical: see ziram
Zirberk: see ziram
Ziride: see ziram
Zithiol: see malathion
Zitox: see ziram
zoocoumarin: see warfarin
Zorial: see fluometuron
Zotox: see arsenic acid
Zrylam: see carbaryl
Zyban: see maneb
Zyban: see thiophanate methyl
Zytron: see DMPA

CAS NUMBERS

10-64-7: see 4-chloroaniline
50-00-0: see formaldehyde
50-29-3: see DDT
51-03-6: see piperonly butoxide
51-28-5: see dinitrophenol
52-51-7: see bronopol
52-68-6: see trichlorfon
52-85-7: see famphur
54-11-5: see nicotine
55-38-9: see fenthion
56-23-5: see carbon tetrachloride
56-35-9: see tributyltin oxide
56-38-2: see parathion
56-72-4: see coumaphos

57-14-7: see unsymmetrical 1,1-dimethylhydrazine
57-24-9: see strychnine
57-74-9: see chlordane
57-92-1: see streptomycin
58-36-6: see OBPA
58-89-9: see lindane
60-15-5: see dimethoate
60-57-1: see dieldrin
61-82-5: see amitrole
62-73-7: see dichlorvos
62-74-8: see sodium fluoroacetate
62-75-9: see dimethylnitrosamine
63-25-2: see carbaryl
66-76-2: see Dicumarol
66-81-9: see cycloheximide
67-66-3: see chloroform
67-68-5: see dimethyl sulfoxide
67-72-1: see hexachloroethane
70-30-4: see hexachlorophene
70-38-2: see dimethrin
70-43-9: see barthrin
71-43-2: see benzene
72-43-5: see methoxychlor
72-54-8: see DDD
72-55-9: see DDE
72-56-0: see ethylan
74-83-9: see methyl bromide
74-90-8: see hydrogen cyanide
75-08-1: see ethyl mercaptan
75-09-2: see methylene chloride
75-15-0: see cardon disulfide
75-21-8: see ethylene oxide
75-60-5: see cacodylic acid
75-99-0: see dalapon
76-06-2: see chloropicrin
76-12-0: see chlorofluorocarbons
76-22-2: see camphor
76-44-8: see heptachlor
76-87-9: see fentin hydroxide
77-06-5: see gibberellic acid
78-34-2: see dioxathion
78-48-8: see DEF
78-87-5: see 1,2-dichloropropane
79-01-6: see trichloroethene
79-02-7: see dichloroacetaldehyde
79-09-4: see propionic acid
80-06-8: see chlorfenethol
80-33-1: see chlorfenson
81-81-2: see warfarin
82-66-6: see diphacinone
82-68-8: see PCNB
83-79-4: see rotenone
84-65-1: see anthraquinone
84-74-2: see dibutyl phthalate
85-34-7: see fenac
85-40-5: see delta[4]tetrahydrophthalimide
86-50-0: see azinphosmethyl
86-87-3: see naphthaleneacetic acid

300-76-5: see naled
301-12-2: see oxydemeton-methyl
309-00-2: see aldrin
311-45-5: see paraoxon
314-40-9: see bromacil
315-17-4: see mexacarbate
319-84-6: see benzene hexachloride
319-85-7: see benzene hexachloride
319-86-8: see benzene hexachloride
330-54-1: see diuron
330-55-2: see linuron
333-41-5: see diazinon
420-04-2: see hydrogen cyanamide
434-16-2: see cholecalciferol
467-69-6: see flurecol-butyl
470-90-6: see chlorfenvinphos
485-31-4: see binapacryl
492-80-8: see auramine
495-48-7: see azoxybenzene
502-55-6: see EXD
504-24-5: see Avitrol 200
507-60-8: see red squill
510-15-6: see chlorobenzilate
513-49-5: see 2-butanamine
513-78-0: see cadmium carbonate
522-70-3: see antimycin A3
533-74-4: see dazomet
534-52-1: see dinitrocresol
542-75-6: see dichloropropene
556-22-9: see glyodin
556-61-6: see methyl isothiocyanate
557-30-2: see glyoxime
563-12-2: see ethion
584-79-2: see allethrin
584-79-2: see bioallethrin
592-01-8: see calcium cyanide
608-73-1: see benzene hexachloride
608-93-5: see pentachlorobenzene
621-64-7: see N-nitrosodipropylamine
623-91-6: see diethyl fumarate
640-19-7: see fluoroacetamide
671-04-5: see carbanolate
722-20-8: see endrin
732-11-6: see phosmet
741-58-2: see bensulide
759-94-4: see EPTC
773-06-0: see AMS
786-19-6: see carbophenothion
814-91-5: see copper oxalate
834-12-8: see ametryn
886-50-0: see terbutryn
900-95-8: see triphenyltin acetate
919-44-8: see monocrotophos
944-22-9: see fonofos
950-37-8: see methidathion
957-51-7: see diphenamid
973-21-7: see dinobuton
991-81-5: see chlormequat chloride

1014-69-3: see desmetryn
1018-64-2: see cadmium chloride
1024-57-3: see heptachlor epoxide
1071-83-6: see glyphosate
1085-98-9: see dichlofluanid
1113-02-6: see omethoate
1124-33-0: see Avitrol 100
1134-23-2: see cycloate
1194-65-6: see dichlobenil
1198-55-6: see tetrachlorocatechol
1214-39-7: see 6-benzyladenine
1249-84-9: see azacosterol
1302-45-0: see aluminum phosphide
1303-96-4: see borax
1303-28-2: see arsenic pentoxide
1306-19-0: see cadmium oxide
1314-84-7: see zinc phosphide
1317-39-1: see copper oxide
1319-77-3: see cresylic acid
1327-53-3: see arsenic trioxide
1330-20-7: see xylene
1332-40-7: see copper oxychloride
1344-81-6: see calcium polyphosphide
1338-02-9: see copper napthenate
1397-94-0: see antimycin A
1420-04-8: see niclosamide
1420-07-1: see dinoterb
1461-22-9: see tributyltin chloride complex
1563-66-2: see carbofuran
1582-09-8: see trifluraline
1596-84-5: see daminozide
1610-18-0: see prometon
1634-78-2: see malaoxon
1646-87-3: see aldicarb sulfoxide
1646-88-4: see aldicarb sulfone
1689-83-4: see ioxynil
1689-84-5: see bromoxynil
1702-17-6: see clopyralid
1746-18-2: see monolinuron
1746-01-6: see TCDD
1754-58-1: see diamidfos
1757-18-2: see Akton
1762-95-4: see ammonium thiocyanate
1836-77-7: see CNP
1861-32-1: see DCPA
1861-40-1: see benefin
1897-45-6: see chlorothalonil
1910-42-5: see paraquat dichloride
1912-24-9: see atrazine
1918-00-9: see dicamba
1918-02-1: see picloram
1918-11-2: see terbucarb
1918-16-7: see propachlor
1929-77-7: see vernolate
1929-82-4: see nitrapyrin
1929-88-0: see benzthiazuron
1982-47-4: see chloroxuron
1982-49-6: see siduron

8032-32-4: see mineral spirits
8047-13-0: see ryania
8052-41-3: see stoddard solvents
8065-48-3: see demeton
8065-36-9: see bufencarb
9003-13-8: see butoxy polypropylene glycol
9006-42-2: see metiram
10034-04-8: see calcium chloride
10035-10-6: see hydrogen bromide
10043-35-3: see boric acid
10048-95-0: see sodium arsenate &IIIé
10101-41-4: see calcium sulfate dihydrate
10124-36-4: see cadmium sulfate
10265-92-6: see acephate-met
10311-84-9: see dialifor
10380-28-6: see oxine-copper
10402-15-0: see copper citrate
10453-86-8: see resmethrin
10605-21-7: see carbendazim
12001-20-6: see cadmium-calcium-copper-zinz-
   chromate complex
12001-03-8: see copper acetoarsenite
12057-74-8: see magnesium phosphide
12069-69-1: see basic copper carbonate
12122-67-7: see zineb
12427-38-2: see maneb
13067-93-1: see cyanophenphos
13071-79-9: see terbufos
13121-70-5: see cyhexatin
13171-21-6: see phosphamidon
13194-48-4: see ethoprop
13356-08-6: see fenbutatin-oxide
13364-45-7: see chlorbromuron
13366-73-9: see photodieldrin
13684-56-5: see desmedipham
13684-63-4: see phenmedipham
14047-09-7: see TCAB
14255-88-0: see fenazaflor
14484-64-1: see ferbam
14491-59-9: see credazine
14816-18-3: see phoxim
14816-20-7: see chlorphoxim
15096-52-3: see cryolite
15263-53-3: see cartap
15457 05-3: see fluorodifen
15652-38-7: see decafentin
15662-33-6: see ryania
15922-78-8: see soʒi.m omadine
15972-60-8: see alachlor
16672-87-0: see ethephone
16752-77-5: see methomyl
16828-95-8: see copper ammonium complex
17109-49-8: see edifenphos
17804-35-2: see benomyl
18181-70-9: see idofenphos
18181-80-1: see bromopropylate
19044-88-3: see oryzalin
19666-30-9: see oxadiazon

20427-59-2: see copper hydroxide
20762-60-1: see potassium azide
21087-64-9: see metribuzin
21609-90-5: see leptophos
21725-46-2: see cyanazine
22224-92-6: see fenamiphos
22781-23-3: see bendiocarb
22936-86-3: see cyprazine
23135-22-0: see oxamyl
23184-66-9: see butachlor
23422-53-9: see formetanate hydrochloride
23505-41-1: see pirimiphos-ethyl
23560-59-0: see heptenophos
23564-05-8: see thiophanate methyl
23564-06-9: see thiophanate ethyl
23947-60-6: see ethirimol
23950-58-5: see pronamide
24017-47-8: see triazophos
24019-80-1: see nuclear polyhedrosis virus
24934-91-6: see chlormephos
25167-83-3: see tetrachlorophenol
25311-71-1: see isofenphos
25954-13-6: see fosamine ammonium
26002-80-2: see phenothrin
26225-79-6: see ethofumesate
26628-22-8: see sodium azide
26718-65-0: see mevinphos
27314-13-2: see norflurazon
28249-77-6: see benthiocarb
28434-00-6: see S-bioallethrin
28434-01-7: see bioresmethrin
28772-56-7: see bromadiolone
29091-05-2: see dinitramine
29232-93-7: see pirimiphos-methyl
29804-22-6: see disparlure
29973-13-5: see ethiofencarb
30525-89-4: see paraformaldehyde
30560-19-1: see acephate
30622-37-8: see penta-s-triazinetrione
32809-16-8: see procymidone
33089-61-1: see amitraz
33113-08-5: see copper ammonium carbonate
33245-39-5: see fluchloralin
33271-65-7: see cupric hydrazinium sulfate
33629-47-9: see butralin
33820-53-0: see isopropalin
34014-18-1: see tebuthiuron
34256-82-1: see acetochlor
34465-46-8: see hexachlorodibenzo-p-dioxin
34681-10-2: see butoxycarboxim
34987-38-7: see diphenyl
35256-85-0: see butam
35367-38-5: see diflubenzuron
35400-43-2: see sulprofos
35764-59-1: see cismethrin
35822-46-9: see heptachlorodibenzo-p-dioxin
36734-19-7: see iprodione
37764-25-3: see dichlormid

CHARTS OF PESTICIDE CHARACTERISTICS

Chapter Four

Charts of Pesticide Characteristics

Charts are alphabetized under common names.

NAME:	Common Trade and Other Chemical CAS Number	Class of Chemical	Chief Pesticide Use; Status	Persistence	Effects on Mammals		Adverse effects on other non-target species
					Immediate Toxicity (Acute)	Long-Term Toxicity (Chronic)	Physical properties
abamectin Affirm; Agrimek; Avermectin B1; Avid; Avomec; MK-936; Vertimec (10E,14E,16E ,22Z)-(1R,4S ,5'S,6S,6'R, 8R,12S,13S,2 0R,21R,24S)-6'[(S)-sec-butyl]-21,24-dihydroxy-5',11,13,22-tetramethyl-2-oxo-3,7,19-trioxatetracyclo [15.6.1.1^{4,8}.0^{20,24}]pentacosa-10,14,16,22-tetraene-6-spiro-2'-(5',6'- dihydro-2'H-pyran)-12-yl 2,6-didioxy-4-O-(2,6-didioxy-3-O-methyl-a-L-arabino-hexapyranosyl)-3-O-methyl-a-L-arabino-hexapyranoside (i) mixture with (10E,14E,16E,22Z)-(1R,4S,5'S,6S,6'R,1R, 12S,13S,20R,21R,24S)-21,22-dihydroxy-6'-isopropyl-5',11,13,22-tetramethyl-2-oxo-3,7,19-trioxatetracyclo[15.6.1.1^{4,8}.0^{20,24}] pentacosa-10,14,16,22-tetraene-6-spiro-2'-(5',6'-dihydro-2'H-pyrano)-12-yl 2,6-dideoxy-4-O-(2,6-dideoxy-3-O-methyl-a-L-arabino-hexopyranoxyl)-3-O-methyl-a-L-arabino-hexapyranoside (ii) (4:1) CAS # 71751-41-2		biological	insecticide acaricide	mod-pers (1)	oral: very high (2) dermal: low to medium (3) inhalation: ?	?	immediate toxicity: birds: low to medium (3) fish: very high (3) aquatic insects: very high (3) water: slightly soluble oil: insoluble non-volatile
acephate Orthene; Ortho 12420; Ortran O,S-dimethyl acetylphosphoramidothioate; O,S-dimethyl acetic phosphoramidothioate CAS # 30560-19-1		organo-phosphate	insecticide	non-pers (1)	oral: medium to high (2,3) dermal: ? inhalation: medium (1)	suspect carcinogen (4) suspect mutagen (4) fetotoxin (4) "some evidence of hormonal effects" (5)	immediate toxicity: birds: medium to high (6) fish: low (7) crustaceans: low (4) molluscs: medium (4) bees: high (13) plants: low; in plant tissue, metabolizes to acephate met (8-10) long-term toxicity: birds: may affect behavior and breeding success (11,12) water: "very soluble" slightly volatile
contaminant(s):							
O,O,S-trimethyl phosphorothioate					oral: high (1)	delayed toxicity (1)	
methylthioacetate					dermal: medium to high (1) inhalation: medium (1)	suspect mutagen (1) eye damage (1)	
transformation product(s):							
acephate-met (see acephate-met)							

83

NAME: Common Trade and Other Chemical CAS Number	Class of Chemical	Chief Pesticide Use; Status	Persistence	Effects on Mammals		Adverse effects on other non-target species Physical properties
				Immediate Toxicity (Acute)	Long-Term Toxicity (Chronic)	
acephate-met BAY 71628; Baythroid TM (with cyfluthrin); methamidophos; Monitor; Ortho 9006; SRA 5172; Tamaron; Tamaron (with parathion) O,S-dimethyl phosphoramidothioate CAS # 10265-92-6	organo-phosphate	insecticide restricted use: USA	non-pers (1)	oral: very high (2) dermal: very high (2) inhalation: high (3)	hair loss (8) decreased fertility (8)	immediate toxicity: birds: high (4) fish: low (5) bees: high (6) crustaceans: very high (7) water: "readily soluble" slightly volatile
acetochlor Acenit; Harness 2-chloro-N-ethoxymethyl-6'-ethylacet-o-toluidide; chloro-N-(ethoxymethyl)-N-(2-ethyl-6-methylphenyl) acetamide CAS # 34256-82-1	amide	herbicide	?	oral: medium (1) dermal: medium (1) inhalation: low to medium (1)	suspect carcinogen (2)	immediate toxicity: birds: medium (1) fish: high (1) crustacean: medium (1) water: soluble "negligible vapor pressure"
acifluorfen Blazer 2L; Blazer 2S; Galaxy (with bentazon); Storm (with bentazone); Tackle 5-(2-chloro-α,α,α-trifluoro-p-tolyloxy)-2-nitrobenzoic acid; 5-[2-chloro-4-(trifluoromethyl) phenoxy]-2-nitrobenzoic acid CAS # 50594-66-6	phenoxy	herbicide	non-pers to mod-pers (1)	oral: medium (1) dermal: ? inhalation: ?	carcinogen (2) heart, kidney, blood & liver damage (2) delayed fetal development (2)	immediate toxicity: birds: medium to high (1) fish: low to medium (1) crustaceans: low (1) water: very soluble non-volatile combustible
acrolein acrylaldehyde; Aqualin; Aqualin Biocide; Aqualin Slimicide; Magnicide H Herbicide acrylic aldehyde; allyl aldehyde; ethylene aldehyde; α-pentyl-β-phenyl acrylaldehyde; 2-propenal CAS # 107-02-8	aldehyde	fungicide herbicide restricted use, USA	non-pers (1)	oral: very high (1) dermal: high (2) inhalation: "extremely toxic" (2)	suspect mutagen (3-5) teratogen (4) embryotoxin (4)	immediate toxicity: birds: very high (6) fish: very high (7) water: "soluble" highly volatile flammable
acrylonitrile Acritet; Acrylofume; Acrylon; Carbacryl; Miller's Fumigrain; VCN; Ventox vinyl cyanide; cyanoethylene; propenenitrile CAS # 107-13-1	cyanide	fumigant insecticide voluntary cancellation of most uses by producer, USA, 1978	?	oral: high (1) dermal: ? inhalation: ?	carcinogen (2,3) mutagen (4) teratogen (5)	water: "soluble" flammable
Akton Akton; Axiom; SD 9098 O-[2-chloro-1-(2,5-dichlorophenyl) vinyl]O,O-diethyl phosphorothioate CAS # 1757-18-2	organo-phosphate	insecticide	mod-pers (1)	oral: very high (2) dermal: ? inhalation: ?	?	immediate toxicity: birds: medium (3) fish: high to very high (4) bees: medium (5)

NAME: Common Trade and Other Chemical CAS Number	Class of Chemical	Chief Pesticide Use; Status	Persistence	Effects on Mammals		Adverse effects on other non-target species Physical properties
				Immediate Toxicity (Acute)	Long-Term Toxicity (Chronic)	
alachlor Adeochlor; Bronco (with glyphosate); Bullet; Cannon; CP 50144; Lariat; Lasso (with atrazine); Lazo; Partner-Mon-9848; Pillarzo; Rambo (with atrazine) 2-chloro-2',6'-diethyl-*N*-(methoxymethyl)-acetanilide CAS # 15972-60-8	amide	herbicide banned in Canada, Sweden restricted use, USA	non-pers to mod-pers (1)	oral: medium (2) dermal: medium (1) inhalation: ?	carcinogen (3,4) suspect mutagen (5,6) eye damage: (3) kidney and liver damage (3)	immediate toxicity: birds: low to medium (1,7) fish: medium to high (1,8) bees: low to medium (9) water: soluble slightly volatile
aldicarb OMS771; Sentry (with lindane); Temik (with lindane); UC 21149 2-methyl-2-(methylthio) propionaldehyde O-(methylcarbamoyl) oxime CAS # 116-06-3	carbamate	insecticide acaracide nematocide restricted use: USA	non-pers to pers (1,2)	oral: very high (1) dermal: very high (1) inhalation: "toxic" (1)	suspect mutagen (3) may decrease learning behavior (4)	immediate toxicity: birds: high to very high (5,6) fish: high to very high (6,7) bees: very high (8) aquatic insects: high (6) water: soluble highly volatile
transformation product(s): **aldicarb sulfoxide** 2-methyl-2-(methylsulfinyl) propionaldehyde O-(methylcarbamoyl) oxime CAS # 1646-87-3				oral: very high (1)		
aldicarb sulfone 2-methyl-2-(methylsulfonyl) propionaldehyde O-(methylcarbamoyl) oxime CAS # 1646-88-4				oral: very high (1)		
aldrin Aldrex; aldrine (France); Aldrite; Aldrosol; Altox; Compound 118; Drinox; HHDN; Octalene; Seedrin hexachlorohexahydro-*endo-exo*-dimethanonaphthalene; 1,2,3,4,10,10,-hexachloro-1,4, 4a,5,8a-hexahydro-1,4-*endo*-*exo*-5,8-dimethanonaphthalene CAS # 309-00-2	organo-chlorine	insecticide most uses banned: USA, 1974 all products cancelled by 1987 with termiticide use ended	mod-pers to pers (1,2)	oral: high to very high (3) dermal: very high (3) inhalation: ?	cumulative (1,2) carcinogen (1,2,4) suspect teratogen (2,5) liver and kidney damage (2,6)	BIOCIDE immediate toxicity: birds: very high (3,7) fish: very high (7) amphibians: medium (7) aquatic insects: very high (7) crustaceans: very high (7) molluscs: very high (7) bees: very high (8) aquatic worms: medium (9) aquatic plants: medium (7) water: "insoluble" oil: "moderately soluble" slightly volatile flammable to combustible
transformation product(s): **dieldrin** (see dieldrin)						

NAME: Common Trade and Other Chemical CAS Number	Class of Chemical	Chief Pesticide Use; Status	Persistence	Effects on Mammals		Adverse effects on other non-target species Physical properties
				Immediate Toxicity (Acute)	Long-Term Toxicity (Chronic)	
allethrin Allethrin concentrate MGK; Alleviate; allycinerin; Exbiol; FDA 1446; FMC 249; Necarboxylic acid; Neo-Pynamin 5/1/30 (with phenothrin & piperonyl butoxide); NIA 249; OMS 468; pallethrine; Pesguard (with phenothrin & piperonyl butoxide); Pynamin; Pyrescel; Pyresin; Pyrocide; RU 28173; Sumithrin Plus (with phenothrin) (RS)-3-allyl-2-methyl-4-oxocyclopent-2-enyl (1RS)-cis,trans-chrysanthemate; (RS)-allethronyl (1R,cis,trans)chrysanthemate CAS # 584-79-2	pyrethroid	insecticide	?	oral: medium to high (1,2,3) dermal: low to medium (2,4) inhalation: medium to high (3)	suspect mutagen (5) suspect immunotoxin (6)	immediate toxicity: birds: low to medium (7,8) fish: low to very high (9,10) crustaceans: very high (10) aquatic insects: very high (10) volatile to highly volatile
isomer(s):						
d-cis/trans-allethrin Pynamin Forte d-cis/trans-allethrin; (RS)-[1R,cis,trans]-chrysanthemate CAS # 42534-61-2	pyrethroid	insecticide	?	oral: high (1) dermal: ? inhalation: ?	?	immediate toxicity: fish: "very high" (2) combustible
esbiothrin allethrin stereoisomer; Detrans (with deltamethrin); K-O (with deltamethrin & piperonyl butoxide); OMS 3045; RU 27436 (RS)-allethronyl [1R,trans]-chrysanthemate	pyrethroid	insecticide	?	oral: high (1) dermal: "low" (2) inhalation: ?	?	
S-bioallethrin Esbiol; esdepallethrin; OMS 3046; RU 16 121; RU 3054 d-allethronyl d-trans-allethrin; (+)-allethronyl (+)-trans-allethrin CAS # 28434-00-6	pyrethroid	insecticide	?	oral: medium to high (1,2) dermal: ? inhalation: medium to high (1)	?	immediate toxicity: birds: low (1) fish: medium to very high (1) water: medium volatile combustible
bioallethrin allethrin stereoisomer; depallethrin; Kefil (with permethrin & piperonyl butoxide); RU 11705; Vapona Flykiller (with permethrin & piperonyl butoxide) d-trans-allethrin; (+)-trans-allethrin; (RS)-allethronyl [1R,trans]-chrysanthemate CAS # 584-79-2	pyrethroid	insecticide	?	oral: medium to high (1,2) dermal: ? inhalation: medium (1)	suspect teratogen (3)	immediate toxicity: fish: very high (2) crustaceans: low (2) water: slightly soluble oil: slightly volatile to volatile combustible
allyl alcohol Hopkins Allyl Alcohol 2-propene-1-ol; vinyl carbinol; 1-propenol-3 CAS # 107-18-6	alcohol	soil fumigant insecticide restricted use, USA, 1976	?	oral: high (1) dermal: very high (1) inhalation: very high (2)	suspect mutagen (3,4) liver damage (5,6)	water: soluble highly volatile flammable
transformation product(s): **acrolein** (see acrolein)						

NAME:	Common Trade and Other Chemical CAS Number	Class of Chemical	Chief Pesticide Use; Status	Persistence	Effects on Mammals		Adverse effects on other non-target species Physical properties
					Immediate Toxicity (Acute)	Long-Term Toxicity (Chronic)	
aluminum phosphide	Celphos (India); Delicia (E. Germany); Phostoxin aluminum phosphide CAS # 1302-45-0	metal/ mineral aluminum	insecticide fumigant restricted use: USA	?	oral: very high (2) dermal: ? inhalation: ?	gastrointestinal damage (1) liver, heart, and kidney damage (1)	water: "slightly soluble"
transformation product(s):							
phosphine gas	CAS # 7803-51-2				oral: very high (1) dermal: medium (1) inhalation: "poisonous" (2)		water: "slightly soluble" oil: "insoluble" combustible, "can ignite spontaneously in cold air; explosive"
ametryn	A 1093; Ametrex; ametryne; Evik; G-34162; Gesapax; Gesapax combi (with atrazine); Gesapax H (with 2,4-D); Trinatox D 2-(ethylamino)-4-(isopropyl amino)-6-(methylthio)-s-triazine; N-ethyl-N'-(1-methyl ethyl)-6-(methylthio)-1,3,5-triazine-2,4-diamine CAS # 834-12-8	triazine	herbicide	non-pers to mod-pers (1)	oral: medium (2) dermal: low to medium (2) inhalation: ?	?	immediate toxicity: birds: low (1) fish: medium (1) crustaceans: high (3) bees: low (4) molluscs: medium to high (1) water: soluble oil: "soluble" slightly volatile "nonflammable"
aminocarb	A363; BAY 44646; Matacil 4-(dimethylamino)-m-tolyl methylcarbamate CAS # 2032-59-9	carbamate	insecticide	non-pers (1)	oral: very high (2) dermal: high (3) inhalation: medium (4)	suspect mutagen (5)	immediate toxicity: birds: very high (4) fish: high (6) crustaceans: low (6) bees: high (7) water: "slightly soluble" slightly soluble
amitraz	Azadieno; Azaform; BAAM; BTS 27419; Estrella; JA 119; Mitaban; Taktic; Triatox; Triazid; U-36059; Zipak (with bifenthrin) N'-(2,4-dimethylphenyl)-N-((2,4-dimethylphenyl(imino) methyl)-N-methylmethamimidamide; N-methyl-N'-2,4-xylyl-N-(N-2,4-xylyl-formimidoyl)formamidine; N',N'-[(methylimino)dimethylidyne] bis[2,4-xylidine] CAS # 33089-61-1	miscel-laneous	insecticide acaricide cancelled USA	?	oral: medium (1) dermal: medium to high (1) inhalation: ?	suspect carcinogen (2,3)	immediate toxicity: birds: medium (1) fish: medium to high (1,4) bees: low to medium (5) water: insoluble oil: "soluble" slightly volatile
transformation product(s):							
N'-[2,4-xylyl]N-methyl formamidine					oral: high (1)	decreased fertility and viability in young (1)	water: "sparingly soluble" slightly volatile

NAME: Common Trade and Other Chemical CAS Number	Class of Chemical	Chief Pesticide Use; Status	Persistence	Effects on Mammals		Adverse effects on other non-target species Physical properties
				Immediate Toxicity (Acute)	Long-Term Toxicity (Chronic)	
amitrole 3,A-T; Altazin (with atrazine); Amerol; Amino Triazole; aminotriazole; Amitrole-T; Amizol; AT; ATA; Azolan; Azole; Chempar Amitrole; Cytrol; Diurol; Domatol; Herbizole; Ustinex; Vorox; Weedazol; Weedazol T; X-All 3-amino-1,2,4-triazole; 3-amino-s-triazole CAS # 61-82-5	triazole	herbicide restricted use, USA	non-pers to mod-pers (1)	oral: low (2) dermal: low to medium (1) inhalation: ?	carcinogen (3,4) suspect mutagen (3) fetotoxin (9) liver damage (9) goiters (5)	immediate toxicity: birds: low to medium (6) fish: low (7) crustaceans: low (7) bees: low to medium (8) water: "soluble" oil: "insoluble" slightly volatile
ammonium arsenate	inorganic					
ammonium thiocyanate Amitrile T.L. (with amitrole); Amitrole-T (with amitrole); ammonium rhodanide; ammonium sulfocyanate; ammonium sulfocyanide; Amthio; Cytrol Amitrole-T (with amitrole); Giror (with amitrole & paraquat); Radazone TL (with amitrole) CAS # 1762-95-4	cyanide	herbicide soil sterilant	non-pers to mod-pers (1)	oral: medium (2) dermal: ? inhalation: ?	may metabolize slowly in the body to cyanide (3)	water: "very soluble" "noncombustible"
amobam Amobam; Chem-O-Bam diammonium ethylene bisdithiocarbamate CAS # 3566-10-7	thiocar-bamate	fungicide	?	oral: high (1) dermal: ? inhalation: ?	?	water: "very soluble"
transformation product(s):						
ethylene thiourea ETU 2-imidazolidinethione CAS # 96-45-7				oral: medium (1)	carcinogen (2,3) suspect mutagen (4) teratogen (4,5,6) increased fluid in skull (5,7) goitrogenic (3)	
AMS Amcide; Ammat; Ammate X; Ammate X-NI; ammonium amidosulphate; ammonium sulphamidate; Ikurin; Sulfamate; sulfamate ammonium amidosulphate; ammonium sulphamidate; ammonium sulfamate CAS # 773-06-0	metal\mineral sulfur	herbicide	non-pers to mod-pers (1)	oral: medium (2) dermal: ? inhalation: ?	?	immediate toxicity: birds: medium (2) fish: low (1) bees: low to medium (3) water: very soluble non-volatile "nonflammable"
anilazine B-622; Direz; Dyrene; Kemate; Triasyn 2,4-dichloro-6-(o-chloro anilino)-s-triazine CAS # 101-05-3	triazine	fungicide	non-pers (1)	oral: low to high (1) dermal: medium (1) inhalation: ?	?	BIOCIDE immediate toxicity: birds: low (2) fish: very high (1) crustaceans: high (2) bees: low (3) phytoplankton: high (2) plants: toxic to some (4) water: "insoluble"

NAME: Common Trade and Other Chemical CAS Number	Class of Chemical	Chief Pesticide Use; Status	Persistence	Effects on Mammals		Adverse effects on other non-target species Physical properties
				Immediate Toxicity (Acute)	Long-Term Toxicity (Chronic)	
anthraquinone Anthracene; anthradione; Corbit; Heolite; Morkit; Quinolate AC, Fs, Quinolate AC Kara (with oxine-copper); Quinolate Triple Kara (with oxine-copper & lindane); Quinolate V 4 X AC, FS, DS (with oxine-copper & carboxin) 9,10-anthraquinone; 9,10-anthracenedione CAS # 84-65-1	quinone	bird repellent	?	oral: medium (1) dermal: low to medium (2) inhalation: ?	suspect mutagen (3,4)	immediate toxicity: birds: low (5) water: "insoluble" combustible
antimycin A antimycin A1; antipiricullin; Fintrol; virosin 3-[[3-(formylamino)-2-hydroxybenzoyl] amino]-8-hexyl-2,6-dimethyl- 4,9-dioxo-1,5-dioxonan-7-yl 3-methylbutanoate CAS # 1397-94-0	antibiotic	fungicide piscicide	non-pers (1)	oral: ? dermal: ? inhalation: ?	?	immediate toxicity: amphibians: low (3) reptiles: low (3) crustaceans: low to medium (2) plankton: low (3) water: "insoluble"
antimycin A3 Blastmycin [2R-(2R*,3S*,6S*,7R*,8R*)]-8-butyl- 3-[[3-(formylamino)-2-hydroxybenzoyl] amino]-2,6-dimethyl- 4,9-dioxo-1,5-dioxonan-7-yl-3-metylbutanoate CAS # 522-70-3	antibiotic	fungicide piscicide	non-pers (1)	oral: very high (2) dermal: ? inhalation: ?	?	water: "insoluble"
ANTU Alrato; Antu; Anturat; Bantu; Chemical 109; Dirax; Kill Kantz; Killer Katz; Kripid; krysid (USSR); Naphtox; Rat-tu; Rattrack; Rattract; Smeesana; U5227 α-napthyl thiourea; 1-(1-napthyl)-2-thiourea; α-naphthylthiocarbamide; 1-naphthalenylthiourea CAS # 86-88-4	miscel- laneous	rodenticide registration voluntarily cancelled by manufacturer , USA	?	oral: very high (1) dermal: ? inhalation: ?	formerly carcinogenic napthylamines suspected as impurities (2)	immediate toxicity: birds: medium (2) water: soluble
contaminant(s): B naphthylamine					carcinogen (1)	
Aramite 88R; Acacide; Aracide; Aramite-15W; Aratron; CES; Niagaramite; Ortho-mite 2(p-tert-butylphenoxy)-1-methylethyl- 2'-chloroethyl sulfite CAS # 140-57-8	organic	acaricide insecticide registration voluntarily cancelled by manufacturer USA	non-pers (1)	oral: medium (2) dermal: ? inhalation: ?	carcinogen (3)	BIOCIDE immediate toxicity: birds: very high (4) fish: very high (5) crustaceans: very high (5) bees: low to medium (6) aquatic insects: very high (5) water: "insoluble" oil: "soluble"

NAME: Common Trade and Other Chemical CAS Number	Class of Chemical	Chief Pesticide Use; Status	Persistence	Effects on Mammals		Adverse effects on other non-target species Physical properties
				Immediate Toxicity (Acute)	Long-Term Toxicity (Chronic)	

Arosurf
 	
Arosurf
 	
poly(oxy-1,2-ethanediyl), a-isooctadecyl-ω-hydroxy
 	
CAS # 52292-17-8
	miscel-laneous	insecticide surfactant	non-pers (1)	oral: low (1)
				
				dermal: ?
				
				inhalation: low (1)
					?	immediate toxicity:
						  fish: low (1)
						  crustaceans: high (1)
						
						combustible

arsenic colloidal arsenic; metallic arsenic Characteristics as a class are given here; variations are indicted for each compound below CAS # 7440-38-2	metal/mineral arsenic	insecticide herbicide rodenticide	pers (1)	oral: high (2) dermal: ? inhalation: ?	cumulative (3) carcinogen (4-6) mutagen (3) teratogen (3,7)	water: "insoluble"
compound(s):						
ammonium arsenate CAS # 53404-17-4	inorganic					
ammonium methanearsonate AMA; Ansar 157; Super Crab-E-Rad A.M.A. monoammonium methanearsonate; monoammonium methylarsonate; methanearsonic acid, monoammonium salt; methylarsonic acid, monoammonium salt; MAMA; MMA; amine methanearsonate* (also called AMA) *combination of dodecylammonium methanearsonate and octylammonium methanearsonate CAS # 2321-53-1	organic	herbicide		oral: medium (1,2)		
arsenic acid Crab Grass Killer; Hi-Yield Dessicant H-10; Zotox arsenic acid; o-arsenic acid CAS # 7778-39-4	inorganic	herbicide		oral: high to very high (1)	suspect mutagen (2)	
arsenic pentoxide arsenic acid anhydride; arsenic anhydride; arsenic oxide; arsenic penthoxide; arsenic [V] oxide CAS # 1303-28-2	inorganic	herbicide fungicide wood preservative			suspect mutagen (1,2) testicular damage (3)	water: soluble
arsenic trioxide Arsenolite; Arsodent; Clandelite; white arsenic arsenic oxide; arsenic (III) oxide; arsenic sesquioxide; arsenious acid;	inorganic	herbicide insecticide rodenticide cancelled USA, 1977		oral: high (1)	carcinogen (2) suspect mutagen (3) fetotoxin (3,4)	

NAME: Common Trade and Other Chemical CAS Number	Class of Chemical	Chief Pesticide Use; Status	Persistence	Effects on Mammals		Adverse effects on other non-target species Physical properties
				Immediate Toxicity (Acute)	Long-Term Toxicity (Chronic)	
arsenic trioxide (*cont.*) arsenious trioxide; arsenolite; arsenous acid; arsenous acid, anhydride; arsenous oxide arsenous oxide, anhydride; arsentrioxide CAS # 1327-53-3						
cacodylic acid Cotton Aide HC; Dilic; DMAA; Moncide; Montar; Phytar 138; Rad-E-Cate 25; Silvisar dimethylhydroxyarsine oxide; dimethylarsinic acid; dimethylarsenic acid; CAS # 75-60-5	organic			oral: medium		
calcium acid methanearsenate CAMA; Super Crab-E-Rad-Calar; Super Dal-E-Rad methanearsonic acid, calcium salt; calcium acid methanearsenate; calcium acid methyl arsenate; calcium methanearsenate CAS # 5902-95-4	organic	herbicide		oral: medium (1)		"nonflammable"
calcium arsenate Chip-Cal Granular; Cucumber Dust; FLAC; KALO; Kilmag; Pencal; Security; Spra-Cal arsenic acid, calcium salt; calcium arsenate; calcium o-arsenate; tricalcium arsenate CAS # 7778-44-1	inorganic	insecticide cancelled USA	pers (1)	oral: medium to very high (2,3) dermal: medium (4)		immediate toxicity: bees: medium (5) water: insoluble to soluble
calcium arsenite	inorganic	insecticide				water: "slightly soluble"
calcium propanearsenate propanearsonic acid, calcium salt; calcium propanearsonate; calcium propyl arsonate CAS # 126-94-3	organic	herbicide		oral: high (1)		
chromated copper arsenate CCA; Osmose; Wolmanized; WoodPlus copper chromated arsenate	inorganic	fungicide insecticide restricted use, USA	"arsenic can be released from pressure treated wood" (1) "sealant...did not reduce the dislodgeable arsenic levels" from wood. (1)			

(*continued on next page*)

NAME: Common Trade and Other Chemical CAS Number	Class of Chemical	Chief Pesticide Use; Status	Persistence	Effects on Mammals		Adverse effects on other non-target species Physical properties
				Immediate Toxicity (Acute)	Long-Term Toxicity (Chronic)	
copper acetoarsenite Emerald green; French green; mitis green; Paris green; Schweinfurt green CAS # 12001-03-8	organic	cancelled USA, 1977		oral: very high (1)		immediate toxicity: amphibians: very high (1)
cupric arsenite Scheele's green arsonic acid copper (2+) salt; arsenious acid copper (2+) salt	inorganic	fungicide insecticide rodenticide				water: "practically insoluble"
DSMA Arrhenal; Arsinyl; Di-Tac; disodium methanearsonate; DMA; DMA 100; Methar 30; Sodar disodium methanearsoate; disodium methylarsonate; methanearsonic acid, disodium salt; methylarsonic acid, disodium salt; disodium acid methanearsonate CAS # 144-21-8	organic	herbicide	mod-pers (1)	oral: medium (2,3) dermal: very high (4) inhalation: low (2)		
lead arsenate arsenate of lead; dibasic lead arsonate; Gypsine; lead acid arsonate; Soprabel; Talbot arsenic acid, lead salt; acid lead arsonate; acid lead o-arsenate CAS # 7784-40-9	inorganic	insecticide; fungicide cancelled, USA	pers (1)	oral: medium to high (1)		water: "slightly soluble"
methyl arsonic acid MAA methanearsonic acid; methylarsenic acid; monomethylarsinic acid CAS # 124-58-3	organic					
MSMA Check Mate; Herb-All; Merge 823; monosodium methanearsonate; Quadmec (with 2,4-D & dicamba & mecoprop); Silvisar; Target MSMA; Trans-Vert monosodium methanearsonate; monosodium methylarsonate; methanearsonic acid, monosodium salt; methylarsonic acid, monosodium salt; monosodium acid methanearsonate CAS # 2163-80-6	organic	herbicide	mod-pers to pers (1,2)	oral: medium to very high (3,4) dermal: medium to very high (4,5) inhalation: low (3)	toxic hepatitis (6)	immediate toxicity: fish: low to medium (3) long-term toxicity: may accumulate in plants (7) water: very soluble
OBPA 10,10'-oxybis-10H-phenox arsine CAS # 58-36-6	organic	antibiotic restricted USA, 1979		oral: very high (1) dermal: very high (1) inhalation: low (1)	cumulative (1)	immediate toxicity: birds: low (1) fish: very high (1) crustaceans: very high (1) water: slightly soluble highly volatile

NAME: Common Trade and Other Chemical CAS Number	Class of Chemical	Chief Pesticide Use; Status	Persistence	Effects on Mammals		Adverse effects on other non-target species Physical properties
				Immediate Toxicity (Acute)	Long-Term Toxicity (Chronic)	
sodium arsenate [I] Fatsco Ant Poison; Sweeney's Ant-Go sodium arsenate; sodium m-arsenate; sodium o-arsenate CAS # 7631-89-2	inorganic	wood preservative all non wood preservative uses cancelled, USA 1986			teratogen (1) embryotoxin (2) liver and kidney effects (3)	water: "very soluble"
sodium arsenate [II] disodium arsenate; disodium arsenic acid; disodium hydrogen arsenate; disodium hydrogen o-arsenate; disodium monohydrogen arsenate; sodium acid arsenate; dibasic sodium arsenate; sodium arsenate dibasic anhydrous CAS # 7778-43-0	inorganic	wood preservation all nonwood preservative cancelled, uses USA, 1986			carcinogen (1) suspect mutagen (2,3)	water: "very soluble"
sodium arsenate [III] arsenic acid, disodium salt, heptahydrate; dibasic sodium arsenate; disodium arsenate, heptahydrate; sodium acid arsenate, heptahydrate; sodium arsenate; dibasic, heptahydrate; disodium arsenate heptahydrate; sodium arseniate CAS # 10048-95-0	inorganic	wood preservative all non wood preservation uses cancelled, USA, 1986			suspect mutagen (1,2) teratogen (3,4) embryotoxin (5)	
sodium arsenite Atlas 'A'; Chem Pels; Chem-Sen 56; Kill-All; Penite; Prodalumnol Double sodium arsenite; sodium m-arsenite; arsenous acid, sodium salt; arsenious acid, monosodium salt; arsenious acid; sodium salt CAS # 7784-46-5	inorganic	fungicide herbicide insecticide cancelled, most uses, USA, 1978	pers (1)	oral: very high (2) dermal: very high (3) inhalation: ?	suspect mutagen (4,5) teratogen (6) fetotoxin (6) embryotoxin (7)	immediate toxicity: birds: high to very high (2) crustaceans: very high (9) fish: high to very high (9) molluscs: very high (9) water: "very soluble"
asulam Asilan; Asulfox F; Asulox; Asulox 40; Candex (with atrazine); Dialam (with diuron); Graslam (with mecoprop & MCPA); Jonnix; Krater (with diuron); MB 9057; Slam (with dalapon); Talent (with paraquat); Target (with dalapon); Tartan (with diuron) methyl sulfanilylcarbamate; methyl [(4-aminophenyl)sulfonyl]carbamate CAS # 3337-71-1	carbamate	herbicide	non-pers (1)	oral: low (1) dermal: low to medium (1) inhalation: low to medium (1)	suspect carcinogen (2)	immediate toxicity: birds: low to medium (1) fish: low (3) crustaceans: low (2) bees: low to medium (4) water: slightly soluble non-volatile "nonflammable"

NAME: Common Trade and Other Chemical CAS Number	Class of Chemical	Chief Pesticide Use; Status	Persistence	Effects on Mammals		Adverse effects on other non-target species Physical properties
				Immediate Toxicity (Acute)	Long-Term Toxicity (Chronic)	
atrazine AAtrex; Altacide Extra (with sodium chlorate); Atlazin (with amitrole); Atradex; Atradex 50; Atranex; Bellater (with cyanazine); Bicep (with metolachlor); Extrazine (with cyanazine); Fogard; Fogard L; Geigy 30,027; Gesaprim; Gesaprim D (with 2,4-D); Lasso (with alachlor); Primatol; Sutan+; Vectal; Vorox Granulat 371; Weedex A 2-chloro-4-ethylamino-6-isopropylamino-s-triazine CAS # 1912-24-9	triazine	herbicide restricted use, USA, 1990	mod-pers to pers (1,2)	oral: low to medium (3,4) dermal: medium (3) inhalation: "low" (4)	carcinogen (5,6) mutagen (7,8) immunotoxin (9) adrenal damage (10)	immediate toxicity: fish: low to high (4) crustaceans: low to medium (4) bees: medium (2) molluscs: high (1) aquatic insects: high to very high (13) long-term toxicity: soil invertebrates: may reduce populations (11) amphibians: may impair reproduction (13) water: slightly soluble slightly volatile
transformation product(s): N-nitrosoatrazine						
auramine 4,4-carbonimidoylbis[N,N-dimethylbenzenamine]monohydrochloride CAS # 492-80-8	miscel-laneous	fungicide	?	oral: ? dermal: ? inhalation: ?	carcinogen (1,2) suspect mutagen (3)	?
Avitrol 100 Avitrol 100 4-nitropyridine N-oxide CAS # 1124-33-0	miscel-laneous	avicide	?	oral: ? dermal: ? inhalation: ?	suspect mutagen (1,2)	immediate toxicity: birds: very high (3)
Avitrol 200 Avitrol 200 4-aminopyridine; 4-aminopyridine hydrochloride CAS # 504-24-5	miscel-laneous	avicide restricted, USA	mod-pers to pers (1)	oral: very high (1) dermal: high (1) inhalation: ?	?	immediate toxicity: birds: very high (2) fish: very high (1) water: very soluble
azacosterol azacholesterol; Azasterol; diazasterol; ornitrol; SC-12937 azacosterol hydrochloride; 20,25-diazacholesterol dihydrochloride; 17-β-(dimethylamino)propyl) methylamino)adrost-5-en-3-β-ol dihydrochloride CAS # 1249-84-9	miscel-laneous	bird sterilant	?	oral: medium to high (1) dermal: ? inhalation: ?	?	water: "moderately soluble" oil: "insoluble"
azinphos-ethyl Acifon; Azinos; azinphos-ethyl; BAY 16259; Bionex; Cotnion-Ethyl; Crysthion; Ethyl Guthion; ethylazinphos; Gusathion A; R1513; triazotion (USSR) O,O-diethyl 3-(4-oxo-1,2,3-benzotriazin-3 (4H)-yl)methyl phosphorodithioate CAS # 2642-71-9	organo-phosphate	insecticide not registered, USA	?	oral: very high (1) dermal: high (1) inhalation: ?	?	immediate toxicity: birds: "toxic" (1) fish: "toxic" (1) non-volatile to slightly volatile "flammable"

NAME: Common Trade and Other Chemical CAS Number	Class of Chemical	Chief Pesticide Use; Status	Persistence	Effects on Mammals		Adverse effects on other non-target species Physical properties
				Immediate Toxicity (Acute)	Long-Term Toxicity (Chronic)	
azinphos-methyl azinphos-methyl; BAY 17147; Carfene; Cotnion-methyl; Gusathion; Guthion; Methyl Guthion; methylazinphos; metil-triazotion (USSR) O,O-dimethyl S[4-oxo-1,2,3-benzotriazin-3(4H)ylmethyl]phosphorodithioate CAS # 86-50-0	organo-phosphate	insecticide acaricide restricted use, USA, 1977	non-pers (1)	oral: high to very high (2,3) dermal: high to very high (2,3) inhalation: ?	suspect carcinogen (4,5) suspect mutagen (6,7)	immediate toxicity: birds: high (3) fish: very high (3) crustaceans: very high (3) aquatic insects: very high (3) slightly volatile combustible
transformation product(s):						
oxygen analogue of azinphosmethyl			appears one week after spraying with azinphos-methyl (1)			
aziprotryne aziprotryn; Brasoran; Brassoron; C07019; Mesoranil; Mezuron 2-azido-4-isopropylamino-6-methylthio-s-triazine; 4-azido-N-(1-methylethyl)-6-(methylthio)-1,3,5-triazin-2-amine CAS # 4658-28-0	triazine	herbicide not registered, USA	non-pers to mod-pers (1)	oral: low to medium (2) dermal: low to medium (2) inhalation: ?	?	water: slightly soluble slightly volatile
azobenzene diphenyldiimide; diphenyldiazene; azobenzide; azobenzol CAS # 103-33-3	miscel-laneous	acaricide fumigant withdrawn from market by manufacturer USA	?	oral: medium (1) dermal: ? inhalation: ?	carcinogen (2,3) mutagen (4,5) liver damage (6) spleen damage (4)	water: "practically insoluble"
transformation product(s):						
azoxybenzene Azomyte diphenyldiazene 1-oxide; azobenzenoxide; azoxybenzide; azosydibenzene CAS # 495-48-7	miscel-laneous	acaricide		oral: low to medium (1)		water: "insoluble"
Bacillus thuringiensis (Berliner) B.T.	biological	insecticide	non-pers (1)	oral, dermal, inhalation: "non-toxic" (2,3)	?	?
varieties:						
Bacillus thuringiensis var. aizawei Certan Bacillus thuringiensis, variety aizawei		larvacide for wax moth				
Bacillus thuringiensis var. israelensis Bactimos; BMC; Skeetal; Teknar; Vectobac Bacillus turingiensis, variety israelensis		larvacide for mosquitoes and some other flies				water: "insoluble"

(continued on next page)

NAME: Common Trade and Other Chemical CAS Number	Class of Chemical	Chief Pesticide Use; Status	Persistence	Effects on Mammals		Adverse effects on other non-target species Physical properties
				Immediate Toxicity (Acute)	Long-Term Toxicity (Chronic)	
Bacillus thuringiensis var. kurstaki Agritol; Attack; Bactospeine; Bactur; Bakthane; Biotrol; BTV; Bug Time; Cekubacillina; Dipel; Foray; Javelin; Larvatrol; Leptox; Novabac; Thuricide; Tribactur *Bacillus thuringiensis, variety kurstaki*		larvacide for moths				water: "insoluble"
Bacillus thuringiensis var. san diego M-One *Bacillus thuringiensis, variety san diego*		larvacide for some beetles				
Bacillus thuringiensis var. tenebrionis Trident *Bacillus thuringiensis, variety tenebrionis*		larvacide for some beetles				
barban barbamate (So. Africa); barbane (France); Carbyen; Caryne; CBN; chlorinat (USSR); Fisons B25; Neoban 4-chloro-2-butynyl *m*-chlorocarbanilate; 4-chloro-2-butyl *N*-(3-chlorophenyl) carbamate CAS # 101-27-9	carbamate	herbicide	mod-pers (1)	oral: medium (2) dermal: ? inhalation: low (2)	?	immediate toxicity: fish: medium (1) bees: low to medium (3) water: slightly soluble to very soluble slightly volatile flammable
barium metaborate	metal/mineral	fungicide	?	oral: ? dermal: ? inhalation: ?	?	water: soluble
barthrin 6-chloropiperonyl chrysanthemate; 6-chloropiperonyl-2,2-dimethyl-3-(2-methylpropenyl) cyclopropanecarboxylate CAS # 70-43-9	pyrethroid	insecticide	?	oral: low (1) dermal: ? inhalation: ?	?	oil: "soluble in kerosene"
benalaxyl Galben C. (with copper oxychloride); Galben F (with folpet); Galben M (with mancozeb); Galben RF (with copper sulfate and folpet); Galben Z (with zineb); Galben; Tairel; Tairel C. (with copper oxychloride); Tairel F (with folpet); Tairel M (with mancozeb); Tairel Z (with zineb) methyl *N*-phenylacetyl-*N*-2, 6-xylyl-ᴅʟ-alaninate CAS # 71626-11-4	miscellaneous	fungicide not registred for use in U.S.	?	oral: medium (1) dermal: ? inhalation: low to medium	?	immediate toxicity: birds: low (1) fish: medium (1) water: slightly soluble slightly volatile combustible

NAME: Common Trade and Other Chemical CAS Number	Class of Chemical	Chief Pesticide Use; Status	Persistence	Effects on Mammals		Adverse effects on other non-target species Physical properties
				Immediate Toxicity (Acute)	Long-Term Toxicity (Chronic)	
benazolin Asset; Banzsalox; Benazolox (with clopyralid); Bencornox; Benopan; Bensecal; Benzar; Cresopur; Galipan; Galtak; Grassland Weedkiller; Herbazolin; Keropur; Legumex Estra; Ley-Cornox (with 2,4-DB & MCPA); Leymin; Springclene 2; Tri-Cornox (with dicamba & dichlorprop); Tri-Cornox Special 4-chloro-2-oxo-3-benzothiazoline acetic acid; 4-chloro-2-oxobenzothiazolin-3-yl-acetic acid; 4-chloro-2-oxo-3(2H)-benzothiazol acetic acid CAS # 3813-05-6	miscel-laneous	herbicide	?	oral: low to medium (1,2) dermal: ? inhalation: ?	?	immediate toxicity: fish: medium (3) crustaceans: medium (3) bees: "non-toxic" (3) water: insoluble "nonflammable"
bendiocarb Dycarb; Ficam; Garvox; Multamat; Niomil; Seedox; Tatoo; Turcam 2,2-dimethyl-1,3-benzodioxol-4-yl methylcarbamate CAS # 22781-23-3	carbamate	insecticide	non-pers to mod-pers (1)	oral: very high (1) dermal: high (1) inhalation: ?	cataracts (1)	immediate toxicity: birds: very high (2) fish: high (1) bees: "highly toxic" (3) plants: toxic to some (4) water: very soluble oil: soluble slightly volatile
contaminant(s):						
dibenzodioxin or dibenzofuran impurities	dibenzo-dioxin/ dibenzo-furan					
transformation product(s):						
methyl isocyanate (the Bhopal chemical)				oral: high (1) dermal: medium to high (2) inhalation: medium (1)	suspect mutagen (3) suspect fetotoxin (3) lung damage (3)	water: "sparingly soluble" highly volatile
benefin Balan; Balfin; Benefex; benfluralin; bethrodine; Binnell; El 110; Quilan N-butyl-N-ethyl-α,α,α-trifluoro-2, 6-dinitro-p-toluidine CAS # 1861-40-1	dinitro-aniline	herbicide	mod-pers (1)	oral: low to medium (2) dermal: low (2) inhalation: medium to high (2)	?	immediate toxicity: birds: low (2) fish: "toxic" (2) bees: "relatively nontoxic" (3) water: insoluble slightly volatile "nonflammable"
benomyl Benlate; DuPont 1991; Tersan 1991; Turf Fungicide methyl 1-(butylcarbamoyl)-2-benzimidazole carbamate CAS # 17804-35-2	benzimi-dazole	fungicide restricted, USA, 1982	mod-pers (1)	oral: low (2) dermal: low to medium (3) inhalation: ?	suspect carcinogen (4,5) mutagen (4,5) teratogen (4,5) liver and testes damage (5,6) reduced sperm (6) blood damage (5)	immediate toxicity: birds: high (7) fish: high to very high (8) earthworms: high (9) crustaceans: low (9) bees: low (10) long-term toxicity: plants: mutagen water: "insoluble" oil: "insoluble"

(continued on next page)

NAME: Common Trade and Other Chemical CAS Number	Class of Chemical	Chief Pesticide Use; Status	Persistence	Effects on Mammals		Adverse effects on other non-target species Physical properties
				Immediate Toxicity (Acute)	Long-Term Toxicity (Chronic)	
transformation product(s): **thiophanate methyl** (see thiophanate methyl)						
carbendazim (see carbendazim)						
bensulide Betasan; Disan; Exporsan; Prefar; Presan; R-4461 S-2-benzenesulfonamidoethyl O,O-di-isopropyl phophorodithioate; O,O-di-isopropyl S-2-phenylsulfonaminoethyl phosphorodithioate CAS # 741-58-2	organo-phosphate	herbicide	mod-pers (1)	oral: medium to high (2) dermal: medium (1) inhalation: ?	?	immediate toxicity: fish: high (2) bees: low to medium (3) water: slightly soluble combustible
bentazon Basagran DP (with dichlorprop); Basagran Ultra (with ioxynil & dichlorprop); Basagran-M60; bendioxide (So. Africa); bentazone; Galaxy (with acifluorfen); Laddok (with atrazine); Storm (with acifluorfen); Triagran (with MCPA & dichlorprop); Ultima (with dichlorprop); Vega (with cyanazine & dichlorprop) 3-isopropyl-1H-1,2,3-benzothioadiazin-4-(3H)-one 2,2-dioxide CAS # 2505-89-0	miscel-laneous	herbicide	non-pers (1)	oral: medium to high (1,2) dermal: medium (1) inhalation: ?	teratogen (3)	immediate toxicity: birds: medium (1) fish: low (1) bees: "harmless" (1) water: soluble oil: very soluble non-volatile combustible
benthiocarb B-3015; Bolero; Saturn; Saturno; thiobencarb S-[(4-chlorophenyl)methyl)] diethylcarbamothioate CAS # 28249-77-6	carbamate	herbicide	non-pers to mod-pers (1,2)	oral: medium (1) dermal: ? inhalation: ?	?	immediate toxicity: birds: low (1) fish: medium to high (3) crustaceans: medium (2) water: slightly soluble slightly volatile flammable
benzalkonium chloride Barquat MB-50; Barquat MB-80; BTC; Hyamine 3500; Zephiram alkyl dimethyl benzylammonium chloride CAS # 8001-54-5	quaternary ammo-nium	antibiotic algacide fungicide	?	oral: high (1) dermal: ? inhalation: ?	mutagen (3)	immediate toxicity: birds: "slightly toxic" (2) fish: "moderately to highly toxic" (2) flammable
benzene benzol; benzole; benzolene; bicarburet of hydrogen; carbon oil; coal naphtha; cyclohexatriene; minderal naptha; motor benzol; phene; phenyl hydride; pyrobenzol; pyrobenzole benzene CAS # 71-43-2	aromatic hydro-carbon	fumigant insecticide solvent pesticide uses cancelled, USA	?	oral: low to high (1,2) dermal: ? inhalation: "very high" (2)	carcinogen (4,5,6) mutagen (4,7) teratogen (8,9) blood damage (5,10) bone damage (2)	water: soluble flammable

NAME: Common Trade and Other Chemical CAS Number	Class of Chemical	Chief Pesticide Use; Status	Persistence	Effects on Mammals		Adverse effects on other non-target species Physical properties
				Immediate Toxicity (Acute)	Long-Term Toxicity (Chronic)	
transformation product(s): **phenol** (see phenol)						
1,2,4-benzenetriol 1,2,4-benzenetriol					suspect mutagen (1)	
benzene hexachloride 666 (Denmark); Benzahex; Benzex; BHC; DHB; Dol; Dolmix; FBHC; Gexane; HCCH; HCH (Europe); Heapoudre; Hexablanc; hexachloran (USSR); Hexafor; Hexamul; Hexyclan; Kotol; Lindacol; Soprocide 1,2,3,4,5,6-hexachlorocyclohexane CAS # 608-73-1 MX8007-42-9 319-84-6 (α isomer) 319-85-7 (β isomer) 319-86-8 (Δ isomer) 58-89-9 (γ isomer) 6108-10-7 (ε isomer)	organo-chlorine	insecticide "not produced or registered in USA"	pers (1)	oral: high (2) dermal: ? inhalation: ?	cumulative (3,4) carcinogen (4,5) mutagen (6) fetotoxin (7) liver damage (8) reproductive effects (7) bone marrow damage (aplastic anemia) (9)	immediate toxicity: birds: medium to high (10) fish: very high (11) amphibians: medium (12) crustaceans: very high (11) aquatic insects: very high (11) aquatic worms: high (14) bees: very high (13) plants: high (12) long-term toxicity: plants: mutagen (15) water: "insoluble" slightly volatile combustible
transformation product(s): **2,4,6-trichlorophenol** (see 2,4,6-trichlorophenol) **2,4,5-trichlorophenol** (see 2,4,5-trichlorophenol)						
benzthiazuron BAY 60618; Gatinon; Gatnon; S 22012 1-benzothiazol-2-yl-3-methylurea; N-(2-benzothiazolyl)-N′-methyl urea CAS # 1929-88-0	urea	herbicide not registered for use in USA	?	oral: medium (1) dermal: ? inhalation: ?	?	water: slightly soluble slightly volatile
6-benzyladenine BAP; Paturyl 6-benzyladenine; 6-benzylamino-purine CAS # 1214-39-7	miscel-laneous	growth regulator	?	oral: very high (1) dermal: ? inhalation: ?	suspect mutagen (2)	immediate toxicity: fish: "toxic" (1) bees: "toxic" (1) water: insoluble
bifenox Mowdown (with 2,4-D) methyl 5-(2,4-dichlorophenoxy)-2-nitrobenzoate; 5-(2,4-dichlorophenoxy)-2-nitro-benzoic acid, methyl ester CAS # 42576-02-3	phenoxy	herbicide	non-pers to mod-pers (1)	oral: low (1) dermal: low to medium (1) inhalation: low (2)	?	immediate toxicity: birds: low (1) fish: high (3) crustaceans: medium (4) aquatic insects: high (4) water: insoluble oil: soluble slightly volatile combustible

NAME: Common Trade and Other Chemical CAS Number	Class of Chemical	Chief Pesticide Use; Status	Persistence	Effects on Mammals		Adverse effects on other non-target species Physical properties
				Immediate Toxicity (Acute)	Long-Term Toxicity (Chronic)	
bifenthrin biphenate; Brigade; FMC 54800; OMS 3024; Talstar; Torant (with clofentezine); Zipak (with amitraz) 2-methylbiphenyl-3-ylmethyl (Z)-(1RS,3RS)-3-)2-chloro-3,3-trifluoroprop-1-enyl)-2,2-dimethylcyclopropanecarboxylate CAS # 82657-04-3	pyrethroid	insecticide acaricide	mod-pers (1)	oral: high to very high (1,2) dermal: ? inhalation: ?	suspect carcinogen (1) suspect mutagen (1)	immediate toxicity: birds: medium (1,2) fish: very high (1,2) crustaceans: very high (2) aquatic invertebrates: very high (1) water: insoluble slightly volatile combustible
binapacryl Acricide; Ambox; Dapacril; dinoseb methacrylate; Endosan; FMC 9044; HOE 2784; Morocide; NIA 9044 2-sec-butyl-4,6-dinitrophenyl-3-methyl-2-butenoate CAS # 485-31-4	phenol	insecticide acaricide fungicide	?	oral: medium to high (1,2) dermal: high (2) inhalation: ?	?	immediate toxicity: fish: high (3) bees: medium (4) water: "practically insoluble"
bithionol Actamer; Bithin; Lorothidol; TBP; Vancide BL or BN; XL-7 2,2'-thiobis(4,6-dichlorophenol) CAS # 97-18-7	organo-chlorine	fungicide cancelled, USA, 1968	?	oral: low to medium (1) dermal: ? inhalation: ?	?	water: "practically insoluble" oil: soluble non-volatile
blasticidin S BLA-S (S)-4-[[3-amino-5-[(aminoiminomethyl)methylamino]-1-oxopentyl]amino]-1-(4-amino-2-oxo-1-(2H)-pyrimidinyl)-1,2,3,4-tetradeoxy-d-erythro-hex-2-enopyranuronic acid CAS # 2079-00-7	antibiotic	fungicide used only in Japan	?	oral: high to very high (1,2) dermal: ? inhalation: ?	?	immediate toxicity: fish: "slightly toxic" (3) water: "insoluble to soluble" "nonflammable"
Bomyl Bomyl; Fly Bait Grits; GC 3707; Swat dimethyl 3-hydroxyglutconate dimethyl-phosphate; dimethyl 3-[(dimethoxyphosphinyl)oxy]-2-pentendioate CAS # 122-10-1	organo-phosphate	insecticide acaricide restricted use, USA	non-pers (1)	oral: very high (2) "highly toxic to humans and animals by ingestion, inhalation, or skin contact" (3)	?	immediate toxicity: birds: very high (4) bees: very high (5) water: "insoluble" oil: "insoluble" highly volatile
borax Agriben (with bromacil); Borascu; Borate; Borocil; Borospray; Gerstley; Neobor; Polybor 3; Pyrobor; Trona; Tronabor; V-Bor sodium tetraborate decahydrate; sodium biborate; sodium pyroborate; sodium tetraborate anhydrous CAS # 1303-96-4	mineral: borate	fungicide herbicide insecticide	mod-pers (1)	oral: low to medium (1) dermal: ? inhalation: ?	?	immediate toxicity: fish: low (2) water: very soluble "nonflammable" accumulates in soils

NAME:	Common Trade and Other Chemical CAS Number	Class of Chemical	Chief Pesticide Use; Status	Persistence	Effects on Mammals		Adverse effects on other non-target species Physical properties
					Immediate Toxicity (Acute)	Long-Term Toxicity (Chronic)	
Bordeaux mixture Bor-Dax; Copper Hydro Bordo; FT 2M (with mancozeb); Fungi-Bordo mixture of calcium hydroxide and copper sulfate CAS # 8011-63-0	inorganic	fungicide	?	oral: ? dermal: ? inhalation: ?	?	water: "insoluble"
boric acid Borofax; borsaure; NCI-C566417; P.F. Harris Famous Tablets; Roach Prufe boracic acid; orthoboric acid CAS # 10043-35-3	inorganic	insecticide fungicide herbicide	?	oral: low to medium (1) dermal: "well absorbed through denuded or abraded skin" (3) inhalation: ?	testes damage (2) gastrointestinal, skin, kidney damage (3)	?
brodifacoum brodifakum (Czech); Havoc; Klerat; Lim-N8; Matikus; PP 581; Ratak Plus; Ropax; Talon; Volid; WBA 8119 3-(4-(4'-bromo(1,1'-biphenyl)-4-yl)-1,2,3,4-tetrahydro-1-naphthalenyl-4-hydroxy-2H-1-benzopyran-2-one; 3-[3-(4'-bromobiphenyl-4-yl)-1,2,3,4-tetrahydro-1-naphthyl]-4-hydroxycoumarin CAS # 56073-10-0	coumarin	rodenticide	mod-pers (1)	oral: very high (2) dermal: very high (2) inhalation: ?	?	immediate toxicity: birds: very high (1) fish: active ingredient (very high) Talon (low) (1) long-term toxicity: "keep away from children, domestic animals, and wildlife" (2) water: "insoluble" oil: "insoluble"
bromacil Agriben (with borax); Bromox; Brush-Off; Croptex Onyx; Cynogan; Du-Dusit (with dichlobenil); Fasco Gransil-X; Foremost Weed-Zapper; Galar (with diuron); Habco Bromex; Herbicide 976; HK-80 Weed Killer; Hysan "600" Weed Killer; Hyvar X; Puritan 3925; Rokar X; Terra-Var; Urox HX; Zep Formula 777; Zilch Liquid Weed Killer 5-bromo-3-sec-butyl-6-methyluracil; 5-bromo-6-methyl-3-(1-methylpropyl)uracil CAS # 314-40-9	uracil	herbicide	mod-pers (1)	oral: low (1) dermal: ? inhalation: "low" (1)	?	immediate toxicity: birds: low (3) fish: low (2) bees: low (4) water: soluble "combustible"
bromadiolone Apobas; bromakil; Bromatrol; Bromone; Bromorat; Canadien 2000; Deadline; Deturi Ratones; Lanirat; Maki; Musal; Ratinus; Rodine-C; Slaymore; Spyant Ratones; Super Asecho; Super-Caid; Super-Rozol; Temus; Topidion 3-[3-(4'-bromo[1,1'-biphenyl]-4-yl)-3-hydroxy-1-phenylpropyl]-4-hydroxy-2H-1-benzopyran-2-one; 3-[-[p-(p-bromophenyl)-hydroxyphenethyl]-benzyl-4-hydroxycoumarin] CAS # 28772-56-7	coumarin	rodenticide	?	oral: very high (1) dermal: ? inhalation: ?	?	immediate toxicity: fish: high (2) water: "insoluble" "nonflammable"

NAME: Common Trade and Other Chemical CAS Number	Class of Chemical	Chief Pesticide Use; Status	Persistence	Effects on Mammals		Adverse effects on other non-target species Physical properties
				Immediate Toxicity (Acute)	Long-Term Toxicity (Chronic)	
bromethalin EL-614; Vengeance N-methyl-2,4-dinitro- N-(2,4,6-tribromophenyl)-6- (trifluoromethyl)-benzenamine CAS # 63333-35-7	miscel-laneous	rodenticide	?	oral: high to very high (1,2) dermal: ? inhalation: high (1)	?	immediate toxicity: birds: very high (2) fish: very high (2) aquatic insects: very high (2) water: "not soluble"
bromophos Brofene; bromofos; Brophene; CELA-S-1942; Kilsect; Netal; Nexion; Omexan; OMS 658 O-(4-bromo-2,5-dichlorophenyl)- O,O-dimethylphosphorothioate CAS # 2104-96-3	organo-phosphate	insecticide acaricide	non-pers to mod-pers (1,2)	oral: low to medium (3,4) dermal: medium to high (4) inhalation: ?	mutagen (5)	immediate toxicity: birds: low (4) fish: high to very high (3,6) bees: "toxic" (3) water: slightly soluble volatile
bromophos-ethyl bromofos-ethyl; CELA S-2225; Filariol; Nexagan; OMS 659; Paraban O-(4-bromo-2,5-dichlorophenyl)- O,O-diethylphosphorothioate CAS # 4824-78-6	organo-phosphate	insecticide acaricide not registered for use in USA	?	oral: high (1) dermal: high (2) inhalation: ?	?	immediate toxicity: birds: medium to high (3) water: slightly soluble slightly volatile
bromopropylate GS 19851; Neoron; phenisobromolate; qAcarol isopropyl 4,4'-dibromobenzilate; 1-methylethyl 4-bromo-α-(4-bromophenyl)- α-hydroxybenzeneacetate CAS # 18181-80-1	miscel-laneous	acaricide	non-pers (1)	oral: low (2) dermal: ? inhalation: ?	?	immediate toxicity: fish: medium to high (2) bees: low to medium (3) water: slightly soluble non-volatile
bromoxynil Actril S (with ioxynil & dichlorprop & MCPA); Brominal; Brominil; Buctril; Buctril 20; Buctril 21; Buctril ME4; butichlorfos (USSR); Chipco Buctril; Chipco Crab-kleen; Dantril (with ioxynil & dichlorprop & MCPA); Harness; MB 10064; Nu-Lawn Weeder; One Shot (with diclofop-methyl & MCPA); Oxytril P (with dichlorprop & ioxynil); Partner; Tetroxone M (with dichlorprop & ioxynil & MCPA); Torch 3F (with atrazine); Trio (with 2,4-D & propanil) 3,5-dibromo-4-hydroxybenzonitrile CAS # 1689-84-5	benzo-nitrile	herbicide restricted USA	non-pers (1)	oral: high (2) dermal: medium (2) inhalation: ?	teratogen (3)	immediate toxicity: birds: medium to high (2) fish: very high (2) bees: low to medium (4) aquatic insects: very high (2) water: soluble oil: soluble slightly volatile combustible
bronopol Bronocot; Bronosol; Bronotak 2-bromo-2-nitropropan-1,3-diol; β-bromo-β-nitromethyleneglycol CAS # 52-51-7	miscel-laneous	bactericide antibiotic	?	oral: high (1) dermal: ? inhalation: ?	?	water: very soluble oil: "insoluble" slightly volatile
bufencarb Bux; Bux-Ten Granular; metalkamate; Ortho 5353; RE-5353 mixture: m-(ethylpropyl)phenyl methylcarbamate and m-(l-methylbutyl)phenyl methylcarbamate CAS # 8065-36-9	carbamate	insecticide cancelled USA	non-pers (1)	oral: high (2) dermal: high (2) inhalation: ?	?	immediate toxicity: birds: low to very high (3) fish: high to very high (2) bees: high (4) water: slightly soluble slightly volatile

NAME: Common Trade and Other Chemical CAS Number	Class of Chemical	Chief Pesticide Use; Status	Persistence	Effects on Mammals		Adverse effects on other non-target species Physical properties
				Immediate Toxicity (Acute)	Long-Term Toxicity (Chronic)	
butachlor Butanex; Butanox; CP 53619; Lambast; Machete; Rasayanchlor N-(butoxymethyl)-2-chloro-2',6-diethylacetanilide; N-(butoxymethyl)-2-chloro-N-(2,6-diethylphenyl) acetamide CAS # 23184-66-9	amide	herbicide not registered: USA	non-pers (1)	oral: medium (1,2) dermal: medium (2) inhalation: medium (1)	suspect mutagen (3)	immediate toxicity: fish: medium to high (1) birds: low to medium (1) water: slightly soluble slightly volatile combustible
butam Comodor; S-15544; Tebutam; Tebutame 2,2-dimethyl-N-(1-methylethyl)-N-(phenylmethyl)propanamide CAS # 35256-85-0	amide	herbicide	non-pers (1)	oral: low to medium (1,2) dermal: ? inhalation: ?	?	oil: "insoluble" flammable
2-butanamine 2-AB; Betafume; Deccotane; Frucote; Tutane 2-butanamine; sec-butylamine; 2-aminobutane CAS # 513-49-5	miscel-laneous	fungicide	?	oral: high (1) dermal: ? inhalation: ?	suspect carcinogen (2,3)	immediate toxicity: birds: high (1) fish: low (4) bees: low to medium (5) water: "soluble" highly volatile flammable
transformation product(s): **N-nitrosomethyl-n-butylamine**					carcinogen (1,2)	
butoxycarboxim Plant Pin 3-(methylsulphonyl)-2-butanone O-[(methylammino)carbaryl]oxime CAS # 34681-10-2	carbamate	insecticide	non-pers to mod-pers (1)	oral: high (1) dermal: ? inhalation: ?	?	immediate toxicity: fish: low (2) water: "soluble"
butoxy polypropylene glycol Crag Fly Repellent; Stabilene Fly Repellent polypropylene glycol monobutylether CAS # 9003-13-8		insect repellent acaricide	?	oral: low (1) dermal: ? inhalation: ?	?	water: "slightly soluble" volatile
butralin A-820; Amchem 70-25; Amex 820; Sector; Tamex N-sec-butyl-4-tert-butyl-2,6-dinitrobenzamine; 4-(1,1-dimethylethyl)-N-(1-methylpropyl)-2,6-dinitrobenzamine; N-sec-butyl-2,6-dinitroanaline CAS # 33629-47-9	dinitro-aniline	herbicide not registered: USA	mod-pers (1)	oral: low to medium (2,3) dermal: medium (4) inhalation: low (5)	?	immediate toxicity: fish: high (4) water: slightly soluble slightly volatile flammable
butylate R-1910; Stauffer R-1910; Sutan GR (with atrazine); Sutan Plus; Sutazin (with atrazine); Sutazine + 18-6G S-ethyl diisobutylthiocarbamate; S-ethyl bis(2-methylpropyl)carbamothiote CAS # 2008-41-5	thiocar-bamate	herbicide	non-pers (1)	oral: low to medium (1) dermal: low to medium (1)	?	immediate toxicity: fish: medium (1) bees: low to medium (3) aquatic invertebrates: medium (2) water: slightly soluble volatile combustible

NAME: Common Trade and Other Chemical CAS Number	Class of Chemical	Chief Pesticide Use; Status	Persistence	Effects on Mammals		Adverse effects on other non-target species / Physical properties
				Immediate Toxicity (Acute)	Long-Term Toxicity (Chronic)	
4,4-bipyridyl	miscel-laneous			oral: high (1) dermal: ? inhalation: ?	pulmonary toxin (1)	
cadmium Characteristics as a class are given here; variations are indicated for each compound below CAS # 7440-43-9	metal/ mineral cadmium	fungicide restricted use in USA	perm	oral: "?" dermal: ? inhalation: very high (1)	cumulative (2,3) kidney damage (4,5) reduced number and growth of young (6)	immediate toxicity: fish: very high (3) crustaceans: medium (8) molluscs: low to medium (8) water: "insoluble"
compound(s): **cadmium carbonate** CAS # 513-78-0	organic	fungicide				water: "soluble"
cadmium chloride Caddy; Vi-Cad CAS # 1018-64-2	inorganic	fungicide		oral: high to very high (1,2)	carcinogen (3) suspect mutagen (4) teratogen (5) suspect fetotoxin (6) cumulative (13) suspect neurotoxin (7) kidney and liver damage (8,9) testes damage (10) immunotoxin (11,12)	immediate toxicity: crustaceans: low to high (16)
cadmium oxide CAS # 1306-19-0	inorganic	nematocide			carcinogen (1)	water: "insoluble"
cadmium sebacate cadmium decanedioate CAS # 4476-04-4	organic	fungicide		oral: high (1) dermal: high (1)		
cadmium succinate Cadminate; Kroma-Clor; Ultra-Clor CAS # 141-00-4	organic	fungicide		oral: medium to high (1,2)		water: very soluble
cadmium sulfate CAS # 10124-36-4	inorganic	fungicide nematocide			carcinogen (1) suspect mutagen (1,2) teratogen (3)	water: "soluble"
calcium Characteristics are indicted for each compound below CAS # 7440-70-2	metal/ mineral calcium		perm			
compound(s): **calcium chlorate** calcium chlorate CAS # 5902-95-4	inorganic	herbicide insecticide		oral: medium (1)	heart damage (2)	
calcium cyanamide Cyanamid	cyanide	herbicide fungicide		oral: medium (1)		

NAME: Common Trade and Other Chemical CAS Number	Class of Chemical	Chief Pesticide Use; Status	Persistence	Effects on Mammals		Adverse effects on other non-target species Physical properties
				Immediate Toxicity (Acute)	Long-Term Toxicity (Chronic)	
calcium cyanide A-Dust; Cyanogas; Degesch Calcium Cyanide A-Dust CAS # 592-01-8	cyanide	fumigant insecticide rodenticide restricted use, USA		oral: very high (1)		
calcium hypochlorite chloride of lime CAS # 7778-54-3	inorganic	fungicide antibiotic algacide				water: soluble
calcium phosphide	inorganic	rodenticide				
calcium polysulfide lime sulphur CAS # 1344-81-6	metal/ mineral sulfur	fungicide				
calcium propionate	organic	fungicide				water: "soluble"
calcium sulfate anhydrous gypsum; Plaster of Paris CAS # 7778-18-9	metal/ mineral sulfur	"inert" carrier				
camphor (1R)-1,7,7-trimethylbicyclo[2.2.1]heptan-2-one; 2-camphanone CAS # 76-22-2	botanical	insect repellent	?	oral: ? dermal: ? inhalation: ? humans seem more susceptible than other species (1)	liver damage (2)	water: "insoluble" combustible
captafol Bayleton Triple (with triadimefon & carbendazim); Crisfolatan; Difolatan; Difosan; Folcid; Foltapet (with folpet); Haipen; Merpafol; Mycodifol; Ortho 5865; Sanspor; Sufonimide; Sulfemmide cis-N-[(1,1,2,2-tetrachloroethyl)thio]-4-cyclohexene-1,2-dicarboximide CAS # 2425-06-1	phthalate	fungicide cancelled USA, 1986	non-pers (1)	oral: low (2) dermal: medium (3) inhalation: ?	carcinogen (4,5) suspect mutagen (6,7) suspect teratogen (6,8)	immediate toxicity: fish: very high (10) birds: low (9) aquatic insects: very high (10) crustaceans: very high (10) water: slightly soluble
captan Acme Garden Fungicide; Aliette Extra (with fosetyl-al & thiabendazole); Captaf; captane (France); Captanex; D-264 Plus Captan; Drexel Captan; Drexel Captan Plus Molybdenum; Merpan; Orthocide Garden Fungicide; SR 406; Stauufer Captan 80; Voncaptan N-[(trichloromethyl)thio]-4-cyclohexene CAS # 133-06-2	phthalate	fungicide most food crop uses cancelled in USA, 1989	non-pers to mod-pers (1,2)	oral: low (2,3) dermal: ? inhalation: ?	carcinogen (4-6) suspect mutagen (1,7) teratogen (8) fetotoxin (1) immunotoxin (9)	immediate toxicity: birds: low to medium (2) fish: very high (1) earthworms: low to medium (2) bees: low (dermal), high (oral) (1,10) long-term toxicity: plants: increased incidence of gall in cherry trees (11) water: slightly soluble oil: insoluble slightly volatile

(continued on next page)

NAME: Common Trade and Other Chemical CAS Number	Class of Chemical	Chief Pesticide Use; Status	Persistence	Effects on Mammals		Adverse effects on other non-target species Physical properties
				Immediate Toxicity (Acute)	Long-Term Toxicity (Chronic)	
transformation product(s): **delta⁴ tetrahydrophthalimide** CAS # 85-40-5					suspect fetotoxin (1)	
carbanolate Banol; Sok; U-12927 6-chloro-3,4-xylyl methylcarbamate; 2-chloro-4,5-dimethylphenyl methylcarbamate CAS # 671-04-5	carbamate	insecticide	?	oral: very high (1) dermal: ? inhalation: ?	?	immediate toxicity: bees: high (2)
carbaryl Arylam; Bercema NMC50; Carpolin; Crag Sevin; Dicarbam; Dyna-carbyl; Germain's; Hexavin; Karbatox 75; Menaphtham; Nac; Pomex; Ravyon; Sevin; Sok; Tercyl; Tricarnam; UC 7744; Vioxan; Zrylam 1-naphthyl methylcarbamate; methylcarbamate 1-naphthalenol CAS # 63-25-2	carbamate	insecticide acaricide mollusicide	non-per to mod-pers (1,2)	oral: medium to high (3) dermal: ? inhalation: low	suspect carcinogen (5) suspect mutagen (6) teratogen (4) fetotoxin (6) suspect viral enhancer (6) decreased fertility from ovary and testes damage through successive generations (7)	immediate toxicity: birds: low to medium (3) fish: very high (10) bees: "extremely toxic" (6) crustaceans: very high (6) earthworms: "extremely toxic" (10) aquatic worms: high to very high (11) aquatic insects: very high (10) plants: toxic to some; chromosome damage in some (12) long-term toxicity: birds: may affect breeding success (8,9) fish: reduction in sex hormone, may affect reproduction (13); increased vulnerability to predation; affects swimming capacity (14) water: slightly soluble to soluble slightly volatile combustible
transformation product(s): **1-naphthol** α-naphthol CAS # 90-15-3	miscel- laneous			oral: high (1)	liver damage (1)	
nitrosocarbaryl from reaction with nitric acid often present in air, soil, saliva when carbaryl is ingested					carcinogen (1) mutagen (1,2)	
carbendazim BCM; BMC; Carbate; carbendazol; Cekudazim; CTR 6669; Custos; Delsene; Delsene M; Equitdazin; Focal; Hoe 17411; Kemdazin; Lignasan; MBC*; MCB; Pillarstin; Supercarb; Tritocol 2-(methoxycarbonylamino)-benzimidazole; methyl 1H-benzimidazol-2-yl-carbamate; methyl-2-benzimidazole carbamate *MBC also used as a trade name for sodium chlorate CAS # 10605-21-7	benzimi- dazole	fungicide	mod-pers	oral: low to medium dermal: low to medium inhalation: ?	suspect carcinogen suspect mutagen testes damage	immediate toxicity: birds: low fish: low to high earthworms: "very toxic" water: slightly soluble non-volatile

NAME: Common Trade and Other Chemical CAS Number	Class of Chemical	Chief Pesticide Use; Status	Persistence	Effects on Mammals		Adverse effects on other non-target species Physical properties
				Immediate Toxicity (Acute)	Long-Term Toxicity (Chronic)	
carbofuran Barbodan; BAY 78537; Brifur; Bripoxur; Carma (with isofenphos); Chinufur; Curaterr; D 1221; FM 10242; Furacarb; Furadan; Furadan 15G; Kenofuran; NIA 10242; Yaltox 2,3-dihydro-2,2-dimethyl-7-benzofuranyl methylcarbamate CAS # 1563-66-2	carbamate	insecticide nematocide acaricide	non-pers to mod-pers (1,2)	oral: high to very high (2,3) dermal: ? inhalation: ? "carbofuran is high toxic by the oral, dermal, and inhalation routes of exposure" (4)	suspect mutagen (3) immunotoxin (5)	immediate toxicity: birds: very high (3,6) fish: high to very high (2) water: soluble slightly volatile
carbon disulfide carbon bisulfide CAS # 75-15-0	metal\mineral sulfur	fumigant insecticide herbicide fungicide rodenticide soil sterilant nematocide	?	oral: ? dermal: ? inhalation: "highly poisonous" (1)	neurotoxin (2) heart damage (3) fetotoxin (3) liver damage (3) kidney damage (3) thyroid, adrenal changes (3)	water: soluble highly volatile flammable
carbon tetrachloride Granosan (with ethylene dichloride); Necatorina perchloromethane; tetrachloromethane CAS # 56-23-5 "susceptibility to carbon tetrachloride poisoning is enhanced by the contemporaneous use of alcohol" (8)	organo-chlorine	insecticide fumigant cancelled USA, 1986	?	oral: low to medium (1,2) dermal: ? inhalation: low (1)	carcinogen (2,3) fetotoxin (2) liver damage (4,5) eye damage (6) kidney damage (6) testes damage (7)	water: "very slightly soluble" "non-inflammable"
transformation product(s):						
phosgene gas				inhalation: "danger period is usually 6 to 24 hours after exposure" (1)	lung damage (2)	water: "slightly soluble" highly volatile "nonflammable gas"
carbophenothion Acarithion; Dagadip; Endyl; Garrathion; Hexathion; Lethox; Nephocarp; R-1303; Stauffer; Trithion S-[(p-chlorophenylthio)methyl O,O-diethyl phosphorodithioate; O,O-diethyl S-[(p-chlorophenylthio)-methyl]phosphorodithioate CAS # 786-19-6	organo-phosphate	insecticide acaricide	?	oral: very high (1) dermal: high to very high (2) inhalation: ?	?	immediate toxicity: birds: high to very high (3) fish: high to very high (4) crustaceans: very high (4) amphibians: very high (5) bees: high (6) water: slightly soluble volatile
carboxin Arrest 75W; Ceravax; D 735; DCMO; Kemekar; Pro-Gro; Quinolate V4 X AC, FS, DS (with oxine-copper & anthraquinone; Uniroyal D 735; Vitaflo; Vitavax; Vitavax 100 5,6-dihydro-2-methyl-N-phenyl-1,4-oxathiin-3-carboxamide; 2,3-dihydro-5-carboxanilido-6-methyl-1,4-oxathiin; 5,6-dihydro-2-methyl-1,4-oxathiin-3-carboxanilide CAS # 5234-68-4	amide	fungicide	non-pers (1,2)	oral: medium to high (3,4) dermal: low to medium (3) inhalation: low (1)	?	immediate toxicity: birds: low to medium (1,5) fish: high (5) crustaceans: low (5) bees: low (6) aquatic insects: low (5) water: soluble combustible

NAME: Common Trade and Other Chemical CAS Number	Class of Chemical	Chief Pesticide Use; Status	Persistence	Effects on Mammals		Adverse effects on other non-target species Physical properties
				Immediate Toxicity (Acute)	Long-Term Toxicity (Chronic)	
cartap Cadan; Caldan; Padan; Patap; Sanvex; T-1258; Vegetox S,S'-(2-dimethylaminotrimethylene)-bis(thiocarbamate); 1,3-bis(carbamoylthio)-2-(N,N-dimethylamino)propane CAS # 15263-53-3	thiocarbamate	insecticide	?	oral: high (1) dermal: low to high (2) inhalation: ?	?	?
CDAA allidochlor; CP-6343; Limit; Randox N,N-diallyl-2-chloroacetamide; 2-chloro-N,N-di-2-propenyl-acetamide; α-chloro-N,N-diallylacetamide CAS # 93-71-0	amide	herbicide	non-pers (1)	oral: medium to high (1,2) dermal: medium to high (1) inhalation: ?	?	immediate toxicity: fish: medium (1) water: very soluble volatile "nonflammable"
CDEC sulfallate; Vegadex 2-chloroallyl diethyldithiocarbamate CAS # 95-06-7	thiocarbamate	herbicide withdrawn from market by manufacturer in USA & Canada	non-pers (1,2)	oral: medium (3) dermal: ? inhalation: ?	carcinogen (4)	immediate toxicity: birds: low (5) bees: low to medium (6) water: soluble oil: "soluble" volatile combustible
transformation product(s): **2-chloroacrolein**					suspect mutagen (1)	
chloramben Amchem Garden Weeder; Amiben (ammonium salt of chloramben); Amilon-WP; chlorambene (France); Ornamental Weeder 46; Vegiven (methyl salt of chloramben) 3-amino-2,5-dichlorobenzoic acid CAS # 133-90-4	miscellaneous	herbicide	non-pers (1)	oral: low (2) dermal: medium (2) inhalation: low (3)	suspect carcinogen (3,4) fetotoxin (5) liver damage (5)	immediate toxicity: fish: low (3) water: soluble "nonflammable"
chloranil G-25804; G-444-E; Spergon; Vulklor 2,3,5,6-tetrachloro-p-benzoquinone CAS# 118-75-2	quinone	fungicide cancelled USA, 1977	?	oral: medium (1) dermal: ? inhalation: ?	carcinogen (2,3)	water: soluble slightly volatile
chlorbenside Chlorocide; Chlorparacide; Chlorsulphacide; Midox; Mitox p-chlorobenzyl p-chlorophenyl sulfide CAS # 103-17-3	organochlorine	acaricide	non-pers (1)	oral: low to medium (2) dermal: ? inhalation: ?	?	water: soluble oil: very soluble slightly volatile
chlorbromuron Bromex (with naled); C-6313; chlorobromuron (France); CIBA 6313; Maloran 3-(4-bromo-3-chlorophenyl)-1-methoxy-1-methylurea; N'-(4-bromo-3-chlorophenyl)-N-methoxy-N-methylurea CAS # 13364-45-7	urea	herbicide	non-pers (1)	oral: low (1) dermal: medium (1) inhalation: ?	?	?

NAME: Common Trade and Other Chemical CAS Number	Class of Chemical	Chief Pesticide Use; Status	Persistence	Effects on Mammals		Adverse effects on other non-target species Physical properties
				Immediate Toxicity (Acute)	Long-Term Toxicity (Chronic)	
chlordane 1068; Aspon-chlordane; Belt; Chlor Kil; Chlordan; Corodane; Kypchlor; Niran; Octa-Klor; Octachlor; Ortho-Klor; Synklor; Termide; Topiclor 20; Velsicol 1068 1,2,4,5,6,7,8,8-octachlor-2,3, 3a,4,7a-hexahydro-4,7-methanoindane (technical chlordane is a mixture of chlorinated hydrocarbons consisting of isomers of chlordane and closely related compounds and byproducts) CAS # 57-74-9	organo- chlorine	insecticide fire ant control in power plants only use of chlordane/ heptachlor permitted in USA, 1988	mod-pers to pers (1-3)	oral: medium to high (4) dermal: ? inhalation: "hazardous by inhalation" (5)	cumulative (2) carcinogen (6) suspect mutagen (1) liver damage (7) testicular damage (8) affects hormone levels (9) neurotoxin (2) blood damage (10)	immediate toxicity: bird: medium to very high (12) fish: high to very high (4,13) crustaceans: high to very high (4,13) aquatic insects: very high (13) long-term toxicity: tendency to bioaccumulate in certain food chains (14) water: insoluble oil: very soluble slightly volatile
contaminant(s): hydrogen chloride hydrochloric acid hydrogen chloride CAS # 7647-01-0				oral: low (1)		
phosgene gas (see carbon tetrachloride)						
hexachlorocyclopentadiene						
carbon monoxide				can potentiate cardiovascular stress in diseased heart (1,2)	brain damage (3)	
transformation product(s): oxychlordane 1-exo,2-endo,4,5,6,7,8,8-octachloro- 2,3-exo-epoxy-2,3,3a,4,7, 7a-hexahydro-4,7-methanoindene				oral: very high (1) more toxic than chlordane (1)	most cumulative of all chlordane derivatives, often found in human fatty tissue (2)	
nonachlor		also component of technical grade chlordane			often found in human adipose tissue (1)	
heptachlor (see heptachlor)						
chlordecone Black Flag Ant Trap; GC 1189; Kepone; Tat Ant Trap 1,1a,3,3a,4,5,5,5a,5b,6-decachloro- octachloro-1,3,4-metheno- 2H-cyclobutano(cd) pentalen-2-one CAS # 143-50-0	organo- chlorine	insecticide acaricide fungicide cancelled USA, 1977	pers (1)	oral: high (2,3) dermal: high (4) inhalation: ?	cumulative (5,6) carcinogen (7,8) suspect fetotoxin (9,10) liver damage (6) neurotoxin (11) reproductive ability decreased (12) suppresses pituitary cell hormones (13) testicular atrophy (17)	immediate toxicity: birds: high (14) fish: very high (15) bees: medium (16) long-term toxicity: birds: affects reproduction (15) fish: abnormal bone development (16) water: "slightly soluble"

NAME: Common Trade and Other Chemical CAS Number	Class of Chemical	Chief Pesticide Use; Status	Persistence	Effects on Mammals		Adverse effects on other non-target species Physical properties
				Immediate Toxicity (Acute)	Long-Term Toxicity (Chronic)	
chlordimeform Acaron; Bermat; chlorphenamidine; CIBA 8514; EP-333; Fundal 500; Fundal Forte (with formetanate hydrochloride); Fundex; Galecron; Ovatoxion; RS141; Schering 36268; Spanon; Spanone N,N-dimethyl-N'-(2-methyl-4-chlorophenyl)-formamidine; N-(4-chloro-2-methylphenyl)-N,N-dimethyl formamidine CAS # 6164-98-3	amide	insecticide acaricide cancelled USA, 1989	non-pers (1)	oral: high (1) dermal: medium to high (1,2) inhalation: high (3)	carcinogen (1)	immediate toxicity: bees: "nontoxic" (4) fish: medium to high (4) water: soluble slightly volatile
transformation product(s):						
4-chloro-o-toluidine					carcinogen (1)	
N-formyl-chloro-o-toluidine					carcinogen (1)	
chlorfenethol BCPE; DCPC; DCPE; Dimite; DMC; Milbex; Qikron 4,4'-dichloro-α-methylbenzhydrol; 1,1-bis(4-chlorophenyl)ethanol; 4-chloro-α-(4-chlorophenyl)-α-methylbenzenemethanol CAS # 80-06-8	organo-chlorine	acaricide	?	oral: medium (1) dermal: ? inhalation: ?	"effects similar to DDT" (2)	water: "practically insoluble"
chlorfenson C-106; chlorofenizon; CPCBS; difenson; ephirsulphonate; Estonmite; Genite 883; K-6451; Lethalaire G-58; Mitran; Orthotran; Ovex; Ovochlor; Ovotox; Ovotran C-854; PCPCBS; Trichlorfenson 4-chlorphenyl 4-chlorobenzenesulphonate CAS # 80-33-1	miscel-laneous	acaricide cancelled USA	?	oral: medium (1,2) dermal: ? inhalation: ?	?	immediate toxicity: birds: medium (3) water: "insoluble" oil: very soluble
chlorfensulphide Micasin 4-chlorophenyl 2,4,5-trichlorophenylazosulphide CAS # 2274-74-0	metal/mineral sulfur	acaricide	?	oral: medium (1) dermal: ? inhalation: ?	?	water: "practically insoluble" oil: "soluble in petroleum solvents"
chlorfenvinphos Apachlor; Birlane; C 8949; CFV; CGA 26351; Compound 4072; Sapecron; SD 7849; Steladone; Supona; Unitox; Vinylphate 2-chloro-1-(2,4-dichlorophenyl)vinyl diethylphosphate; 2,4-dichloro-α-(chloromethylene)benzyl diethyl phosphate CAS # 470-90-6	organo-phosphate	insecticide acaricide	non-pers to mod-pers (1,2)	oral: high to very high (2,3) dermal: high to very high (2,4) inhalation: ?	?	immediate toxicity: birds: high (5) fish: high (4) water: soluble slightly volatile "nonflammable"
chlorfluazuron CGA 112913; CGA 913; IKI-7899; UC-62644 1-[3,5-dichloro-4-(3-chloro-5-trifluoromethyl-2-pyridiloxy)phenyl]-3-(2,6-difluorobenzoyl) urea CAS # 71422-67-8	urea	insect growth regulator	?	oral: low (1) dermal: ? inhalation: ?	?	immediate toxicity: fish: low (1) water: insoluble

NAME:	Common Trade and Other Chemical CAS Number	Class of Chemical	Chief Pesticide Use; Status	Persistence	Effects on Mammals		Adverse effects on other non-target species Physical properties
					Immediate Toxicity (Acute)	Long-Term Toxicity (Chronic)	
sodium dichloroisocyanurate 1,3-dichloro-s-triazine-2,4,6(1H,3H,5H) trione sodium salt CAS # 2893-78-9							
sodium dichloroisocyanurate dihydrate ACL 56; CDB Clearon; DICD 1,3-dichloro-s-triazine-2,4,6(1H,3H,5H) trione sodium salt dihydrate CAS # 51580-86-0							
trichloroisocyanuric acid ACL 90; CDB 90; Plus; T.I.C.A.; TCCA trichloro-s-triazinetrione; trichloroisocyanurate CAS # 87-90-1							immediate toxicity: birds: medium (1) fish: high (1) crustaceans: high to very high (1)
chlorinated isocyanurates	triazine	fungicide algacide cancelled, most uses, USA	?	oral: "low" (1) dermal: "low" (1) inhalation: ?	kidney and urinary tract stone formation (1)	?
compound(s): **dichloroisocyanuric acid** ACL 70 1,3-dichloro-s-triazine- 2,4,6(1H,3H,5H) trione; dichloro-s-triazinetrione CAS # 2782-57-2							
penta-s-triazinetrione ACL 66; DS [Mono-(1,3,5-trichloro) tetra(1-potassium-3,5-dichloro] penta-s-triazinetrione CAS # 30622-37-8							
potassium dichloroisocyanurate ACL 59; P.D.I.C. 1,3-dichloro-s-triazine-2,4,6(1H,3H,5H) trione potassium salt CAS # 2244-21-5							
chlorflurecol CF 125; chlorflurecol-methyl (USA, Gr. Britain); chlorflurenol (not to be confused with chlorofluorenol); chlorflurenol-methyl; Curbiset; IT 3456; Maintain; Maintain CF 125 methyl 2-chloro-9-hydroxyfluorene-9-carboxylate; methyl 3-chloro-9-hydroxyfluorene-9H-carboxylic acid CAS # 2536-31-4	miscel-laneous	herbicide	non-pers (1)	oral: low to medium (1,2) dermal: medium (1) inhalation: ?	?	immediate toxicity: bird: low (1) fish: medium (1) bee: "nontoxic" (2) water: slightly soluble oil: soluble slightly volatile flammable

NAME: Common Trade and Other Chemical CAS Number	Class of Chemical	Chief Pesticide Use; Status	Persistence	Effects on Mammals		Adverse effects on other non-target species Physical properties
				Immediate Toxicity (Acute)	Long-Term Toxicity (Chronic)	

chlorimuron
 	
Canopy (with metribuzin); Classic; DPX-F6025; Gemini (with linuron)
 	
2-[[[[(4-chloro-6-methoxyprimidin-2-yl)amino]carbonyl]amino]sulfonyl]benzoate
 	
CAS # 90982-32-4
	urea	herbicide	non-pers (1)	oral: medium (1)
				
				dermal: ?
				
				inhalation: low to medium (1)
					?	immediate toxicity:
						  birds: low to medium (1)
						  fish: low (1)
						  crustaceans: low (1)
						
						water: slightly soluble
						
						non-volatile

chlorine
 	
chloor (Durch); Chlor (German); chlore (French); cloro (Italian)
 	
chlorine
 	
CAS # 7782-50-5
	miscel-laneous	algacide antibiotic fungicide	?	oral: ?
				
				dermal: "causes burns" (1)
				
				inhalation: "low" (2)
					erosion of teeth (2)	BIOCIDE
						
						immediate toxicity:
						  fish: very high (3)
						  crustaceans: very high (3)
						  aquatic insects: very high (3)
						  aquatic worms: high (3)
						  molluscs: very high (3)
						  amphibians: medium (3)
						  phytoplankton: high to very high (3)
						long-term toxicity:
						  crustaceans: decreased reproduction (3)
						  fish: disequilibrium (4,5)
						  "reacts explosively or forms explosive compounds with many common substances"
						
						water: soluble
						
						highly volatile

contaminant(s): **hexachlorobenzene** (see hexachlorobenzene)						
transformation product(s): **chloroform** (see chloroform)						

chlormephos
 	
Dotan; MC 2188
 	
S-(chloromethyl) O,O-diethyl phosphorodithioate
 	
CAS # 24934-91-6
	organo-phosphate	insecticide	non-pers (1,2)	oral: very high (1)
				
				dermal: very high (1)
				
				inhalation: ?
					?	immediate toxicity:
						  bees: "toxic" (1)
						  birds: high (1)
						  fish: high (3)
						
						water: slightly soluble
						
						volatile

chlormequat
 	
2-chloroethyl-N,N,N-trimethylammonium ion
 	
CAS # 7003-89-6
	bypiridyl	growth regulator	"persists in soil less than one season" (1)	oral: medium (2)
				
				dermal: ?
				
				inhalation: low to medium (2)
					?	immediate toxicity:
						  birds: medium (2)
						  fish: low (2)
						  molluscs: low (2)
						  crustaceans: medium (2)
						
						"non-volatile"

salt(s): **chlormequat chloride** Arotex-extra; Barleyquat-B; Bettaquat-B; CCC; CeCeCe; chlorocholine chloride; Cycocel; Cycocel-Extra; Cycogan Extra; Cycogen; Ethanaminium; Farmacel; Hico CCC; Hormocel; Lihocin; Minc; Nainit; Padquat; Titan 2-chloroethyl-N,N,N-trimethylammonium chloride CAS # 991-81-5						

NAME: Common Trade and Other Chemical CAS Number	Class of Chemical	Chief Pesticide Use; Status	Persistence	Effects on Mammals		Adverse effects on other non-target species Physical properties
				Immediate Toxicity (Acute)	Long-Term Toxicity (Chronic)	

chlorobenzilate
 	
Acaraben; Akar; Benz-O-Chlor; Benzilan; Compound 338; Folbex; G 23992; Geigy 338; Kop-Mite
 	
ethyl 4,4'-dichlorobenzilate
 	
CAS # 510-15-6
	organo-chlorine	acaricide
		
		restricted USA, 1979
			non-pers (1)	oral: medium (2)
				
				dermal: ?
				
				inhalation: low (1)
					carcinogen (3,4) testes damage (1,4)	immediate toxicity:
						  birds: low (1)
						  fish: high (1)
						  bees: low (5)
						
						water: slightly soluble
						
						slightly volatile

chlorofluorocarbons Freon Genetron 1,2-dichloro-1,1,2,2-tetrafluorethane CAS # 76-12-0	miscel-laneous	aerosol propellant almost all pesticide uses banned, USA	?	oral: ? dermal: ? inhalation: low (1)	see piperonyl butoxide (1)	long-term toxicity: reduction in protective ozone layer in earth's stratosphere, producing a global increase in ultraviolet radiation at the earth's surface; increase in skin cancers and mutation rates; may also cause climate changes
transformation product(s): **chlorine** (see chlorine) **phosgene gas** (see carbon tetrachloride)						
chloroform trichloromethane; formyl trichloride CAS # 67-66-3	organo-chlorine	acaricide insecticide fumigant	?	oral: medium to very high (1,2) dermal: ? inhalation: very high (3)	carcinogen (2,4,5) mutagen (6) fetotoxin (2,4) neurotoxin (7,8) liver damage (3,7) kidney damage (2) lung,thyroid and bladder damage (5)	water: soluble highly volatile "nonflammable"
transformation product(s): **phosgene gas** (see carbon tetrachloride)						

chloroneb
 	
chloronebe (France); Demosan (with thiram); Soil Fungicide 1823; Teremec; Terraneb; Tersan SP Turf Fungicide
 	
1,4-dichloro-2,5-dimethoxybenzene
	organo-chlorine	fungicide	mod pers to pers (1)	oral: low to medium (1,2)
				
				dermal: ?
				
				inhalation: low (1)
					kidney and liver damage (1) blood cell, spleen, and bone marrow damage (1)	immediate toxicity:
						  birds: low (1)
						  fish: medium (1)
						  crustaceans: medium (1)
						
						water: slightly soluble
						
						volatile

transformation product(s):
 	
TCDD
 	
2,3,7,8-tetrachlorodibenzo-p-dioxin
 	
CAS # 1746-01-6
	dibenzo-dioxin		mod-pers to perm (1,2)	oral: very high (2,3)
				
				dermal: very high (2)
				
				inhalation: ?
					cumulative (4,5) carcinogen (4,6) suspect mutagen (7) teratogen (8,9) chloracne (2,7) thymic atrophy (10,11) hirsutism (7) affects vitamin A Balance in liver & kidney (12,13)	immediate toxicity:
						  birds: very high (14)
						  amphibians: very high (2)
						
						water: insoluble
						
						oil: slightly soluble
						
						slightly volatile

NAME: Common Trade and Other Chemical CAS Number	Class of Chemical	Chief Pesticide Use; Status	Persistence	Effects on Mammals		Adverse effects on other non-target species Physical properties
				Immediate Toxicity (Acute)	Long-Term Toxicity (Chronic)	
chlorophacinone Actosin C; Afnor; Caid; Delta; Drat; Lepit; Liphadione; LM 91; Microzul; Muriol; Partox; Quick; Ramucide; Ranac; Ratomet; Raviac; Ravicac; Rozol; Topitox 2-(a-p-chlorophenyl-a-phenylacetyl)-indane-1, 3-dione; 2-((p-chlorophenyl)phenylacetyl)- 1,3-indandione; 2-[(4-chlorophenyl)phenylacetyl]- 1,3-indandione CAS # 3691-35-8	indan- dione	rodenticide	?	oral: very high (1) dermal: high to very high (2) inhalation: ?	?	immediate toxicity: birds: high (2) water: "sparingly soluble" oil: "soluble" "nonflammable"
chloropicrin Acquinite; Aquinite; Chlor-O-Pic; Dojyopicrin; Dolochlor; Dowfume MC 33; G-25; Larvacide; Niklor; Nitrochloroform; Pic-Clor; Picfume; Picride; PS; S-1; Tear Gas; Terr-O-Gas (with methyl bromide); Triclor; Vomiting GAs; Vorlex 201 trichloronitromethane CAS # 76-06-2	organo- chlorine	fumigant	?	oral: high (1) dermal: ? inhalation: high (1)	suspect mutagen (2) anemia (3) irregular heartbeat (3) recurrent asthma (3)	water: slightly soluble highly volatile "nonflammable"
chloropropylate Acaralate; Chlormite; G 24163; Rospin isopropyl 4,4'-dichlorobenzilate CAS # 5836-10-2	organo- chlorine	acaricide discon- tinued, USA	?	oral: medium (1) dermal: low to medium (2) inhalation: ?	?	immediate toxicity: fish: low (4) bees: low to medium (5) water: slightly soluble slightly volatile
chlorothalonil Blazon; Bravo; Bravo C/M; Bravocarb (with carbendazim); chlorthalonil; Clortocar Ramato; Clortosip (with copper oxychloride & maneb); Dacobre; Daconil 2787; Exotherm; Exotherm Termil; Forturf; Nopocide; Termil tetrachloroisophthalonitrile CAS # 1897-45-6	benzo- nitrile	fungicide	mod-pers (1,2)	oral: low (3) dermal: low to medium (3) inhalation: high (3)	carcinogen (4,5) hyperexcitability (5) skin damage (6) eye damage (7) kidney damage (5)	immediate toxicity: birds: low to medium (1) fish: very high (1) bees: low (9) "aquatic organisms": very high (1) plants: toxic to some (10) water: insoluble oil: slightly soluble slightly volatile
transformation product(s): **4-hydroxy-2,5,6- trichloroisophtalonitrile**					anemia (1)	immediate toxicity: birds: medium (1)
chloroxuron Aitkens Lawn Sand Plus (with ferrous sulfate); Ashlade D-Moss (with ferrous sulfate); C-1983; chloroxifenidim; Norex; Tenoram 3-[p-(p-chlorophenoxy)phenyl]-1,1- dimethylurea CAS # 1982-47-4	urea	herbicide	mod-pers (1)	oral: low (1) dermal: medium (2) inhalation: ?	?	immediate toxicity: birds: low to medium (3) fish: low to medium (1) bees: low (4) water: slightly soluble non-volatile nonflammable

NAME: Common Trade and Other Chemical CAS Number	Class of Chemical	Chief Pesticide Use; Status	Persistence	Effects on Mammals		Adverse effects on other non-target species Physical properties
				Immediate Toxicity (Acute)	Long-Term Toxicity (Chronic)	
chlorphoxim BAY SRA 7747; Baythion C 2-chloro-*a*-[(diethoxyphosphinothioyloxyimino)-phenylacetonitrile; *O*-diethyl phosphorothioate CAS # 14816-20-7	organo-phosphate	insecticide; acaricide	?	oral: low to medium (1) dermal: low to high (1) inhalation: ?	?	water: slightly soluble slightly volatile
chlorpropham Beet Kleen (with fenuron & propham); Chloro-IPC; Chlorpropham; chlorprophame (France); CICP; CIPC; Faso Wy-Hoe; Furloe; Herbon Yellow (with fenuron); Jack Wilson Chloro 51; Morlex (with fenuron & ethofumesate & propham); Preweed; Sprout Nip; Spud-Nic; T-H Klean Drop; Taterpex; Triherbide-CIPC; Y-3 isopropyl *m*-chlorocarbanilate; isopropyl *N*-(3-chlorophenyl)carbamate CAS # 101-21-3	carbamate	herbicide	mod-pers (1)	oral: medium (2) dermal: medium (1) inhalation: low (1)	decreased body weight (3)	immediate toxicity: birds: low to medium (3)
chlorpyrifos Dowco 179; Dursban; Lepister (with flucythrinate); Lorsban; Pyrinex; Salut (with dimethoate) *O,O*-diethyl *O*-3,5,6-trichloro-2-pyridyl phosphorothioate CAS # 2921-88-2	organo-phosphate	insecticide	mod-pers (1)	oral: medium to high (1,2) dermal: medium to high (3,4) inhalation: high (6)	cumulative (13,14) fetotoxin (6) delayed neurotoxin (7) bulls: sterility and impotence (8)	BIOCIDE immediate toxicity: birds: high to very high (1,9) molluscs: very high (1) fish: very high (1) amphibians: low to high (2) crustaceans: very high (1) bees: very high (10) aquatic insects: very high (1) long-term toxicity: birds: leg weakness (1); delayed neurotoxicity (11) fish: affects growth (1) crustaceans: affects reproduction & equilibrium (1) plants: toxic to some (12) water: slightly soluble slightly volatile
contaminant(s): **sulfoTEPP** (see sulfoTEPP)						
transformation product(s): **pyridinol** 3,5,6-trichloro-2-pyridinol						slightly volatile
chlorsulfuron DPX-W4189; Finesse; Glean; Glean 20DF; Glean T; Glean TP; Telar 2-chloro-*N*-[(4-methoxy-1,3,5-triazin-2-yl)aminocarbonyl]benzenesulfonamide CAS # 64902-72-3	urea	herbicide	non-pers to mod-pers (1)	oral: low (2) dermal: medium (2) inhalation: low to medium (1)	?	immediate toxicity: birds: low to medium (1) fish: low (1) water: soluble "slightly volatile" "nonflammable"

NAME: Common Trade and Other Chemical CAS Number	Class of Chemical	Chief Pesticide Use; Status	Persistence	Effects on Mammals		Adverse effects on other non-target species / Physical properties
				Immediate Toxicity (Acute)	Long-Term Toxicity (Chronic)	
cholecalciferol Quintox; Rampage; Vitamin D3 9,10-Seocholestra-5,7,10(19)-t rein-3 betaol; activated 7-dehydro-colesterol; oleovitamin D₃ CAS # 434-16-2	miscel- laneous	rodenticide	?	oral: very high (1) dermal: ? inhalation: ?	arterial damage (2)	immediate toxicity: birds: "low" (3) water: "insoluble" oil: "soluble"
transformation product(s): **calcitriol**					teratogen (1) fetotoxin (1)	
cinmethylin Argold; Cinch exo-1-methyl-4-(1-methylethyl)-2-[(2-methylphenyl)methoxy]-7-oxabicyclo[2.2.1]heptane CAS # 87818-31-3	miscel- laneous	herbicide	mod-pers (1)	oral: medium (2) dermal: low to medium (1) inhalation: ?	?	water: slightly soluble slightly volatile combustible
citronella Ceylon citronella oil; Java citronello oil; oil of citronell Ceylon: mixture of 60% geraniol, 15% citronellal, 10-15% camphene & diphenthene; Java: mixture of 25-50% citronellal, 25-45% geraniol CAS # MX8000-29-1	oil	insect repellent	?	?	?	water: "not soluble" to "slightly soluble" combustible
clofentezine Acaristop; Apollo; Apolo; NC-21314; Panatac; Torero (with tau-fluvalinate); Viktor (with fenpropathrin) 3,6-bis(2-chlorophenyl)-1,2,4,5-tetrazine CAS # 74115-24-5	miscel- laneous	acaricide	mod-pers (1)	oral: low to medium (2) dermal: low to medium (2) inhalation: ?	?	immediate toxicity: birds: low (3) fish: low to high (3,4) water: insoluble non-volatile to slightly volatile
clomazone Command; Commence (with trifluralin); Dimethazone; Fenoxan; FMC-57020; Gamit; Merit 2-[(2-chlorophenyl)methyl]-4,4-dimethyl-3-isoxazolidinone CAS # 81777-89-1	miscel- laneous	herbicide	non-pers to mod-pers (1,2)	oral: medium (1,3) dermal: low to medium (3) inhalation: medium (3)	?	immediate toxicity: birds: low to medium (1,4) fish: low to medium (1,4) aquatic insects: medium (1) crustaceans: medium to high (1) molluscs: medium (5) water: soluble volatile combustible
clopyralid 3,6-DPA; Benazolox (with benazolin); Curtail (with 2,4-D); Dowco 290; Format; Kerb Mix Matrikerb (with pronamide); Lontrel (with 2,4-D); Lontrel Plus (with mecoprop & dichlorprop & MCPA); Mayclene (with dichlorprop & MCPA); Reclaim; Stinger 3,6-dichloro-2-pyridine carboxylic acid; 3,6-dichloropicolinic acid CAS # 1702-17-6	miscel- laneous	herbicide	non-pers to mod-pers (1)	oral: medium to high (1) dermal: low to medium (1) inhalation: ?	?	immediate toxicity: fish: low (2) slightly volatile flammable to combustible

NAME: Common Trade and Other Chemical CAS Number	Class of Chemical	Chief Pesticide Use; Status	Persistence	Effects on Mammals		Adverse effects on other non-target species Physical properties
				Immediate Toxicity (Acute)	Long-Term Toxicity (Chronic)	
CNP chlornitrofen; MC 1478; MO; Mo-338 2,4,6-trichlorophenyl-4-nitrophenyl ether CAS # 1836-77-7	phenoxy	herbicide	?	oral: low (1) dermal: low to medium (1) inhalation: ?	?	immediate toxicity: fish: low (1) water: insoluble
copper Characteristics as a class are given here; variations are indicated for each compound below CAS # 7440-50-8	metal/ mineral copper		perm	oral: ? dermal: ? inhalation: ?	?	?
compound(s):						
basic copper carbonate Bremen blue; Bremen gree; Cobredon; Copophos; Kop Karb; Malachite; Nutra-Spray Basic Copper Carbonate cupric carbonate; copper carbonate hydroxide; cupric subcarbonate CAS # 12069-69-1	organic	fungicide		oral: high (1)		immediate toxicity: fish: "toxic" (1) water: "insoluble"
basic copper chloride	inorganic			oral: medium (1)	"irreversible eye damage" (2)	immediate toxicity: fish: high (3) water: "insoluble" "corrosive"
basic cupric acetate cupric subacetate CAS # 142-71-2	organic	fungicide		oral: medium (1)		water: "soluble"
Bordeaux mixture Bor-Dax; Copper Hydro Bordo; FT 2M (with mancozeb); Fungi-Bordo mixture of calcium hydroxide and copper sulfate (see these separate listings) CAS # 8011-63-0	copper	fungicide				water: "insoluble"
chromated copper arsenate (see arsenic)						
copper ammonium carbonate Cal Cop 1-; Copper-Count-N; For-Cop-80NC; K-Cop Sol Kop 10; Taytox CAS # 33113-08-5	organic	fungicide antibiotic	"long lasting once applied to the crop" (1)			immediate toxicity: fish: "toxic to fish" (1) water: "readily soluble" to very soluble
copper ammonium sulfate	inorganic					immediate toxicity: fish: medium to very high (1)
copper bis(3-phenylsalicylate) bis(2-hydroxyl[1,1'-biphenyl]-3-carboxylato-O,O)copper ; bis(3-phenylsalicylato)copper; copper-3-phanyl salicylate CAS # 5328-04-1	organic	fungicide		oral: medium (1)		water: "practically insoluble" oil: very soluble "non-volatile"
copper carbonate	organic	fungicide				water: "practically insoluble"

(continued on next page)

NAME: Common Trade and Other Chemical CAS Number	Class of Chemical	Chief Pesticide Use; Status	Persistence	Effects on Mammals		Adverse effects on other non-target species Physical properties
				Immediate Toxicity (Acute)	Long-Term Toxicity (Chronic)	
copper chelate	organic	herbicide algacide		oral: medium (1)		immediate toxicity: fish: medium to high (1) crustaceans: low (1) water: "soluble" oil: "insoluble" highly volatile "nonflammable"
copper citrate Cuprocitrol 2-hydroxyl-2,3-propanetricarbo xylic acid copper salt; cupric citrate CAS # 10402-15-0	organic	algacide antibiotic				
copper ethylenediamine complex Komeen	organic	herbicide				immediate toxicity: fish: low to high (1) water: "soluble" "nonflammable"
copper hydroxide Blue Shield; Champion; Comac Parasol; Criscobre; Cudrox; Cuidrox; Cupravit Blue; Kocide; Kocide 101; Kocide 404S; Parasol copper (II) hydroxide; cupric hydroxide; hydrated cupric oxide CAS # 20427-59-2	inorganic	antibiotic fungicide		oral: medium (1)	"causes irreversible eye damage" (2)	immediate toxicity: birds: low to medium (1) fish: low to high (1) water: slightly soluble
copper linoleate Citcop; Cop-O-cide; Copoloid; Gro-Tone Liquid; Knew; TC-90 copper salts of fatty and rosin acid. 20 - 25% copper abietate, 8 - 12% copper linoleate and copper oleate	organic	fungicide voluntary cancellation by producer, USA		"relatively low toxicity" (1)		immediate toxicity: fish: "toxic to fish" (1) water: insoluble to slightly soluble oil: "soluble"
copper napthenate Cop-R-Nap; Copper-Cure; Coppernate; Cuprinol Brown; Cuprinol Green; Curpinol; NCN; Tryosan; Tuscopper copper salt of napthenic acid CAS # 1338-02-9	organic	fungicide insecticide		oral: low to very high (1,2) dermal: low to medium (3)		water: "soluble" oil: very soluble ? volatile "flammable"
copper nitrate CAS # 3251-23-8	inorganic	fungicide				
copper oxalate CAS # 814-91-5	organic	bird repellent				water: "practically insoluble"
copper oxide Caocobre; Copox; Copper-Sandoz; Copper-Sardez; Cuprocide; Fungi-Rhap; Kuprite; Nordox; Oleocuivre; Perecot; Perenox; Triangle*; Yellow Cuprocide brown copper oxide; cuprous oxide; dicopper oxide *see Bordeaux mixture CAS # 1317-39-1	inorganic	fungicide developed to replace Bordeaux mixture		oral: medium to high (1,2)		immediate toxicity: fish: low (2) bees: "harmless to bees" (3) water: "insoluble"

| NAME: | Common Trade and Other Chemical CAS Number | Class of Chemical | Chief Pesticide Use; Status | Persistence | Effects on Mammals | | Adverse effects on other non-target species Physical properties |
					Immediate Toxicity (Acute)	Long-Term Toxicity (Chronic)	
copper oxychloride Atacamite; BASF-Grunkupfer; Blitox; Colloidox; Cop Tox; Cophamate; Cos; Coxysan; Cupravit; Cuprin; Cuprocaffaro; Cuprosan Blue; Ferncot; Fytolan; Kauritil; Kilex; Micropcop; Perecol; Recop; Rhodioacuivre basic copper chloride; basic cupric chloride; copper (II) chloride oxide hydrate; dicopper chloride trihydroxide; oxychlorure de cuivre CAS # 1332-40-7		inorganic	fungicide		oral: medium (1)	"irreversible eye damage" (2)	immediate toxicity: fish: high (3) water: "insoluble"
copper oxychloride sulfate Capro 57; Copro 53; Coxysul; CS-56 CAS # 8012-69-9		inorganic	fungicide		"causes substantial but temporary eye injury" (1)		immediate toxicity: bees: low to medium (2)
copper sulfate Blue Copperas; Blue Vitriol; Bluestone; Copace-E; Copperfine-Zinc; Foltimil (with folpet); Neutrocrop; Nu-Cop; Phyton 27; Spray-Cop; Triangle* basic copper sulfate; copper sulfate monohydrate; copper sulfate pentahydrate *See Bordeaux mixture CAS # 7758-98-7		inorganic	fungicide herbicide algacide		oral: medium to high (1,2) dermal: low to medium (2) inhalation: low to high (2)	mutagen (4,5) "bioaccumulates" (3)	immediate toxicity: fish: low to very high (1,2) bees: "toxic" crustaceans: low (2) molluscs: medium (2)
copper sulfate monohydrate		inorganic	fungicide				immediate toxicity: fish: "toxic to fish" (1)
copper tea complex Algae-Rhap CU; K-Pool; K-Tea copper triethanolamine complex		organic	algacide herbicide				immediate toxicity: fish: medium (1) water: soluble
copper zinc chromate Crag Fungicide 658; Experimental Fungicide 658		inorganic	fungicide antibiotic				water: "practically insoluble"
cupric hydrazinium sulfate Mathieson 466; Omazene; Omazine copper (II) dihydrazinium disulfate; bis(hydrazine)bis(hydrogensulf ato)copper; copper dihydrazine disulfate CAS # 33271-65-7		organic	fungicide				
oxine-copper Bioquin; Copper 8; Copper Oxinate; Cunilate 2472; Cuproquin; Dokirin; Dormycin; Fruitdo; Milmer 1; Quinolate; Quinolate 15; Quinolate 20; Quinolate AC, Fs, Quinolate AC Kara (with anthraquinone); Quinolate MG SAFI (with endosulfan & lindane); Quinolate Triple Kara (with anthraquinon & lindane); Quinolate V 4 X AC, FS, DS (with anthraquinone & carboxin); Quinolate V 4 X Triple (with lindane); Tomo-oxiran CAS # 10380-28-6		organic	fungicide wood preservative		oral: low to medium (1)	suspect mutagen (2)	immediate toxicity: fish: high (3) water: "insoluble" "non-volatile"

(continued on next page)

NAME: Common Trade and Other Chemical CAS Number	Class of Chemical	Chief Pesticide Use; Status	Persistence	Effects on Mammals		Adverse effects on other non-target species Physical properties
				Immediate Toxicity (Acute)	Long-Term Toxicity (Chronic)	
zinc coposil BSZ; Citrusperse; Cop-O-Zinc a combination of basic copper sulfate and basic zinc sulfate or zinc oxide	inorganic	fungicide used as a safener with lead arsenate				immediate toxicity: fish: "toxic to fish" (1)
coumafuryl Foumarin; Fumasol; Furmarin; furmarin (U.K., New Zealand); Kill-Ko-Rat; Krumkil; Lurat; Mouse Blues; Rat-A-Way; Ratafin; tomarin (Turkey) 3-(α-(2-acetonylfurfury)- 4-hydroxycoumarin; 3-[1-(2-furanyl)-3-oxobutyl]- hydroxy-2H-1-benzopyran-2-one; 4-hydroxy-3-[3-oxo-1-(2-furyl)butyl] coumarin CAS # 117-52-2	coumarin	rodenticide	?	oral: very high (1) dermal: ? inhalation: ?	?	water: "slightly soluble"
salt(s): **sodium salt of coumafuryl (Fumasol)**						water: "very soluble"
coumaphos Agridip; Asuntol; BAY 21/199; Baymix; Co-Ral; diolice; Meldane; Muscatox; Perizin; Resistox; Umbethion O,O-diethyl O-3-chloro-4-methyl-2-oxo-2H-1-benzopyran-7-yl phosphorothioate CAS # 56-72-4	organo-phosphate	insecticide "do not use before or after the application of natural or synthetic pyrethrins or compounds used to synergize them"	?	oral: very high (1) dermal: high (1) inhalation: high (1)	?	immediate toxicity: birds: very high (3) fish: very high (4) aquatic insects: very high (4) crustaceans: very high (4) long-term toxicity: birds: delayed neurotoxicity (5) water: slightly soluble oil: "slightly soluble" slightly volatile
4-CPA Fruitone; Marks 4-CPA; PCPA; Tomato Fix; Tomato Hold; Tomatotone (4-chlorophenoxy)acetic acid; (p-chlorophenoxy)acetic acid; parachlorophenoxyacetic acid CAS # 122-88-3	organo-chlorine	plant growth regulator	non-pers (1)	oral: ? dermal: ? inhalation: ? "chemically related to 2,4-D and exhibits similar acute toxicity in laboratory animals" (2)	?	immediate toxicity: fish: low (1) water: "very soluble"
transformation product(s): **phenol** (see phenol)						
credazine H-722; Kusakira; SW-6701; SW-6721 (3-(2-methylphenoxy)pyridazine ; 3-o-tolyloxypyridazine CAS # 14491-59-9	miscel-laneous	herbicide	"mod-pers" (1)	oral: medium (1) dermal: low to medium (1) inhalation: ?	?	immediate toxicity: fish: low (1) water: soluble

NAME: Common Trade and Other Chemical CAS Number	Class of Chemical	Chief Pesticide Use; Status	Persistence	Effects on Mammals		Adverse effects on other non-target species Physical properties
				Immediate Toxicity (Acute)	Long-Term Toxicity (Chronic)	
creosote (coal tar) brick oil; coal tar creosote; coal tar oil; creosote oil; creosotum; cresylic creosote; dead oil; naphthalene oil; tar oil complex mixture of less volatile hydrocarbons, with up to 3% phenolic constituents, obtained from destructive distillation of coal. Over 300 components identified; composition highly variable, depending on coal source and process; most abundant constituent is napthalene: see napthalene CAS # 8001-58-9	phenol	wood preservative restricted USA, 1986	?	oral: ? dermal: ? inhalation: ?	carcinogen (1,2) suspect mutagen (2) lung and liver damage (3)	water: "practically insouble" combustible
creosote (wood tar) beachwood creosote; Creasote; wood creosote mixture: contains chiefly guaiacol (o-methoxyphenol; CAS # 90051) and creosol (2-methoxy-4-methylphenol; CAS # 93-56-1) with little or no phenol or cresols obtained from destructive distillation of wood CAS # 8021-39-4	phenol	wood preservative restricted USA, 1986	?	oral: medium to very high (1) dermal: very high (1) inhalation: medium (1)	suspect mutagen (2)	long-term toxicity: fish: carcinogen (3) water: "insoluble" oil: "soluble"
cresylic acid cresols; tricresol mixture of cresols (methylphenols) obtained from coal tar; usually contains a few percents phenol, and other aromatic compounds depending on grade CAS # 1319-77-3	phenol	insecticide fungicide antibiotic	?	oral: medium (1) dermal: ? inhalation: ?	?	immediate toxicity: fish: low (2) highly volatile combustible
crotoxyphos Cio-Vap; Ciodrin; Cypon EC; Decrotox; Duo-Kill 1-phenylethyl 3-(dimethoxyphosphinoyloxy)isocrotonate; 1-phenylethyl 3-(dimethoxyphosphinyloxy)isocrotonate; dimethyl(E)-1-methyl-2- (1-phenylethoxycarbonyl) vinyl phosphate; (E)-1-phenylethyl 3-[(dimethoxyphosphinyl)oxy]-2-butenoate; a-methylbenzyl(a)-3-hydroxycrotonate ester with dimethyl phosphate CAS # 7700-17-6	organo- phosphate	insecticide	?	oral: high (1,2) dermal: high to very high (1,2) inhalation: ?	?	immediate toxicity: birds: medium to high (3) fish: medium to very high (4) crustaceans: very high (4) bees: high (5) water: soluble slightly volatile
crufomate Dowco 132; Kempak; Montrel; Ruelene 4-tert-butyl-2-chlorophenyl methyl methylphosphoramidate CAS # 299-86-5	organo- phosphate	insecticide	?	oral: medium to high (1,2) dermal: ? inhalation: ?	?	immediate toxicity: birds: high (3) water: "practically insoluble" oil: "practically insoluble"
cryolite Koyoside; Kryocide; Prokil sodium fluoaluminate; sodium aluminofluoride CAS # 15096-52-3	metal/ mineral	insecticide	?	oral: low (1) dermal: very high (2) inhalation: ?	fluoride accumulation in bones (3)	BIOCIDE immediate toxicity: birds: low to medium (3) fish: low (4) crustaceans: medium (4) bees: low (5) plant: damages some crops, especially peaches (6) water: soluble

NAME:	Common Trade and Other Chemical CAS Number	Class of Chemical	Chief Pesticide Use; Status	Persistence	Effects on Mammals		Adverse effects on other non-target species Physical properties
					Immediate Toxicity (Acute)	Long-Term Toxicity (Chronic)	
cyanazine Bellater (with atrazine); Bladex; Bladotyl (with mecoprop); Blagal (with MCPA); Blazine (with atrazine); Conquest (with atrazine); DW 3418; Extrazine (with atrazine); Fortrol; Holtox (with atrazine); Scogal; SD 15417; Vega (with bentazon & dichlorprop); WL 19805 2-(4-chloro-6-ethyamino-s-triazin-2-ylamino)-2-methylpropionitrile CAS # 21725-46-2		triazine	herbicide restricted USA, 1988	non-pers (1)	oral: high (1) dermal: low to medium (1) inhalation: low to medium (2)	teratogen (3)	immediate toxicity: birds: low to high (4) fish: low to medium (5) crustaceans: high (5) bees: low (6) water: soluble non-volatile "nonflammable"
cyanophenphos CYP; S-4087; Surecide O-p-cyanophenyl O-ethyl phenyl phosphonothioate; O-(4-cyanophenyl) O-ethyl phenyl phosphonothioate CAS # 13067-93-1		organo-phosphate	insecticide	?	oral: high to very high (1) dermal: ? inhalation: ?	liver damage (2)	immediate toxicity: bees: "toxic" (3) long-term toxicity: birds: delayed neurotoxin (4) water: slightly soluble slightly volatile
cyanophos Cyanox; CYAP; Cynock; S-4084 O-4-cyanophenyl O,O-dimethyl phosphorothioate; 4-dimethoxyphosphinothioyloxyl benzonitrile; O-(4-cyanophenyl) O,O-dimethyl phosphorothioate; O,O-dimethyl phosphorothioate O-ester with p-hydroxybenzonitrile CAS # 2636-26-2		organo-phosphate	insecticide	?	oral: medium (1) dermal: ? inhalation: ?	?	immediate toxicity: fish: medium (1) bees: "toxic to honeybees" (2) long-term toxicity: birds: delayed neurotoxicity (3) water: slightly soluble slightly volatile
cycloate Eurex; hexylthiocarbam; R-2063; Ro-Neet; Ronit S-ethylcyclohexylethylthiocarbamate; S-ethylcyclohexylethylcarbamothioate; S-ethyl-N-ethylthiocyclohexanecarbamate CAS # 1134-23-2		thiocar-bamate	herbicide	non-pers (1)	oral: medium (1) dermal: low to medium (2) inhalation: ?	?	immediate toxicity: birds: low to medium (3) fish: medium (1) bees: low to medium (4) water: slightly soluble volatile combustible
cyclohexane hexahydrobenzene; hexamethylene; hexanapthene CAS # 110-82-7		miscel-laneous	solvent	?	oral: ? dermal: ? inhalation: very high (1)	?	immediate toxicity: fungus: "slight fungicidal action" water: slightly soluble highly volatile flammable

NAME: Common Trade and Other Chemical CAS Number	Class of Chemical	Chief Pesticide Use; Status	Persistence	Effects on Mammals		Adverse effects on other non-target species Physical properties
				Immediate Toxicity (Acute)	Long-Term Toxicity (Chronic)	
cyclohexanone Anone; Hytrol O; Nadone cyclohexanone; ketohexamethylene; pimelic ketone CAS # 108-94-1	miscel-laneous	solvent	?	oral: medium (1) dermal: ? inhalation: ?	suspect mutagen (1)	water: very soluble slightly volatile combustible
cycloheximide Acti-Aid; Acti-Dione; Acti-Dione BR Concentration; Acti-Dione PM; Acti-Dione RZ (with pentachloronitrobenzene); Acti-Dione TGF; Acti-Dione Thiram (with thiram); Actispray; Hizarocin; Kaken 3[2-(3,5-dimethyl-2-oxocyclohexyl)-2-hydroxyethyl]-glutarimide; naramycin CAS # 66-81-9	amide	fungicide growth regulator Acti-dione* voluntarily withdrawn from registration in Canada	non-pers (1)	oral: medium to very high (2,3) dermal: very high (1) inhalation: ?	suspect mutagen (4,5) suspect teratogen (6,7) immunotoxin (10)	immediate toxicity: birds: high to very high (1,8) fish: high (1) bees: low (9)
cycluron 3-cyclo-octyl-1,1-dimethylurea CAS # 2163-69-1	urea	herbicide	?	oral: medium (1) dermal: ? inhalation: ?	?	water: soluble
cyfluthrin Baygon Spray (with dichlorvos & propoxur); Baythroid F (with omethoate); Baythroid TM (with acephate-met); Insectipen; Muscatox (with phoxim) (RS)-α-cyano-4-fluoro-3-[phenoxybenzyl (1RS)-cis,trans-3-(2,2-dichlorovinyl)-2,2-dimethylcyclopropanecarboxylate CAS # 68359-37-5	pyrethroid	insecticide	non-pers to mod-pers (1)	oral: low to very high (2,3) dermal: ? inhalation: ?	?	immediate toxicity: fish: high to very high (1,4) crustacean: very high (5) bees: high (6) water: insoluble to slightly soluble non-volatile to volatile flammable
λ-cyhalothrin Icone; Karate; OMS 3021; PP 321 λ-cyhalothrin; α-cyano-3-phenoxybenzyl 3-(2-chloro-3,3,3-trifluropropenyl)-2,2-dimethylcyclopropanecarboxylate (1:1 mixture of (Z)-(1S,3S)-R-ester and (Z)-(1R,3R) S-ester CAS # 91465-08-6	pyrethroid	insecticide	non-pers (1)	oral: high (1) dermal: ? inhalation: very high (1)	?	immediate toxicity: fish: very high (2) water: insoluble non-volatile
cymoxanil Bayleton AN, Curzatem (with triadimefon); Fytospore (with mancozeb) 1-(2-cyano-2-methoxymino-acetyl)-3-ethylurea	urea	fungicide not registered for use in U.S.	non-pers (1)	oral: medium (1) dermal: ? inhalation: ?	?	immediate toxicity: fish: medium (1) water: soluble
cypermethrin Ammo; Arrivo; Barricade; Cymbush; Cymperator; Cypercopal; cypermethrine; Cyrux; Demon; Fenom; Flectron; Folcord; Imperator; Kafil Super; Nurelle; Polytrin; Ripcord; Siperin; Toppel; Ustaad; WL 43467 (RS)-α-cyano-3-phenoxybenzyl (1RS-cis,trans-3-9 2,2-dichlorovinyl)-2,2-dimethyl-cyclopropanecarboxylate CAS # 52315-07-8	pyrethroid	insecticide	non-pers (1)	oral: medium to high (2,3) dermal: ? inhalation: ?	mutagen (4,5) immunotoxin (6)	immediate toxicity: birds: low (7,8) fish: high to very high (1,9) crustaceans: very high (10) bees: "toxic" (2) water: insoluble oil: "lipophilic" non-volatile

NAME: Common Trade and Other Chemical CAS Number	Class of Chemical	Chief Pesticide Use; Status	Persistence	Effects on Mammals		Adverse effects on other non-target species Physical properties
				Immediate Toxicity (Acute)	Long-Term Toxicity (Chronic)	
cyprazine Outfox; Prefox (with ethiolate); S-6115; S-9115 2-chloro-4-(cyclopropylamino)-6-(isopropylamino)-s-triazine; 6-chloro-N-cyclopropyl-N'-(1-methylethyl)-1,3,5-triazine-2,4-diamine CAS # 22936-86-3	triazine	herbicide cancelled USA	?	oral: low to medium (1,2) dermal: medium (2) inhalation: ?	lung damage (3)	immediate toxicity: birds: medium (6) fish: medium (4) water: slightly soluble combustible
cyromazine Larvadex; Trigard N-cyclopropyl-1,3,5-triazine-2,4,6-triamine CAS # 66215-27-8	triazine	insect growth regulator	?	oral: medium (1) dermal: ? inhalation: medium (1)	suspect fetotoxin (2,3)	immediate toxicity: fish: low (1,4) water: very soluble slightly volatile
transformation product(s): **melamine** CAS # 108-78-1				oral: medium (1)	carcinogen (1) bladder stones (1)	
cythioate AC 26691; American Cyanamid Cl 26691; Cyflee; Proban* * Proban is also a trade name for an herbicide, 1-(3,4-dichlorophenyl)-3-isopro-pyl-3-(2-propenyl)urea, not listed in this guide. O,O-dimethyl O-p-sulfamoylphenyl phosphorothioate; p-hydroxybenzenesulfonamide, O-ester with O,O-dimethyl phosphorothioate CAS # 115-93-5	organo-phosphate	insecticide	pers (1)	oral: high (2) dermal: ? inhalation: ?	?	volatility: "negligible"
2,4-D Agricorn D; Agrotect; Amidox; Cloroxone; College Brand Weed Killer; Ded-Weed Aero Ester; Demise; Dicotox; Dinoxol; Dymec; Esteron 44; Fersone; Green Cross Amine 80; Hormotox; Lawn-Keep; Lithane; Miracle; Niagara Am Sol; Plantgard; Raid Weed Killer; Weedone 2,4-dichlorophenoxyacetic acid CAS # 94-75-7	phenoxy	herbicide restricted USA	non-pers to mod-pers (1,2)	oral: medium to high (2,3) dermal: high (4) inhalation: medium to high (5)	carcinogen (6,7) suspect mutagen (8,9) teratogen (10,11) suspect fetotoxin (12) anorexia (12) immunotoxin (13) toxic injury to liver, kidney, & central nervous system (19)	immediate toxicity: birds: low to high (5,14) fish: low to very high (5,15) amphibians: low to medium (5) crustaceans: low to very high (5) molluscs: medium (11) non-target insect: low to high (11) bees: low to medium (16) soil organisms: low (17) long-term toxicity: birds: can affect egg production (17) fish: cumulative (17) amphibians: inhibits frog egg development (17) crustaceans: may significantly reduce population (17) molluscs: reduction in population; cumulative (17) plants: leaf malformation (16) soil organisms: may inhibit growth (17) can favor growth of insects and pathogens (18)

NAME: Common Trade and Other Chemical CAS Number	Class of Chemical	Chief Pesticide Use; Status	Persistence	Effects on Mammals		Adverse effects on other non-target species Physical properties
				Immediate Toxicity (Acute)	Long-Term Toxicity (Chronic)	
2,4-D (*cont.*)						water: "insoluble" to soluble" oil: "insoluble" highly volatile
transformation product(s):						
2,7-dichlorodibenzo-p-dioxin	dibenzo-dioxin				carcinogen (1)	slightly volatile
1,3,7-trichlorodibenzo-p-dioxin (see dibenzodioxin class)	dibenzo-dioxin					
1,3,6,8-tetrachlorodibenzo-p-dioxin 1,3,6,8-TCDD	dibenzo-dioxin					water: insoluble
1,3,7,9-tetrachlorodibenzo-dioxin	dibenzo-dioxin					
TCDD (see chloroneb)						
2,4-dichlorophenol						
dalapon Basfapon; BH Dalapon; Dalapon-Na; Destral (with 2,4-D & diuron); Dowpon; Gramevin; Lignum (with atrazine); Radapon; Slam (with asulam); Target (with asulam); Unipon 2,2-dichloropropionic acid (and sodium salt) CAS # 75-99-0	miscel-laneous	herbicide	non-pers (1)	oral: low to medium (2) dermal: low to medium (3) inhalation: low (2)	?	immediate toxicity: birds: low (2) fish: low (4) bees: low to medium (5) crustaceans: medium (4) aquatic insects: low (4) water: very soluble volatility "negligible" "nonflammable"
daminozide Alar; Alar-85; Aminozide; B-995; B-nine; DSMA; Kylar; Kylar-85; SAD-85 butanedioic acid mono (2,2-dimethyl hydrazide); succinic acid 2,2-dimethyl hydrazide CAS # 1596-84-5	miscel-laneous	plant growth regulator cancelled for food crop uses, at request of producer, USA, Nov. 1989	"does not persist in soil" (3)	oral: low (1,2) dermal: ? inhalation: low 91)	carcinogen (3,4)	immediate toxicity: birds: low (1,2) fish: low (1) crustaceans: low (5) bees: low (6) water: "soluble" to "slightly soluble" slightly volatile
transformation product(s):						
unsymmetrical 1,1-dimethylhydrazine dimazine; UMDH asymmetrical dimethyl hydrazine; 1,1-dimethylhydrazine; N,N-dimethylhydrazine CAS # 57-14-7	miscel-laneous	is also rocket-fuel		oral: high (1) dermal: "corrosive to skin" (1)	carcinogen (2-4) mutagen (5,6)	highly volatile
dazomet Basamid; Crag Fungicide 974; Crag Nemacide; DMTT; Micro-Fume; Mylone; N-521; Preservit	miscel-laneous	herbicide fungicide nematocide	non-pers to mod-pers (1,2)	oral: medium to high (3) dermal: ?	liver and kidney damage (6)	immediate toxicity: fish: "toxic to fish" (4) bee: low to medium (5)

(continued on next page)

NAME: Common Trade and Other Chemical CAS Number	Class of Chemical	Chief Pesticide Use; Status	Persistence	Effects on Mammals		Adverse effects on other non-target species / Physical properties
				Immediate Toxicity (Acute)	Long-Term Toxicity (Chronic)	
dazomet (*cont.*) tetrahydro-3,5-dimethyl-2H-l,3, 5-thiadiazine-2-thione; 3,5-dimethyl-2H-l,3,5-thiadiazine-2-thione CAS # 533-74-4				inhalation: medium (3)		water: soluble volatile "nonflammable"
transformation product(s): **carbon disulfide** (see carbon disulfide)						
methyl isothiocyanate (see methyl isothiocyanate)						
formaldehyde (see formaldehyde)						
hydrogen sulfide (see calcium polysulfide)						
2,4-DB Butoxone (with 2,4-D & MCPA); Butoxone SB; Butyrac 118; Butyrac ester; Embutox; Ley-Cornox (with benazolin & MCPA); MB 2878 4-(2,4-dichlorophenoxy) butyric acid CAS # 94-82-6	phenoxy	herbicide	non-pers (1)	oral: medium (1) dermal: low to medium (2)	mutagen (2)	immediate toxicity: fish: medium (2) bees: low to medium (3) water: slightly soluble oil: "highly soluble"
salt(s): **2,4-DB sodium** CAS # 10433-59-7				oral: medium to high (1)		
transformation product(s): **2,4-D** (see 2,4-D)						
DBCP BBC 12; CHA-KEM-CO; dibromochloropropane; Fumazone; Garden Fume; Green Light; Nemafume; Nemagon; Oxy DBCP 1,2-dibromo-3-chloropropane CAS # 96-12-8	organo-chlorine	nematocide fumigant cancelled, most uses, USA, 1977	?	oral: high (1) dermal: high (1) inhalation: ?	carcinogen (2,3) suspect mutagen (4) testicular damage (5,6) kidney & liver damage (6)	immediate toxicity: birds: high (7) molluscs: "high" (7) earthworms: "high" (7) bees: low (8) water: soluble oil: "miscible in hydrocarbon oils" highly volatile combustible
DCPA chlorthal dimethyl; DAC 893; Dacthal; Decimate (with propachlor) dimethyl tetrachloroterephthalate; dimethyl 2,3,5,6-tetrachloro-1,4-benzenedicarboxylate CAS # 1861-32-1	phthalate	herbicide	mod-pers (1)	oral: medium (1,2) dermal: medium (1) inhalation: ?	suspect carcinogen (3) suspect mutagen (3) suspect teratogen (3)	immediate toxicity: birds: low to medium (1) bees: low (4) slightly volatile "nonflammable"
contaminant(s): **TCDD** (see chloroneb)						
hexachlorobenzene (see hexachlorobenzene)						

NAME: Common Trade and Other Chemical CAS Number	Class of Chemical	Chief Pesticide Use; Status	Persistence	Effects on Mammals		Adverse effects on other non-target species Physical properties
				Immediate Toxicity (Acute)	Long-Term Toxicity (Chronic)	
D-D 1,2-Dichloropropane with 1,3-dichloropropene CAS # 8003-19-8	organo-chlorine	fungicide herbicide insecticide nematocide soil fumigant	non-pers (1)	oral: high (2) dermal: high (2) inhalation: ?	suspect carcinogen (1,3) suspect mutagen (1,4) liver and kidney damage (3,5)	immediate toxicity: fish: very high (1) crustaceans: very high (1) water: soluble highly volatile flammable
DDD Rhothane; TDE dichloro diphenyl dichloroethane; 2,2-bis(p-chloropheny)-1,1-dichloroethane CAS # 72-54-8	organo-chlorine	insecticide cancelled USA, 1971	pers; non-pers on plant surfaces (1)	oral: medium to high (2) dermal: ? inhalation: ?	cumulative (3) suspect carcinogen (4,5) mutagen (6) atrophy of adrenal cortex (5)	immediate toxicity: birds: low to high (7) fish: very high (8) amphibians: very high (5) crustaceans: very high (8) bees: low (9) aquatic insects: very high (8) water: "practically insoluble"
DDE dichlorodiphenyl dichloroethylene; 1,1-dichloro-2,2-bis (p-chlorophenyl) ethylene CAS # 72-55-9	organo-chlorine	insecticide	pers (1)	oral: medium (2) dermal: ? inhalation: ?	cumulative (3) carcinogen (4,5)	immediate toxicity: fish: very high (6) bees: low to medium (7) long-term toxicity: birds: embryo mortality in ducks; cumulative (8,9); eggshell thinning (10) water: insoluble oil: "fat soluble"
DDT Anofex; Arkotine; Chlorophenothane; Dedelo; Diamekta 50%; Didimac; Dinocide; Genitox; Gesapon; Gesarex; Gyron; Ixodex; Kopsol; Neocidol; Pentachlorin; Pentech; Rukseam; Santobane; Tech DDT; Zerdane dichloro diphenyl trichloroethane; 1,1,1,-trichloro-2,2,-bis(p-chlorophenyl) ethane CAS # 50-29-3	organo-chlorine	insecticide cancelled USA, 1972	pers (1)	oral: high (2) dermal: medium to high (3) inhalation: ?	cumulative (3) carcinogen (4) mutagen (5) fetotoxin (5) embryotoxin (3) decreased fertility (3) hormone changes (6) aplastic anemia (6) liver damage (7) immunotoxin (8)	BIOCIDE immediate toxicity: birds: low to medium (9) fish: very high (10) amphibians: low to medium (9) crustaceans: very high (10) bees: high (11) aquatic insect: very high (10) aquatic worm: low (12) long-term toxicity: birds: diminished reproduction; eggshell thinning; cumulative (9,13) fish: affects reproduction (1) snakes: much more toxic to egg-laying than to viviparous snakes (14) reduced photosynthesis by marine phytoplankton (15) bioaccumulates (16) molluscs: reduces shell growth (17) water: insoluble oil: "very soluble" slightly volatile

(continued on next page)

NAME: Common Trade and Other Chemical CAS Number	Class of Chemical	Chief Pesticide Use; Status	Persistence	Effects on Mammals		Adverse effects on other non-target species Physical properties
				Immediate Toxicity (Acute)	Long-Term Toxicity (Chronic)	
transformation product(s):						
dicofol (see dicofol)						
DDE (see DDE)						
DDD (see DDD)						
DEET Autan; Blockade; Chemform; Delphene; DET; Detamide; Dieltamid; diethyl toluamide; Flypel; M-DET; Metaldephene; MGK Diethyltoluamide; Off; Repel *N,N*-diethyl-*m*-toluamide; *N,N*-diethyl-3-methylbenzamide CAS # 134-62-3	amide	insect repellent	?	oral: medium (1,2) dermal: medium (2) inhalation: ?	neurotoxin (3)	water: "insoluble"
DEF butyphos; Chemagro B-1776; DEF defoliant; DeGreen; E-Z-off D; Fos-Fall "A"; Ortho Phosphate Defoliant; SST *S,S,S*-tributylphosphorotrithioate CAS # 78-48-8	organo-phosphate	herbicide defoliant	non-pers (1)	oral: high (2) dermal: high to very high (3) inhalation: ?	suspect delayed neurotoxin (4)	BIOCIDE immediate toxicity: birds: medium to high (5) fish: very high (6) amphibians: high (7) aquatic insects: very high (6) crustaceans: very high (6) bees: low to medium (8) long-term toxicity: birds: delayed neurotoxcity (9) fish: feeding behavior inhibited and swimming capacity affected (10) "high potential for phytotoxicity" (11) water: "insoluble"
deltamethrin Butoss; Cislin; Crackdown; decamethrin; Decis; Decis Dan (with endosulfan & fenitrothion & profenfos); Delsekte; deltamethrine; Deltaphos (with triazophos); Detrans (with esbiothrin); K-O (with esbiothrin & piperonyl butoxide); K-Obiol (with piperonyl butoxide); K-Otek; K-Othrin; Kothrin; NRDC 161; OMS 1998; RU 22974 [1*R*-[1α(S*),3α]]-cyano(3-phenoxyphenyl) methyl 3-(2,2-dibromoethenyl)-2,2-dimethylcyclopropanecarboxylate CAS # 52918-63-5	pyrethroid	insecticide	?	oral: medium to very high (1,2) dermal: ? inhalation: ?	?	water: insoluble non-volatile
demeton BAY 10756; Demox; E-1059; mercaptofos (USSR); Systemox; Systox mixture of *O,O*-diethyl *O*-2-(ethylthio)ethyl phosphorothioate and *O,O*-diethyl *S*-2-(ethylthio)ethyl phosphorothioate CAS # 8065-48-3	organo-phosphate	insecticide acaricide	non-pers (1)	oral: very high (2) dermal: very high (3) inhalation: ? readily absorbed through the skin (4)	mutagen (5,6) suspect teratogen (7) embryotoxin (7)	immediate toxicity: birds: very high (8) fish: very high (9) amphibians: medium (8) crustaceans: very high (9) bees: high (10) toxic to some plants (4) water: soluble slightly volatile

NAME: Common Trade and Other Chemical CAS Number	Class of Chemical	Chief Pesticide Use; Status	Persistence	Effects on Mammals		Adverse effects on other non-target species Physical properties
				Immediate Toxicity (Acute)	Long-Term Toxicity (Chronic)	
demeton-methyl BAY 15203; BAY 21/116; Duratox; Met-Systox; methyl demeton; Methylsystox O,O-dimethyl O(and S)-2-(ethylthio)ethyl phosphorothioate CAS # MX8022-00-2	organo-phosphate	insecticide acaricide	?	oral: very high (1) dermal: high to very high (1) inhalation: very high (2)	?	immediate toxicity: fish: low to medium (2) long-term toxicity: birds: teratogenic (3) phytotoxic to some plants (4) water: soluble slightly volatile
desmedipham Betanal 475; Betanal AM; Betanex; Bethanol-475; EP-475; SN 38107 ethyl-m-hydroxycarbanilate carbanilate (ester); 3-ethoxycarbonylaminophenyl-N-phenylcarbamate; ethyl [3-[[(phenylamino)carbonyl]oxy]-phenyl]carbamate CAS # 13684-56-5	carbamate	herbicide	non-pers (1)	oral: low to medium (2,3) dermal: medium (4) inhalation: ?	?	immediate toxicity: fish: medium to high (2) bees: low to medium (5) water: slightly soluble non-volatile combustible
desmetryn desmetryne; G 34360; Samuron; Semeron; Topusyn 2-isopropylamino-4-methylamino-6-methylthio-s-triazine CAS # 1014-69-3	triazine	herbicide	non-pers (1)	oral: medium (1) dermal: ? inhalation: ?	?	immediate toxicity: "negligible toxicity to wildlife" (2) water: soluble slightly volatile
dialifor dialifos (except USA); dialiphor; Hercules 14503; Torak S-(2-chloro-1-phthalimid oethyl) (O,O-diethyl phosphorodithioate; S-[2-chloro-1-(1,3-dihydro-1,3-dioxo-2H-isoindol-2-yl)ethyl] O,O-diethyl phosphorodithioate CAS # 10311-84-9	organo-phosphate	insecticide acaricide	non-pers (1)	oral: high to very high (2,3) dermal: very high (3)	suspect teratogen (4)	immediate toxicity: bees: medium (5) water: "insoluble"
diallate 2,3-DCDT; Avadex; CP 15336; DATC S-2,3-dichloroallyl diisopropylthiocarbamate; S-2,3-dichloroallyl diisopropylthiolcarbamate CAS # 2303-16-4	thiocar-bamate	herbicide restricted USA, 1982	non-pers (1)	oral: medium to high (1,2) dermal: medium (1) inhalation: ?	carcinogen (3,4) suspect mutagen (3) "testicular and ovarian effects" (5) delayed neurotoxicity (3)	immediate toxicity: fish: medium (1) crustacean: medium (1) bees: low to medium (6) water: slightly soluble volatile
transformation product(s): **2-chloroacrolein**					suspect mutagen (1)	
diamidfos Dowco-169; Nellite phenyl N,N'-dimethyl phosphorodiamidate CAS # 1754-58-1	organo-phosphate	nematocide	?	oral: high (1) dermal: very high (1) inhalation: ?	?	immediate toxicity: birds: very high (2) water: very soluble
diatomaceous earth Celatom; Celite; Fuller's Earth; infusorial earth; Kenite; kieselguhr; Perma Guard*; silica; Super-Cel	metal/mineral silica	insecticide desiccant	perm	oral: ? dermal: ? inhalation: ?	lung damage (1)	immediate toxicity: bees: low (2)

(continued on next page)

NAME: Common Trade and Other Chemical CAS Number	Class of Chemical	Chief Pesticide Use; Status	Persistence	Effects on Mammals		Adverse effects on other non-target species Physical properties
				Immediate Toxicity (Acute)	Long-Term Toxicity (Chronic)	
diatomaceous earth (*cont.*) siliceous earth (composed of shells of diatoms); silicon dioxide *combined with pyrethrins CAS # 61790-53-2						
diazinon AG 500; Alfa-tox; Basudin; Dazzel; Diazajet; Diazatol; Diazide; Diazinon; Diazitol; Diazol; dimpylate; Dipofene; G 24480; Gardentox; Knox Out 2FM; Neocidol; Nipsan; Sarolex; Spectracide O,O-diethyl O-(2-isopropyl-6-methyl-4-pyrimidinyl) phosphorothioate CAS # 333-41-5	organo-phosphate	insecticide nematocide banned from use on golf courses and turf farms in USA	non-pers (1,2)	oral: medium to high (3,4) dermal: low to high (4) inhalation: medium (3)	suspect mutagen (5,6) fetotoxin (7) suspect neurotoxin (8) allergic dermatitis (9) conjunctivitis (9) immunotoxin (10)	BIOCIDE immediate toxicity: birds: very high (1,11) fish: very high (12) amphibians: very high (11) crustaceans: very high (12) bees: very high (13) aquatic insects: very high (14) aquatic worm: high (9) plants: toxic to some (15) long-term toxicity: birds: teratogen (2) water: slightly soluble oil: very soluble volatile combustible
contaminant(s): **isodiazinon**					porphyria (1)	
transformation product(s): **sulfoTEPP** (see sulfoTEPP)						
TEPP (see TEPP)						
dibutyl phthalate DBP di-*n*-butyl phthalate CAS # 84-74-2	phthalate	insect repellent	?	oral: low (1) dermal: ? inhalation: ?	suspect mutagen (2) suspect teratogen (3) testicular atrophy (4)	water: soluble volatile flammable
dicamba Banex; Banlene Solo (with ioxynil & dichlorporp); Banvel D; Banvel K (with 2,4-D); Banvell II; Bio Lawn Weedkiller (with 2,4-D & ioxynil); Brush Buster; Dianat; Docklene (with MCPA-sodium); Endox (with mecoprop); Fallow Master (with glyphosate); Fettel (with mecoprop & triclopyr); Herrisol (with MCPA-sodium); Lawnsman (with dichlorprop & MCPA); MDBA; Mediben; Nettle-Ban (with 2,4,5-T & 2,4-D); Super Trimec (with 2,4-D & dichlorprop); Trooper; Weedmaster (with 2,4-D) 3,6-dichloro-o-anisic acid; 2-methoxy-3,6-dichlorobenzoic acid	organo-chlorine	herbicide	non-pers to mod-pers (1,2)	oral: medium (3,4) dermal: ? inhalation: ?	?	immediate toxicity: birds: medium (2) fish: low (3,5) crustaceans: low to medium (2,5) water: soluble slightly volatile "nonflammable"

NAME: Common Trade and Other Chemical CAS Number	Class of Chemical	Chief Pesticide Use; Status	Persistence	Effects on Mammals		Adverse effects on other non-target species Physical properties
				Immediate Toxicity (Acute)	Long-Term Toxicity (Chronic)	
dicamba *(cont.)* CAS # 1918-00-9						
transformation product(s): **2,7-dichlorodibenzo-p-dioxin** (see 2,4-D)						
dimethylnitrosamine CAS # 62-75-9				oral: very high (1,2)	carcinogen (3,4) mutagen (1,4)	water: "soluble" "volatile" "nonflammable"
dicapthon AC 4124; American Cyanamid 4124; Captec; Di-Captan; Isomeric Chlorthion; OMS-214 *O*-2-chloro-4-nitrophenyl *O,O*-dimethyl phosphorothioate; *p*-nitro-*o*-chlorophenyl dimethyl thionophosphate CAS # 2463-84-5	organo-phosphate	insecticide	non-pers to mod-pers (1)	oral: high (2) dermal: high (3) inhalation: ?	?	water: "practically insoluble"
dichlobenil Barrier 2G; Barrier 50W; Casoron; Casoron G; Decabane; Du-Dusit (with bromacil); Dyclomec 4G; Dyclomec G2; Forte (with simazine); Fydulan G (with dalpon-sodium); Fydulex G (with dalpon-sodium); Fydusit (with bromacil); Norosac 4G; Norosac10 G; Prefix D; Silbenil 2,6-dichlorobenzonitrile CAS # 1194-65-6	benzo-nitrile	herbicide	mod-pers (1)	oral: medium (1,2) dermal: high (1) inhalation: ?	suspect carcinogen (3)	immediate toxicity: fish: medium (1,2) crustaceans: medium (1) bees: low (4) water: slightly soluble slightly volatile "nonflammable"
transformation product(s): **2,4-dichlorobenzamide** BAM					suspect neurotoxin (1)	
dichlofenthion Bromex; dichlorofenthion; diclofention; Hex-Nema; Mobilawn; Nemacide; Tri-VC-13; VC-13 *O*-2,4-dichlorophenyl *O,O*-diethyl phosphorothioate CAS # 97-17-6	organo-phosphate	insecticide nematocide	?	oral: high (1,2) dermal: medium to high (3) inhalation: ?	?	immediate toxicity: birds: high to very high (4) fish: very high (5) crustaceans: very high (5) aquatic insects: very high (5) water: insoluble oil: "miscible in kerosene"
dichlofluanid BAY 47531; Cupro-Euparene (with copper oxychloride); dichlofluanide (France); Elvaren; Elvaron; Eparen; Euparen; Euparene; Euparin; KUE 13032c *N'*-dichlorofluoromethylthio-*N,N*-dimethyl-*N'*-phenylsulfamide; 1,1-dichloro-*N*-[(dimethylamino)sulfonyl]-fluoro-*N*-phenylmethanesulfenamide CAS # 1085-98-9	amide	fungicide	?	oral: low to medium (1,2) dermal: low to medium (2) inhalation: low to high (3)	?	immediate toxicity: fish: medium to high (3) bees: low to medium (4) water: slightly soluble slightly volatile

NAME: Common Trade and Other Chemical CAS Number	Class of Chemical	Chief Pesticide Use; Status	Persistence	Effects on Mammals		Adverse effects on other non-target species Physical properties
				Immediate Toxicity (Acute)	Long-Term Toxicity (Chronic)	
dichlone Algistat; Compound 604; Kolo-100 (with sulfur); Phygon; Phygon Paste; Phygon Seed Protectant; Phygon XL; Sanquinon dichloronapthoquinone; 2,3-dichloro-1,4-napthoquinone CAS # 117-80-6	quinone	fungicide algacide	?	oral: medium (1) dermal: ? inhalation: ?	liver damage (2)	immediate toxicity: fish: high to very high (3) crustaceans: medium to very high (4) "insect predators": "high" (3) water: insoluble
dichloran AL-50; Allisan; Botran; CNA; DCNA; dicloran; ditranil; Kiwi Luster; Resinan; U-2069 2,6-dichloro-4-nitroaniline CAS # 99-30-9	dinitro-aniline	fungicide	non-pers (1)	oral: low to medium (1) dermal: ? inhalation: low (1)	suspect mutagen (2) liver & kidney damage (3) cataracts (6)	immediate toxicity: birds: low to medium (4) fish: low to high (1) crustaceans: medium (1) bees: low (5) water: slightly soluble slightly volatile
dichlormid Capsolane (with EPTC); Eradicane E (with EPTC); Eradicane G (with EPTC); Surpass (with vernolate); Sutan+ (with butylate); Sutan+ 10G (with butylate); Sutar 85E (with butylate); Sutazine (with butylate & atrazine) N,N-dially-2,2-dichloroacetamide CAS # 37764-25-3	amide	herbicide safener	non-pers (1)	oral: medium (2) dermal: low to medium (2) inhalation: medium (2)		immediate toxicity: birds: low (2) fish: low (2) water: soluble oil: very soluble slightly volatile combustible
dichloroethyl ether Chlorex bis (2-chloroethyl) ether; 2,2'-dichloroethyl ether CAS # 111-44-4	miscel-laneous	fumigant insecticide	?	oral: high (1,2) dermal: ? inhalation: ? inhalation: "May cause liver damage if inhaled" (2)	carcinogen (3)	highly volatile flammable to combustible
dichlorophen antiphen; DDDM; DDM; dichlorophene (France); G-4; Prenetol GDC (alkaline solution); Preventol GD; Super Moxxtox 2,2'-methylene bis(4-chlorophenol); di-(5-chloro-2-hydroxyphenyl)methane CAS # 97-23-4	phenol	fungicide	?	oral: medium (1) dermal: ? inhalation: ?	?	immediate toxicity: fish: high (2) crustaceans: very high (3) water: slightly soluble non-volatile
1,2-dichloropropane D-D (with dichloropropene); Nematox (with dichloropropene); propylene dichloride; Telone (with dichloropropene) 1,2-dichloropropane CAS # 78-87-5	organo-chlorine	fungicide herbicide insecticide soil fumigant	mod-pers to pers (1,2)	oral: medium (1) dermal: medium (1) inhalation: low (1)	carcinogen (3) suspect mutagen (1) liver damage (1)	immediate toxicity: fish: low (1) crustaceans: low (1) water: soluble highly volatile
dichloropropene D-D (with 1,2-dichloropropane); D-D92; DCP; Dedisol C; Dorlone II; Nematox (with 1,2-dichloropropane); Nematox II; Telone (with 1,2-dichloropropane); Telone II 1,3-dichloropropene; 1,3-dichloro-1-propene CAS # 542-75-6	organo-chlorine	soil fumigant nematocide	"non-pers" (1)	oral: high (2) dermal: high (2) inhalation: ?	carcinogen (3,4) suspect mutagen (1,5) liver & kidney damage (1,5)	immediate toxicity: fish: medium to high (6,7) crustaceans: very high (7) bees: low to medium (8) water: soluble highly volatile flammable

NAME: Common Trade and Other Chemical CAS Number	Class of Chemical	Chief Pesticide Use; Status	Persistence	Effects on Mammals		Adverse effects on other non-target species Physical properties
				Immediate Toxicity (Acute)	Long-Term Toxicity (Chronic)	
dichlorprop 2,4-DP; 2-(2,4-DP); Actril S (with bromoxynil & ioxynil & MCPA); Basagran Ultra (with ioxynil & bentazon); Clenecorn (with mecoprop); Cornoxynil (with bromoxynil); Desormone; Farmon (with MCPA); Hedonal DP; Hormatox; Kildip; Lawnsman (with dicamba & MCPA); Lontrel Plus (with mecoprop & clopyralid & MCPA); Minverva (with bromoxynil & ioxynil & MCPA); Polymone; Polytox; Seritox 50 (with MCPA); Tri-Cornox (with dicamba & benazolin); Ultima (with bentazon); Weedone DCP (with 2,4-D) 2-(2,4-dichlorophenoxy)propionic acid; 2-(2,4-dichlorphenoxy)propanoic acid; s-(2,4-dichlorophenoxy)propionic acid CAS # 120-36-5	phenoxy	herbicide	non-pers to mod-pers (1,2)	oral: medium to high (3,4) dermal: high (5) inhalation: low to medium (6)	suspect carcinogen (7) suspect mutagen (8) liver damage (9)	immediate toxicity: fish: low to high (2,10) water: soluble combustible
contaminant(s): **TCDD** (see chloroneb)						
dichlorvos Apavap; Atgard; Aygard V; Benfos; Brevinyl; Brevinyl E50; Canogard; DDVP; Dedevap; Devikol; Divipan; Equigel; Mafu Strip; No-Pest Insecticide Strip; Raid Ant & Roach Killer; Raid Solid Insect Killer; SD 1750; Task; Unifos; Vaponite (in no-pest strips & some flea collars) O,O-dimethyl-2,2-dichlorovinyl phosphate; 2,2-dichlorovinyl dimethyl phosphate CAS # 62-73-7	organo-phosphate	insecticide	non-pers (1,2)	oral: high (3) dermal: high to very high (4) inhalation: high (3,5)	carcinogen (2,6) mutagen (3,7) suspect teratogen (2,3) sperm and other reproductive abnormalities (3) kills human white blood cells (8) inhibits steroid synthesis (9) indications of bone marrow damage and aplastic anemia (10) immunotoxin (11)	immediate toxicity: birds: very high (12) fish: very high (13) bees: very high (1) crustaceans: very high (13) aquatic insects: very high (13) long-term toxicity: birds: delayed neurotoxin (11) water: soluble oil: very soluble highly volatile combustible
diclofop-methyl dichlofop-methyl; dichlorfop-methyl; diclofop; Hoegrass; Hoelon; Illoxan; One Shot (with bromoxynil & MCPA) 2-(4-(2,4-dichlorophenoxy)phenoxy)-propanoic acid methyl ester; methyl 2-(4-(2,4-dichlorophenoxy)phenoxy)-propionate CAS # 51338-27-3	phenoxy	herbicide	non-pers to mod-pers (1,2)	oral: medium (2,3) dermal: medium to very high (2) inhalation: ?	?	immediate toxicity: birds: low to medium (2) fish: medium to high (2) water: slightly soluble to very soluble slightly volatile to highly volatile flammable
dicofol Acarin; Acavers 35 (with methomyl); FW-293; Kelthane; Kethane Mixte (with methyl parathion); Mitigan; Mixte (with methyl parathion); Parasoufre Acaricide (with methyl parathion & sulfur); Tedane Extra (with dinocap & tetradiforn or mancozeb); Tuver Acaricide (with ethion & methyl parathion) 1,1-bis(p-chlorophenyl)-2,2,2-trichloroethanol; 4,4'-dichloro-α-(trichloromethyl)benzhydrol CAS # 115-32-2	organo-chlorine	insecticide acaricide	mod-pers (1)	oral: medium to high (2,3) dermal: medium to very high (2,3) inhalation: ?	carcinogen (4)	immediate toxicity: birds: medium to high (5) fish: very high (6) crustaceans: low (7) bees: low (8) aquatic insects: very high (6) water: "insoluble" oil: "soluble" flammable

(continued on next page)

NAME: Common Trade and Other Chemical CAS Number	Class of Chemical	Chief Pesticide Use; Status	Persistence	Effects on Mammals		Adverse effects on other non-target species Physical properties
				Immediate Toxicity (Acute)	Long-Term Toxicity (Chronic)	
contaminant(s): **DDT** (see DDT)						
DDE (see DDE)						
DDD (see DDD)						
dicrotophos Bidrin; C709; Carbicron; Diapadrin; Ektafos; SD 3562 dimethyl cis-2-dimethyl-carbamoyl-1-methylvinyl phosphate; dimethyl phosphate ester of 3-hydroxy-N,N-dimethyl-cis-crotonamide; 3-dimethoxyphosphinyloxy-N,N-dimethylisocrotonamide CAS # 141-66-2	organo-phosphate	insecticide	non-pers to mod-pers (1,2)	oral: very high (1,3) dermal: high to very high (1) inhalation: high to very high (1)	suspect mutagen (3,4) "no organophosphate type delayed neurotoxicity" (3)	immediate toxicity: birds: very high (1,5) fish: low to medium (1,6) amphibians: medium (5) crustaceans: medium to high (6) aquatic insects: high (7) slightly volatile combustible
transformation product(s): **monocrotophos** (see monocrotophos)						
Dicumarol Dicoumarin; Dicoumarol; Dicumarin; Dicumarol; Melitoxin bishydroxycoumarin; 3,3'-methylenebis (4-hydroxycoumarin) CAS # 66-76-2	coumarin	rodenticide	?	oral: medium (1) dermal: ? inhalation: ?	?	water: "practically insoluble"
dieldrin Alvit; Anter; Compound 497; dieldrine (France); Dielmoth; Dorytox; Dudubitoke; Eldrinol; HEOD (Canada); Hododrex; Illoxol; Kyadrin; Octalox; Pestex; Red Shield hexachloroepoxyoctahydro-endo-exo-dimethanonaphthalene and related compounds; 1,2,3,4,10,10-hexachloro-6,7-epoxy-1,4,4a,5,6,7,8,8a-octahydro 1,4-endo-exo-5, 8-dimethanonaphthalene and related compounds CAS # 60-57-1	organo-chlorine	insecticide cancelled USA, 1971	pers (1)	oral: high to very high (2) dermal: very high (2) inhalation: very high (19)	cumulative (3,4) suspect carcinogen (1,5,6) suspect teratogen (7,8) immunotoxin (9,10) abnormal brain waves, behavior changes (10,11)	BIOCIDE immediate toxicity: birds: high to very high (12) fish: very high (13) amphibians: very high (1) crustaceans: very high (14) molluscs: very high (1) bees: very high (15) aquatic insects: very high (13) aquatic worms: medium (16) plankton: very high (14) long-term toxicity: birds: damages reproduction, eggshell thinning (17,18) water: insoluble oil: very soluble slightly volatile "nonflammable"
transformation product(s): **photodieldrin** 2a,2,2,4,5,5a-hexachlorodecahydro-2,4,6-metheno-2H-cychlopenta [4,5]pentaleno[1,2-]oxirene CAS # 13366-73-9				"more toxic than dieldrin" (1)		

NAME:	Common Trade and Other Chemical CAS Number	Class of Chemical	Chief Pesticide Use; Status	Persistence	Effects on Mammals		Adverse effects on other non-target species Physical properties
					Immediate Toxicity (Acute)	Long-Term Toxicity (Chronic)	

dienochlor
 	
HRS-16; Pentac
 	
bis(pentachloro-2,4-cyclopentadiene-1-yl); decachlorobis(2,4-cyclopentadiene-1-yl)
 	
CAS # 2227-17-0
		organo-chlorine	acaricide	non-pers (1)	oral: medium (2)
					
					dermal: medium (2)
					
					inhalation: high (3)
						?	immediate toxicity:
							  bird: low to medium (4)
							  fish: medium to high (4)
							  crustacean: high (5)
							  plants: low (6)
							
							water: insoluble
							
							slightly volatile
							
							toxic fumes if burned

transformation product(s): **chlorine** (see chlorine)							
hydrogen chloride hydrochloric acid CAS # 7647-01-0					oral: low (1)		
phosgene gas (see carbon tetrachloride)							

diethatyl
 	
Barrix (with ethofumesate); H-22234; Hercules 22234
 	
N-chloroacetyl-N-(2,6-diethylphenyl)glycine
 	
CAS # 38727-55-8
		amide	herbicide	mod-pers (1)	oral: medium (2)
					
					dermal: low to medium (3)
					
					inhalation: ?
						?	immediate toxicity:
							  birds: low (3)
							  fish: medium (1)
							
							water: soluble
							
							slightly volatile
							
							flammable

diethyl fumarate 2-butenedioc acid, diethyl ester CAS # 623-91-6		miscel-laneous		?	oral: medium synergistic with malathion	?	?
difenacoum Compo; Matrak; Neosorexa; Ramor; Rastop; Ratak; Ratrick; Silo; WBA 8107 3-(3-diphenyl-4-yl-1,2,3,4-tetrahydro-1-nephtryl)-4-hydroxycoumarin; 3-(3-(1,1'-biphenyl)-4-yl-1,2, 3,4-tetrahydro-1-napthalenyl)-4-hydroxy-2H-1-benzopyran-2-one CAS # 5607-07-5		coumarin	rodenticide	?	oral: high to very high (1) dermal: ? inhalation: ?	?	water: slightly soluble

difenzoquat
 	
AC 84777 Finaven; Avenge; Mataven; Superaven
 	
1,2-dimethyl-3,5-diphenyl-1 H-pyrazolium methyl sulfate
 	
CAS # 49866-87-7
		bypiridyl	herbicide	mod-pers (1,2)	oral: high to very high (1,2)
					
					dermal: low to medium (2)
					
					inhalation: low (2)
						?	immediate toxicity:
							  birds: "low toxicity to birds" (2)
							  fish: low (1)
							  crustaceans: high (3)
							  bees: "nontoxic to bees" (4)
							
							water: very soluble
							oil: "insoluble"
							
							slightly volatile

transformation product(s): **monomethyl pyrazole**							"volatile"

NAME: Common Trade and Other Chemical CAS Number	Class of Chemical	Chief Pesticide Use; Status	Persistence	Effects on Mammals		Adverse effects on other non-target species Physical properties
				Immediate Toxicity (Acute)	Long-Term Toxicity (Chronic)	
diflubenzuron deflubenzon; diflubenuron; difluron; Dimilin; Dimilin IG; Dimilin W-25; Micromite; OMS 1804; PH 60-40; TH 60-40; Vigilante N-[[(4-chlorophenyl)amino]carbonyl]-2,6-difluorobenzamide CAS # 35367-38-5	urea	insecticide restricted use in USA	non-pers to mod pers (1-3)	oral: low (4) dermal: low (4) inhalation: low (4)	affects oxygen carrying capacity of red blood cells via methemoglobinemia and sulfemoglobinemia (3,5)	immediate toxicity: birds: low (4) fish: low (4) crustaceans: very high (4) long-term toxicity: crustaceans: affects reproduction (6) water: "soluble" oil: "virtually insoluble"
transformation product(s): **4-chloroaniline** CAS # 10-64-7					carcinogen (1) may cause methemoglobinemia and sulfhemoglobinemia (2)	
dihydrosafrole 1,2-methylenedioxy-4-propylbenzene CAS # 94-58-6	methylene dioxy	intermediate in the synthesis of piperonyl butoxide	?	oral: medium (1) dermal: ? inhalation: ?	carcinogen (1,2)	?
dikegulac sodium Atrimmec; Atrinal; Cutlass sodium salt of 2,3:4,6-di-O-isopropylidene-α-L-xylo-2-hexalofuranosonic acid CAS # 52508-35-7	miscellaneous	plant growth regulator	?	oral: low (1) dermal: ? inhalation: ?	?	immediate toxicity: birds: low (2) fish: low (1) bees: low (3) water: very soluble non-volatile combustible
dimefox Hanane; Terra-Sytam bis(dimethylamino) fluorophosphine oxide CAS # 115-26-4	organophosphate	insecticide acaricide	?	oral: very high (1) dermal: very high (1) inhalation: "vapor toxicity hazard is high" (1)	?	highly volatile
dimethirimol Milcurb 5-n-butyl-2-dimethylamino-4-hydroxy-6-methylpyrimidine CAS # 5221-53-4	pyrimidine	fungicide	mod-pers (1)	oral: low to medium (2,3) dermal: ? inhalation: ?	?	immediate toxicity: birds: medium (1) fish: low (2) bees: "nontoxic to bees" (2) water: soluble slightly volatile
dimethoate Bio Long Last (with permethrin); Bio Systemic Insecticide; Cyanotril (with flucythrinate); Cygon; Cykuthoate; Daphene; Devigon; Dimet; Dimethogen; Flutrin (with flucythrinate); fosfamid (USSR); Fostion MM; Mikantop (with fenvalerate); Perfekthion; Rogor; Roxion; Salut (with chlorpyrifos); Trimetion; Turbair Systemic Insecticide; Vitex O,O-dimethyl S-(N-methylcarbamoylmethyl) phosphorodithioate; 2-dimethoxyphosphinothioylthio-N-methylacetamide CAS # 60-15-5	organophosphate	insecticide acaricide cancelled, most products, USA, 1981	non-pers (1,2)	oral: high to very high (2,3) dermal: high to very high (1,2) inhalation: ?	suspect carcinogen (4) mutagen (2,5) suspect teratogen (2,6) blood damage (7) testicular atrophy (4) kidney damage (4) immunotoxin (8)	immediate toxicity: birds: high to very high (1,2) fish: low to high (1,2) crustaceans: high to very high (2,9) bees: very high (10) aquatic insects: very high (2) water: very soluble slightly volatile flammable to combustible

NAME: Common Trade and Other Chemical CAS Number	Class of Chemical	Chief Pesticide Use; Status	Persistence	Effects on Mammals		Adverse effects on other non-target species Physical properties
				Immediate Toxicity (Acute)	Long-Term Toxicity (Chronic)	
transformation product(s): **omethoate** (see omethoate)						
dimethrin dimethrine (France) 2,4-dimethylbenzyl (RS)-cis,trans-2,2-dimethyl-3-(2-methylprop-1-enyl)cyclopropanecarboxylate CAS # 70-38-2	pyrethroid	insecticide	?	oral: high (1) dermal: ? inhalation: ?	?	immediate toxicity: fish: high to very high (2,3)
dimethylamine dimethylamine aqueous solution; DMA CAS # 124-40-3	miscellaneous	insect attractant	?	oral: ? dermal: ? inhalation: ?	?	water: "very soluble"
transformation product(s): **dimethylnitrosamine** CAS # 62-75-9				oral: very high (1,2)	carcinogen (3,4) mutagen (1,4)	water: "soluble" "volatile" "nonflammable"
dimethyl phthalate DMP; NTM dimethyl 1,2-benzenedicarboxylate CAS # 131-11-3	phthalate	insect repellent	?	oral: low (1) dermal: ? inhalation: very high (2)	suspect mutagen (3,4)	water: soluble oil: "soluble" volatile combustible
dimethyl sulfoxide DMSO sulfinylbis[methane] CAS # 67-68-5	metal/ mineral sulfur	solvent	?	oral: low (1) dermal: ? inhalation: ?	suspect mutagen (2)	immediate toxicity: birds: high (3) water: "soluble" combustible
dinex dinitrocyclohexylphenol; DN Dust No.12; DN-111; DN-75; DNOCHP; Dowspray 17; Dynone-II; pedinex (France) 2-cyclohexyl-4,6-dinitrophenol ; dicyclohexylamine salt; 4,6-dinitro-o-cyclohexyl phenol CAS # 131-89-5	phenol	insecticide acaricide	?	oral: high (1) dermal: ? inhalation: ?	?	immediate toxicity: bees: low (2) aquatic worms: very high (3)
dinitramine Cobex; Cobexo; diethamine; dinitroamine; USB-3584 N^3,N^3-diethyl-2,4-dinitro-6-trifluoro-methyl-m-phenylenediamine; N^4,N^4-diethyl-α,α,α-trifluoro-3,5-dinitrotoluene-2,4-diamine; N^3,N^3-2,4-dinitro-6-(trifluoromethyl)-1,3-benzendiamine CAS # 29091-05-2	dinitro-aniline	herbicide	mod-pers (1)	oral: medium (2) dermal: medium (2) inhalation: ?	?	immediate toxicity: birds: low (1) fish: medium to very high (1,3) water: slightly soluble oil: very soluble slightly volatile flammable
dinitrocresol Antinonnin; Capsine; Chemosect DNOC; Dekrysil; Detal; Ditrosol; DNOC; Effusan; Elgetol 30; Gioallolio; Ibertox Pasta; Jackyl S; Krenite; Lipan; Nitrador; Prokarbol; Sandolin A; Sinox; Triacide; Trifrina 4,6-dinitro-o-cresol; 2-methyl-4,6-dinitrophenol CAS # 534-52-1	phenol	defoliant herbicide fungicide insecticide	non-pers (1)	oral: high to very high (2)	cumulative (2,3)	immediate toxicity: birds: very high (4) fish: high to very high (5) crustaceans: high (5) molluscs: high (1) aquatic insects: high (5) water: soluble slightly volatile

NAME: Common Trade and Other Chemical CAS Number	Class of Chemical	Chief Pesticide Use; Status	Persistence	Effects on Mammals		Adverse effects on other non-target species Physical properties
				Immediate Toxicity (Acute)	Long-Term Toxicity (Chronic)	
dinitrophenol Chemox PE; Nitro Kleenup 2,4-dinitrophenol; α-dinitrophenol CAS # 51-28-5	phenol	insecticide acaricide fungicide	?	oral: very high (1) dermal: ? inhalation: ? "readily absorbed through intact skin" (2)	brain, liver and kidney damage (3)	immediate toxicity: birds: very high (4) plants: "phytotoxic to green plants (1)
dinobuton Acarelte; Acarelte Forte; Acrex; Dessin; Dinofen; Drawinol; MC 1053; P-1053; Systol; Sytasol; Talan; UC 19786 1-methylethyl 2-(1-methylpropyl)-4,6-dinitrophenyl carbonate; 2-sec-butyl-4,6-dinitrophenyl isopropyl carbonate CAS # 973-21-7	nitro	acaricide fungicide	?	oral: medium to high (1,2) dermal: ? inhalation: ?	?	immediate toxicity: fish: "toxic" (2) bird: high (3) bees: "toxic" (2) water: "practically insoluble"
dinocap Acarthane (with dicofol); Arathane; Cakucap; Crotothane; Iscothane; Karathane; Mildane; Mildex; Sialite; Tedane Extra (with dicofol & tetradiforn or w/ mancozeb) 2-(1-methylheptyl)-4,6-dinitrophenocrotonate CAS # 39300-45-3	phenol	fungicide acaricide restricted USA, 1989	non-pers to mod-pers (1,2)	oral: medium (1) dermal: ? inhalation: high (1)	teratogen (3,4)	immediate toxicity: fish: very high (5) bees: medium (6) water: insoluble highly volatile
dinoseb Dinitrall; dinitro; Dinitro Weed-Killer; dinitrobutyl phenol; dinosebe (France); DN-289; DNBP; DNOSBP; Dow General Weedkiller; Dow Selective Weedkiller; Elgetol 318; Gebutox; Kilseb; Knoxweed; Nitropone C; Premerge; Sinox General Subitex; Unicrop DNBP; Vertac Dinitro Weedkiller 2-sec-butyl-4,6-dinitrophenol; 4,6-dinitro-o-sec-butylphenol; 2,4-dinitro-6-sec-butylphenol CAS # 88-85-7	phenol	herbicide fungicide insecticide dessicant cancelled USA, 1988	non-pers to mod-pers (1,2)	oral: high to very high (3,4) dermal: high to very high (1,5) inhalation: ?	teratogen (6,7) male sterility (7)	immediate toxicity: fish: medium to high (8) bird: very high (9) water: soluble oil: "soluble"
transformation product(s): **nitrosamines**					suspect carcinogen (1)	

dinoterb
 	
DNTBP; Herbogil; Herbogil Liquid D; Nixone
 	
2-tert-butyl-4,6-dinitrophenol
 	
other salts:
    dinoterb-ammonium
    dinoterb-diolamine
    dinoterb-sodium
    dinoterb-dimethylammonium
 	
CAS # 1420-07-1
	phenol	herbicide	?	oral: very high (1)
				
				dermal: ?
				
				inhalation: ?
					?	immediate toxicity:
						fish: very high (1)
						
						water: slightly soluble

NAME:	Common Trade and Other Chemical CAS Number	Class of Chemical	Chief Pesticide Use; Status	Persistence	Effects on Mammals		Adverse effects on other non-target species Physical properties
					Immediate Toxicity (Acute)	Long-Term Toxicity (Chronic)	
dinoterb acetate MC1108 2-*tert*-butyl-4,6-dinitrophenol acetate CAS # 3204-27-1		phenol	herbicide nematocide	?	oral: high (1) dermal: low to medium (1) inhalation: ?	?	water: "almost insoluble"
dioxabenzofos Salithion 2-methoxy-4*H*-1,2,3-benzo-dioxaphosphorin-2-sulfide CAS # 38260-54-7		organo-phosphate	insecticide	?	oral: high (1) dermal: high (2) inhalation: ?	?	immediate toxicity: fish: medium (1) long-term toxicity: birds: delayed neurotoxicity (3) water: slightly soluble volatile
dioxacarb Elocron; Famid; Flocron; Gamid; Ulvair 2-(1,3-dioxolan-2-yl)phenyl methylcarbamate CAS # 6988-21-2		carbamate	insecticide	non-pers (1)	oral: high (1,2) dermal: medium to high (2) inhalation: very high (1)	?	immediate toxicity: birds: "slightly toxic" (1) fish: low (1) bees: high (3) water: soluble slightly volatile
dioxathion Delnav; Deltic; Hercules AC 528; Navadel; Ruphos 2,3-p-dioxanedithiol S,S-bis(O,O-diethyl phosphorodithioate) CAS # 78-34-2		organo-phosphate	insecticide acaricide	"mod-pers" (1)	oral: very high (2) dermal: high (2) inhalation: high (3)	suspect mutagen (4)	immediate toxicity: birds: high (1) fish: very high (5) crustaceans: very high (5) bees: medium (6) water: "practically insoluble"
diphacinone diphacin (Turkey); Diphacinon; diphenadione; Gold Crest; Kill-Ko Rat Killer; Promar; Ramik 2-(diphenylacetyl)-1*H*-indene-1,3(2*H*)-dione; 2-diphenylacetyl-1,3-indandione CAS # 82-66-6		indan-dione	rodenticide	"maintains a long persistence" (1)	oral: very high (2) dermal: ? inhalation: ?	?	immediate toxicity: birds: medium (2) fish: medium (2) water: slightly soluble non-volatile
diphenamid Dymid; Enide; Rideon N,N-dimethyl-2,2-diphenylacetamide CAS # 957-51-7		amide	herbicide	mod-pers (1)	oral: medium to high (1) dermal: "no toxicity" (1) inhalation: "no toxicity" (1)	?	immediate toxicity: birds: "very low toxicity" (2) fish: low (3) crustaceans: low (3) bees: low (4) water: soluble "nonflammable"
diphenyl biphenyl; Lemonene; Phenador-X; phenylbenzene; PHPH CAS # 34987-38-7		aromatic hydro-carbon	fungicide	?	oral: medium (1) dermal: ? inhalation: ?	kidney, liver, & brain damage (2)	water: "insoluble" highly volatile
diphenylamine Big Dipper; Coraza; Deccoscald 282; DPA; No Scald; Scaldip; Shield DPA N-phenylbenzenamine CAS # 122-39-4		miscel-laneous	fungicide	?	oral: medium to high (1) dermal: ? inhalation: ?	?	water: "insoluble" combustible

NAME: Common Trade and Other Chemical CAS Number	Class of Chemical	Chief Pesticide Use; Status	Persistence	Effects on Mammals		Adverse effects on other non-target species Physical properties
				Immediate Toxicity (Acute)	Long-Term Toxicity (Chronic)	
dipropetryn Cotodon (with metolachlor); Cotofor; GS-16068; Sancap 2-ethylthio-4,6-bis(isopropylamino)-s-triazine; 6-(ethylthio)-N,N'-bis(1-methylethyl)- 1,3,5-triazine-2,4-diamine CAS # 4147-51-7	triazine	herbicide	mod-pers (1)	oral: low to medium (1,2) dermal: low to medium (2) inhalation: low (3)	?	immediate toxicity: fish: medium to high (1) water: slightly soluble slightly volatile "nonflammable"
diquat Actor (with paraquat); Aquacide; Cleansweep (with paraquat); Cyclone (with paraquat); Dextrone (with paraquat dichloride); Groundhog (with paraquat & amitrole & simazine); Midstream; Parable (with paraquat); Pardi Weedol (with paraquat); Pathclear (with paraquat & amitrole & simazine); Preglone (with paraquat); Priglone (with paraquat); Reglex; Reglone; Seccatutto (with paraquat); Soltair (with paraquat & simazine); Spraygrow (with paraquat); Spraytop (with paraquat); Torpedo; Weedtrine-D 6,7-dihydropyrido[1,2-a:2',1'-c] pyrazinediium; 1,1'-ethylene-2,2'-dipyridyldiylium CAS # 2764-72-9	bypiridyl	herbicide dessicant	non-pers (1)	oral: high (2) dermal: high to very high (2,3) inhalation: ?	suspect mutagen (3) suspect teratogen (4) suspect fetotoxin (3) suspect embroytoxin (6) liver damage (5) cataracts (7)	immediate toxicity: birds: medium (8) fish: low (9) crustaceans: low (9) bees: medium (1) water: very soluble non-volatile "nonflammable"
contaminant(s): **ethylene dibromide** (see ethylene dibromide)						
diram ammonium dimethyl- dithiocarbamate CAS # 3226-36-6	carbamate	fungicide	?	oral: medium to high (1) dermal: ? inhalation: ?	?	
disparlure Hercon Disrupt Gypsy Moth; Hercon Luretape Gypsy Moth; Kisparmone; Pherocon GM cis-7,8-epoxy-2-methyloctadecane CAS # 29804-22-6	biological	attractant	?	oral: low (1) dermal: ? inhalation: ?	?	
disulfoton BAY 19639; Disyston; Disyston O (with isofenphos); Dithiodemeton; Dithiosystox; Doubledown (with fonofos); Ekanon; Ekatin TD; Ethimeton; Frumin G; Frumin-Al; Insyst-D; Knave; M-74 (USSR); S276; Solvigran; Solvirex; Twinspan O,O-diethyl S-2-(ethylthio)ethyl phosphorodithioate CAS # 298-04-4	organo- phosphate	insecticide acaricide	non-pers (1,2)	oral: very high (3) dermal: very high (3) inhalation: high (4)	suspect mutagen (5)	immediate toxicity: birds: very high (6) fish: medium to very high (7) crustaceans: very high (7) aquatic insects: very high (7) water: slightly soluble oil: "soluble" volatile combustible
ditalimfos Dowco 199; Laptran; Plondrel O,O-diethyl phthalmide phosphonothioate CAS # 5131-24-8	organo- phosphate	fungicide	?	oral: low to medium (1) dermal: medium to high (1) inhalation: ?	?	immediate toxicity: birds: medium (2) water: soluble slightly volatile

NAME: Common Trade and Other Chemical CAS Number	Class of Chemical	Chief Pesticide Use; Status	Persistence	Effects on Mammals		Adverse effects on other non-target species Physical properties
				Immediate Toxicity (Acute)	Long-Term Toxicity (Chronic)	

diuron
 	
Cekiuron; Dailon; DCMU; Destral (with dalapon & 2,4-D); Di-on; Diurex; DMU; Dynex; Gramixel (with paraquat); Gramuron (with paraquat); Karmex; Krater (with asulam); Marmer; Paracol (with paraquat); Rassapron (with atrazine & amitrole); Surefire (with paraquat); Totacol (with paraquat); Weed-Free g (with bromacil)
 	
3-(3,4-dichlorophenyl)-1,1-dimethylurea
 	
CAS # 330-54-1
	urea	herbicide	mod-pers to pers (1,2)	oral: medium (3)
				
				dermal: low to medium (4)
				
				inhalation: ?
				
				poisoning potential increased with protein-deficient diet (5)
					suspect mutagen (6)
					suspect teratogen (7)
					growth inhibition (5)
					anemia (5)
						BIOCIDE
						
						immediate toxicity:
						  birds: low to medium (8)
						  fish: medium to high (9)
						  crustaceans: medium to high (9)
						  bees: low (10)
						  aquatic insects: high (9)
						  phytoplankton: "very high" (2)
						
						long-term toxicity:
						  fish: gill damage, inhibits reproduction (2)
						  molluscs: shell growth inhibited (2)
						  can reduce oxygen content of ponds (2)
						  water: slightly soluble
						
						oil: soluble
						non-volatile
						"nonflammable"

contaminant(s): **TCAB** TCAB 3,4,3',4'-tetrachloroazobenzene structure is analogous to TCDD (see under 2,4,5-T) CAS # 14047-09-7		also a component of diuron			suspect mutagen (1,2) chloracne & hyperkeratosis (3)	

DMPA
 	
Dow 1329; Dow Crabgrass Killer; Dowco 118; Zytron
 	
O-2,4-dichlorophenyl O'-methyl isopropylphosphoroamidothioate; (1-methylethyl)phosphoramidothoic acid O-(2,4-dichlorphenyl) O-2,4-(dichlorphenyl) O-methyl ester; dimethylolpropionic acid
CAS # 299-85-4
	organo-phosphate	herbicide plant growth regulator	?	oral: medium to high (1,2)
				
				dermal: ?
				
				inhalation: ?
					?	immediate toxicity:
						  birds: medium (2)
						
						long-term toxicity:
						  birds: neurotoxin (3)
						
						water: slightly soluble

dodine
 	
AC 5223; Carpene; Curitan; Cyprex; Cyprex 65W; dodin; dodine acetate; doguadine (France); Efuzin; Melprex; Mexprex; Syllit; Triododine; tsitrex (USSR); Venturol; Vondodine
 	
dodecylguanidine acetate; laurylguanidine acetate
 	
CAS # 2439-10-3
	miscel-laneous	fungicide herbicide	?	oral: medium (1)
				
				dermal: low to high (1)
				
				inhalation: high (2)
					?	immediate toxicity:
						  birds: medium (3)
						  fish: high (3)
						  bees: low to medium (4)
						  insect predators: "toxic to some" (5)
						
						water: soluble
						
						volatile

drazoxolon
 	
Ganocide; Mil-Col; Primicid (with pirimiphos-ethyl); SAIsan F; Sporacol
 	
4-(2-chlorophenylhydrazono)-3-methyl-1,2-oxazol-5(4H)-one; 4-(2-chlorophenylhydrazono)-3-methylisoxazol-5(4H)-one
 	
CAS # 5707-69-7
	miscel-laneous	fungicide	?	oral: high to very high (1,2)
				
				dermal: ?
				
				inhalation: ?
					neurotoxin (2,3)
					heart damage (2)
						immediate toxicity:
						  birds: high to very high (2,4)
						  fish: high (1)
						
						water: "insoluble"
						
						slightly volatile

NAME: Common Trade and Other Chemical CAS Number	Class of Chemical	Chief Pesticide Use; Status	Persistence	Effects on Mammals		Adverse effects on other non-target species Physical properties
				Immediate Toxicity (Acute)	Long-Term Toxicity (Chronic)	
edifenphos BAY 78418; ediphenphos; Hinosan; SRA 7847 O-ethyl-S,S-diphenyl phosphorodithioate CAS # 17109-49-8	organo-phosphate	fungicide	non-pers (1)	oral: high (2) dermal: ? inhalation: high (2)	?	immediate toxicity: birds: medium (2) fish: high (2) water: "practically insoluble" volatile
endod derived from plant species *Phytolacca dodecandra*	botanical	molluscicide	?	oral: medium to high (1) dermal: ? inhalation: ?	?	immediate toxicity: fish: high (2) plants: medium to high (1)
endosulfan benzoepin; Beosit; Chlorthiepin; Cyclodan; Endocel; FMC 5462; Hoe 2671; Insectophene; Kop-Thiodan; Malic; Malix; NIZ 5462; Quinolate MG SAFI (with oxine-copper & lindane); Rogodan (with dimethoate); Thifor; Thimul; thiodan (Iran, USSR); Thionex; Tiovel hexachlorohexhydromethano-2,4, 3-benzodioxathiepin oxide; 6,7,8,9,10,10-hexachloro-1,5,5a,6,9,9a-hexahydro-6,9-methano-2,4,3-benzodioxathiepin 3-oxide; 5-norbornene-2,3-dimethanol-1, 4,5,6,7,7-hexachlorocyclic sulfite CAS # 115-29-7	organo-chlorine	insecticide	mod-pers (1)	oral: very high (2) dermal: very high (2) inhalation: very high (2)	suspect carcinogen (4) mutagen (5,6) kidney damage (4) eye damage (7) suppression of immune responses (8) red blood cell damage (9)	BIOCIDE immediate toxicity: birds: high to very high (10) fish: very high (11) crustaceans: very high (11) bees: very high (12) amphibians: very high (13) molluscs: very high (14) long-term toxicity: fish: ovary damage (15,16) crustaceans: cumulative (18) can inhibit fungal growth (17) damages some plants: some flowers, grapes, and birches (19) water: insoluble slightly volatile "nonflammable"
endothall Accelerate; Aquathol; Des-i-cate; Endothal; endothal (Europe, except Italy); Herbicide 273; Herbicide 282; Hydout; Hydrothal 47; Hydrothol 191; Niagrathal; Ripenthol; Tri-Endothal 3,6-endoxohexahydrophthalic acid; 7-oxabicyclo(2,2,1) heptane-2,3-dicarboxylic acid also the following salts: dipotassium disodium dihydroxyaluminum mono(N,N-dimethyltridecylamine) di(N,N-dimethyltridecylamine) mono(N,N-dimethylcocoamine) di(N,N-dimethylcocoamine) CAS # 129-67-9	phthalate	herbicide	non-pers (1)	oral: high to very high (2,3) dermal: ? inhalation: low to medium (4)	?	immediate toxicity: fish (salts-disodium, dimethyl amine, cocoamine): low to high (5) fish (acid): low to high (6) crustaceans (dipotassium salts): low (7) crustaceans (acid): high (7) aquatic insects (amine salts): medium (5) long-term toxicity: fish (amine salt): damage to liver & testes (8) bioaccumulates in pond bottom arthopods (7) water: very soluble "nonflammable"

NAME: Common Trade and Other Chemical CAS Number	Class of Chemical	Chief Pesticide Use; Status	Persistence	Effects on Mammals		Adverse effects on other non-target species Physical properties
				Immediate Toxicity (Acute)	Long-Term Toxicity (Chronic)	
endrin Accelerate; Agrine; Compound 269; Drinafog; Endrex; Endricol; Endrotox; Enpar; Envel; Hexadrin; Insectrin; Mendrin; Multitox; nendrin (India & So. Africa); Oktonex; OMS-197; Palmarol; SD 3419 hexachloroepoxyoctahydro-endo-endo-dimethanonapthalene; 1,2,3,4,10,10-hexachloro-6,7-epoxy-1,4,4a,5,6,7,8,8a-octa-hydro-1,4-endo,endo-5,8-dimethanonapthalene CAS # 722-20-8	organo-chlorine	insecticide cancelled USA, 1979	pers (1)	oral: very high (2) dermal: very high (2) inhalation: ?	suspect carcinogen (3,4) teratogen (5,6) suspect neurotoxin (7)	BIOCIDE immediate toxicity: birds: very high (8) fish: very high (9) amphibians: high to very high (10) crustaceans: very high (9) bees: high (11) aquatic insects: very high (9) plants: toxic to some plants (1) algae: high to very high (12) long-term toxicity: birds: reproductive damage (13,14) bioaccumulates in fish and molluscs (1) water: "insoluble" slightly volatile "nonflammable"
transformation product(s): **12-ketoendrin**				five times as toxic as endrin (1)		
epichlorohydrin 1-chloro-2,3-epoxypropane CAS # 106-89-8	organo-chlorine	insecticide fumigant	?	oral: high dermal: ? inhalation: ?	carcinogen (1,2) mutagen (3,4) kidney damage (2) decreased sperm motility (5,6)	water: "insoluble" flammable
EPN Meidon 15 Dust (with carbaryl) O-ethyl O-p-nitrophenyl phenyl phosphonothioate; ethyl p-nitrophenyl thionobenzenephosphonate CAS # 2104-64-5	organo-phosphate	insecticide acaricide cancelled USA, 1983	mod-pers (1)	oral: very high (1) dermal: high (1) inhalation: very high (1)	fetotoxin (2) delayed neurotoxin (3)	immediate toxicity: birds: very high (4) fish: high to very high (5) crustaceans: very high (1) bees: very high (6) toxic to some plants (7) water: "slightly soluble" slightly volatile
EPTC Alirox; Anclirox; Capsolane (with dichlormid); Eptam; Eradicane E (with dichlormid); Eradicane G (with dichlormid); Witox S-ethyl dipropylthiolcarbamate; S-ethyl dipropylthiocarbamate CAS # 759-94-4	thiocar-bamate	herbicide	non-pers (1)	oral: medium (2) dermal: low to medium (3) inhalation: ?	?	immediate toxicity: birds: high (4) fish: low to medium (2) crustaceans: low (5) water: soluble volatile flammable
contaminant(s): **N-nitrosodipropylamine** CAS # 621-64-7				oral: high (1)	carcinogen (1)	
transformation product(s): **ethyl mercaptan** CAS # 75-08-1				oral: medium (1) dermal: low (1)		

NAME: Common Trade and Other Chemical CAS Number	Class of Chemical	Chief Pesticide Use; Status	Persistence	Effects on Mammals		Adverse effects on other non-target species / Physical properties
				Immediate Toxicity (Acute)	Long-Term Toxicity (Chronic)	

erbon
 	
Baron; Erbon R; Novege; Novon
 	
2-(2,4,5-trichlorophenoxy)ethyl 2,2,-dichloroproprionate
 	
CAS # 136-25-4
	phenoxy	herbicide
		
		cancelled USA, 1980
			?	oral: medium (1)
				
				dermal: ?
				
				inhalation: ?
					?	immediate toxicity:
						  birds: medium (2)
						  bees: low to medium (3)
						
						water: "insoluble"
						
						oil: "soluble"
						
						"very low volatility"
						"nonflammable"

transformation product(s): **dalapon** (see dalapon)						

ethalfluralin
 	
Edge; EL-161; Grindor (with atrazine); Maizor (with atrazine); Somilan; Sonalan
 	
N-ethyl-N-(2-methyl-2-propenyl)-2,6-dinitro-4-(trifluoromethyl)benzeneamine; 2,6-dinitro-N-ethyl-N-(2-methyl-2-propenyl)-α,α,α-trifluoro-p-toluidine
 	
CAS # 55283-68-6
	dinitro-aniline	herbicide	non-pers (1)	oral: low to high (1)
				
				dermal: low to medium (2)
				
				inhalation: low (2)
					carcinogen (3) teratogen (3)	immediate toxicity:
						  birds: medium (3)
						  fish: very high (1)
						  crustaceans: very high (3)
						
						water: insoluble
						
						slightly volatile
						
						"not flammable"

ethephon
 	
Arvest; Bromaflor; Cagro; Cepha; Cerone; Composan; Etheverse; Ethrel; Flordimex; Florel; Prep; Terpal C (with chlormequat chloride); Terpal M (with chlormequat chloride & mepiquat chloride)
 	
2-(chloroethyl)phosphonic acid
CAS # 16672-87-0
	organo-phosphate	plant growth regulator	non-pers (1)	oral: medium (1,2)
				
				dermal: medium (2)
				
				inhalation: ?
					?	immediate toxicity:
						  birds: medium (1)
						  fish: low (2)
						  bees: low to medium (3)
						  crustaceans: low (4)
						
						water: soluble
						
						oil: "insoluble"
						"nonflammable"

ethiofencarb
 	
Croneton; ethiofencarp; ethiophencarp; Hox 1901
 	
2-ethyl-mercaptomethyl-phenyl-N-methylcarbamate; 2-[(ethylthio)methyl]phenyl methylcarbamate; α-ethylthio-o-tolyl methylcarbamate
CAS # 29973-13-5
	carbamate	insecticide	?	oral: high (1)
				
				dermal: ?
				
				inhalation: high (1)
					?	immediate toxicity:
						  birds: high (1)
						  fish: low to medium (1)
						
						water: slightly soluble
						
						slightly volatile

ethiolate Prefox (with cyprazine); S 6176; S-15076 S-ethyl diethylthiocarbamate; S-ethyl diethylcarbamothioate CAS # 2941-55-1	thiocar-bamate	herbicide	"non-pers" (1)	oral: medium to high (1) dermal: high (1) inhalation: ?	?	immediate toxicity: birds: medium (1) water: soluble volatile combustible

ethion
 	
Acarfor (with dicofol); diethion (France, India, So. Africa); Embathion; Ethanox; Ethiol; Ethodon; FMC 1240; Hylemox; Itopaz; Kwit; NIA 1240; Niagara 1240; Nialate; Rhodiacide; Rhodocide; RP-Thion; Tuver Acaricide (with dicofol & methyl parathion); Vegfru Fosmite
O,O,O',O-tetraethyl S,S'-methylene bisphosphorodithioate; S',S'-methylene O,O,O',O-tetraethyl di(phosphorodithioate); S',S'-methylene bis(O,O-diethyl phosphorodithioate)
CAS # 563-12-2
	organo-phosphate	insecticide acaricide	non-pers to mod-pers (1)	oral: high to very high (3,4)
				
				dermal: high to very high (1,4)
				
				inhalation: medium to high (4,5)
					?	immediate toxicity:
						  birds: low to medium (6)
						  fish: medium to high (7)
						  crustaceans: medium to high (7)
						  bees: medium (8)
						  aquatic insects: medium (7)
						long-term toxicity:
						  birds: teratogen (9)
						
						water: slightly soluble
						oil: "soluble"
						slightly volatile
						"nonflammable"

NAME: Common Trade and Other Chemical CAS Number	Class of Chemical	Chief Pesticide Use; Status	Persistence	Effects on Mammals		Adverse effects on other non-target species Physical properties
				Immediate Toxicity (Acute)	Long-Term Toxicity (Chronic)	
ethiozin BAY SMY 1500; Lektran; Tycor 4-amino-6-*tert*-butyl-3-ethylthio-1,2,4-triazin-5(4*H*)-one; 4-amino-6-(1,1-dimethylethyl)-3-(ethylthio)-1,2,4-triazin5(4*H*)-one; 3-ethylthio-4-amino-6-*tert*-butyl-1,2,4-triazine-5-one CAS # 64529-56-2	triazine	herbicide	?	oral: medium (1) dermal: low to medium (1) inhalation: ?	?	water: soluble slightly volatile "nonflammable"

ethirimol
 	
Milcurb Super; Milgo E; Milstem
 	
5-butyl-2-ethylamino-6-methylp yrimidin-4-ol;
5-butyl-2-(ethylamino)-6-methyl-4(1*H*)-pyrimidinone;
5-butyl-2-(ethylamino)-6-methyl-4-pyrimidinol
 	
CAS # 23947-60-6
	pyrimi-dine	fungicide	mod-pers (1)	oral: low to medium (1)
				
				dermal: ?
				
				inhalation: ?
					?	immediate toxicity:
						  birds: medium (1)
						  fish: low (1)
						
						water: soluble
						
						slightly volatile

ethofumesate
 	
Barrix (with diethatyl-ethyl); Betanal Perfekt (with phenmedipham); Betanal Progress (with phenmedipham & desmedipham); Betanal Tandem (with phenmedipham); Betaron (with phenmedipham); Magnum; Morlex (with chlorpropham & fenuron & propham); NC 8438; Nortron; Tramat
 	
2-ethoxy-2,3-dihydro-3,3-dimethyl-5-benzofuranyl methanosulphonate
 	
CAS # 26225-79-6
	miscel-laneous	herbicide	non-pers	oral: low (1)
				
				dermal: ?
				
				inhalation: ?
					?	immediate toxicity:
						  birds: low to medium (2)
						  fish: low (2)
						  crustaceans: low (2)
						
						water: soluble
						
						slightly volatile

ethoprop
 	
ethoprophos; Jolt; Mocap; prophos; V-C-9-104
 	
0-ethyl S,S-dipropyl phosphorodithioate
 	
CAS # 13194-48-4
	organo-phosphate	insecticide, nematocide	mod-pers (1)	oral: high to very high (1,2)
				
				dermal: very high (2)
				
				inhalation: very high (2)
					suspect mutagen (2)	immediate toxicity:
						  birds: high to very high (3)
						  fish: medium to very high (2)
						  crustaceans: very high (2)
						
						volatile

ethoxyquin ethoxyquine (France); Nix-Scald; Santoquin; Stop-Scald 6-ethoxy-1,2-dihydro-2,2,4-trimethylquinoline; 1,2-dihydro-6-ethoxy-2,2,4-trimethylquinoline CAS # 91-53-2	miscel-laneous	fungicide	?	oral: medium (1) dermal: ? inhalation: ?	suspect carcinogen (2,3) suspect mutagen (4)	combustible
ethyl formate ethyl methanoate CAS # 109-94-4	alcohol	fumigant	?	oral: medium (1) dermal: ? inhalation: low (2)	?	immediate toxicity: bees: low to medium (3) water: slightly soluble flammable
ethyl hexanediol 6-12 Repellent; ethohexadiol; Rutgers 612; Tantoo Bomb 2-ethyl-1,3-hexanediol CAS # 94-96-2	alcohol	insect repellent	?	oral: low to medium (1,2) dermal: medium (2) inhalation: ?	?	water: soluble volatile combustible

NAME: Common Trade and Other Chemical CAS Number	Class of Chemical	Chief Pesticide Use; Status	Persistence	Effects on Mammals		Adverse effects on other non-target species Physical properties
				Immediate Toxicity (Acute)	Long-Term Toxicity (Chronic)	
ethylan Perthane; Q-137; Raid Mothproofer 1,1-dichloro-2,2-bis(p-ethylphenyl)ethane; diethyl diphenyl dichloroethane CAS # 72-56-0	organo-chlorine	insecticide cancelled, most uses, USA, 1980	"mod-pers" (1)	oral: low (2) dermal: ? inhalation: ?	?	immediate toxicity: fish: very high (3) crustaceans: very high (3) bees: high (4) water: "practically insoluble" oil: "soluble"
ethylene dibromide Aadibroom; Agrogas; Bromofume; Carboxide; Cartox; Celmide; DM23 Forte; Dowfume EDB; E-D-Bee; EDB; Edesol; Fumo-Gas; Granosan; Iscobrome D; Nemtosol; Nephis; Soilbrom 40; Soilbrom-90EC; Tradiafume; Unifume 1,2-dibromoethane CAS # 106-93-4	miscel-laneous	fumigant insecticide cancelled, USA, 1989	non-pers to pers (1,2)	oral: high (3) dermal: high (4) inhalation: ?	carcinogen (5) mutagen (2,6) suspect teratogen (7,8) liver, kidney, heart, & spleen damage (9) sperm & egg damage (5) disulfiram enhances toxic effects of EDB (5)	immediate toxicity: birds: high (10) fish: medium (11) long-term toxicity: fish: liver & kidney damage (12) plants: mutagen (13) bioaccumulates (3) water: soluble highly volatile "nonflammable"
ethylene dichloride Borer-sol; Brocide; Chlorasol; Destruxol; Di-Chlor-Mulsion; Dutch Liquid; EDC; ethylene chloride 1,2-dichloroethane; α,β-dichloroethane CAS # 107-06-2	organo-chlorine	insect fumigant	?	oral: medium (1) dermal: ? inhalation: ?	carcinogen (2) suspect mutagen (2,3) liver and kidney damage (2,4,5)	water: soluble highly volatile flammable
ethylene oxide Anprolene; Carboxide; Cartox (W. Germany); ETO; Etox; Oxirane; Oxyfume; Oxyfume 12; T-Gas 1,2-epoxyethane CAS # 75-21-8		fumigant fungicide	?	oral: high (1) dermal: ? inhalation: ?	carcinogen (2,3) suspect mutagen (4,5) nerve damage (6) testicular atrophy (7)	long-term toxicity: plants: mutagen (2) water: "soluble" highly volatile "flammable"
etridiazole AAterra; Ban-rot (with thiophanate-methyl); Dwell; ethazol; ethazole; Koban; Pansoil; Terra-Coat L 205 (with PCNB); Terra-Coat L21; Terradactyl (with chlorothalonil); Terrazole; Truban ethyl 3-trichloromethyl-1,2,4-thiadiazol-5-yl ether; 5-ethoxy-3-trichloromethyl-1,2,4-thiadiazole CAS # 2593-15-9	miscel-laneous	fungicide	mod-pers (1)	oral: low to medium (1,2) dermal: high (3) inhalation: ?	?	immediate toxicity: birds: medium (1) fish: low to medium (1) water: slightly soluble slightly volatile flammable
etrimfos Ekamet; Ekamet ULV; Satisfar O-6-ethoxy-2-ethylpyrimidin-4-yl O,O-dimethyl phosphorothioate; O-(6-ethoxy-2-ethyl-4-pyrimidinyl) O,O-dimethyl phosphorothioate CAS # 38260-54-7	organo-phosphate	insecticide	non-pers (1)	oral: medium to high (2) dermal: medium to high (3) inhalation: low (4)	?	immediate toxicity: fish: medium (4) water: slightly soluble slightly volatile
EXD DEX; Herbisan #5; Sulfasan diethyl dithiobis(thioformate); di[ethoxy(thiocarbonyl)] disulfide CAS # 502-55-6	miscel-laneous	herbicide	"non-pers" (1)	oral: medium (2) dermal: ? inhalation: ?	?	water: insoluble

NAME: Common Trade and Other Chemical CAS Number	Class of Chemical	Chief Pesticide Use; Status	Persistence	Effects on Mammals		Adverse effects on other non-target species Physical properties
				Immediate Toxicity (Acute)	Long-Term Toxicity (Chronic)	

famphur
 	
American Cyanamid 28023; Bo-Ana; Dovip; Famfos; Famophos; Warbex
 	
O,O-dimethyl O-[p(dimethylsulfamoyl)phenyl] phosphorothioate
CAS # 52-85-7
	organo-phosphate	insecticide	?	oral: high to very high (1)
				
				dermal: medium to high (1)
				
				inhalation: ?
					?	immediate toxicity:
						  birds: very high (2)
						  secondary poisoning possible (3,4)

fenac
 	
ACPM-673-A; Fenatrol; Kanepar; TCPA; Tri-Fen; Tri-Fene; Trifene
 	
2,3,6-trichlorophenylacetic acid (and sodium salt)
 	
CAS # 85-34-7
	organo-chlorine	herbicide	pers (1)	oral: medium (1)
				
				dermal: low to medium (1)
				
				inhalation: ?
					?	immediate toxicity:
						  fish: low to medium (2)
						  crustaceans: low (2)
						  aquatic insects: low (2)
						
						water: soluble
						
						volatile
						
						"nonflammable"

fenaminosulf
 	
BAY 22555; BAY 5072; DAPA; Dexon; diazoben; Lesan; Nemacur O (with isofenphos)
 	
sodium [4-(dimethylamino)phenyl]diazene sulfonate;
sodium p-(dimethylamino) benzendiazosulfonate
 	
CAS # 140-56-7
	sulfur	fungicide
		
		registration withdrawn in Canada 1990
			mod-pers (1)	oral: high to very high (2,3)
				
				dermal: very high (2)
				
				inhalation: ?
					mutagen (4,5) kidney damage (6)	immediate toxicity:
						  birds: very high (7)
						  fish: low (8)
						  crustaceans: medium (8)
						  aquatic insects: low to medium (8)
						  bees: low (9)
						
						water: very soluble
						
						oil: "insoluble"
						
						"non-volatile"

fenamiphos
 	
BAY 68138; Inemacury; Nemacur; Nemacur P; phenamiphos
 	
ethyl 3-methyl-4-(methylthio)phenyl (1-methylethyl)phosphoramidate;
ethyl 4-methylthio-m-tolyl isopropylphosphoramidate
CAS # 22224-92-6
	organo-phosphate	nematocide	mod-pers (1)	oral: high to very high (2)
				
				dermal: very high (2)
				
				inhalation: ?
					?	immediate toxicity:
						  birds: very high (2)
						  fish: medium to very high (3)
						  bees: high (4)
						
						water: soluble
						
						slightly volatile
						flammable

fenarimol Bloc; EL-222; Fenzol; Rimidin; Rimidine Plus (with carbendazim & maneb); Rubigan; Transflo 2,4'-dichloro-α-(pyrimidin-5-yl) benzhydryl alcohol; 3-(2-chlorophenyl)-3-(4-chloro-phenyl)-5-pyrimidinemethanol CAS # 60168-88-9	miscel-laneous	fungicide	mod-pers (1)	oral: medium (2) dermal: ? inhalation: ?	suspect carcinogen (3) teratogen (4) decreased male fertility (5)	immediate toxicity: fish: high (1) water: slightly soluble slightly volatile

fenitrothion
 	
Accothion; Agrothion; BAY 41831; Cyfen; Cytel; Danathion; Debucol; Dicontal Neu (with trichlorfon); Docofen; Fenitox; Fenstan; Folithion; MEP; Novathion; Nuvanol; Pesguard ANS (with tetramethrin); S 5660; Sumimix (with fenpropathrin); Verthion
 	
O,O-dimethyl O-(4-nitro-m-tolyl) phosphorothioate
CAS # 122-14-5
	organo-phosphate	insecticide acaricide	non-pers (1)	oral: medium to high (2)
				
				dermal: high (2)
				
				inhalation: ?
					suspect mutagen (3) suspect viral enhancer, implicated in Reye's syndrome (4) behavioral deficits in newborn (5) immunotoxin (6)	immediate toxicity:
						  birds: medium to very high (7)
						  fish: medium (8)
						  crustaceans: very high (8)
						  aquatic insects: very high (8)
						  bees: very high (9)
						  aquatic worms: medium (10)
						
						water: "practically insoluble"
						slightly volatile

(continued on next page)

NAME: Common Trade and Other Chemical CAS Number	Class of Chemical	Chief Pesticide Use; Status	Persistence	Effects on Mammals		Adverse effects on other non-target species Physical properties
				Immediate Toxicity (Acute)	Long-Term Toxicity (Chronic)	
contaminant(s): p-nitro-m-cresol						
O,O,S-trimethyl phosphorothioate				oral: high (1)	delayed toxicity (1)	
O,S,S-trimethyl phosphorodithiote				oral: very high (1)	immunotoxin (1)	
fenoxaprop-ethyl Acclaim; Excel; Furore; Option; Whip (±)-ethyl 2-[4-[(6-chloro-2-benoxazolyl) oxy]phenoxy] propanoate CAS # 66441-23-4	miscel-laneous	herbicide	non-pers (1,2)	oral: low to medium (3,4) dermal: ? inhalation: high (3)	teratogen (5)	immediate toxicity: birds: low to medium (2) fish: medium to high (3) crustaceans: medium (3) water: insoluble non-volatile "flammable"
fenpropathrin Danitol; fenpropathrine; Herald; Kilumal; Meothrin; Ortho Danitol; Rody; S-3206; SD 41706; Sumimik; Sumimix (with fenitrothion); Viktor (with clofentizine); WL 41706; XE 938 (RS)-α-cyano-3-phenoxybenzyl 2,2,3,3-tetramethylcyclopropanecarboxylate CAS # 64257-84-7	pyrethroid	insecticide	non-pers (1,2)	oral: high (3) dermal: ? inhalation: ?	immunotoxin (4)	immediate toxicity: birds: medium (3) fish: very high (3) water: insoluble slightly volatile
fensulfothion BAY 25141; Dasanit; DMSP; S 767; Terracur; Terracur P O,O-diethyl O-[p-methylsulfinyl)phenyl] phosphorothioate CAS # 115-90-2	organo-phosphate	insecticide	mod-pers (1)	oral: very high (2) dermal: very high (1) inhalation: ?	?	immediate toxicity: birds: very high (3) fish: medium to very high (4,5) bees: very high (6) water: soluble
fenthion BAY 29493; Baycid; Bayer 4895; Baytex; DMPT; Ekalux; Entex; Lebaycid; mercaptophos; Quelatox; Queletox; S 1752; Spotton; Talodex; Tiguvon O,O-dimethyl-O-[4-methylthio]-m-tolyl phosphorothioate; phosphorothioic acid O,O-dimethyl O-3-methyl-4-(methylthio)phenyl CAS # 55-38-9	organo-phosphate	insecticide acaricide avicide restricted use, USA	non-pers (1,2)	oral: medium to high (3) dermal: high (4) inhalation: ?	suspect carcinogen (5) delayed neurotoxin (6) suspect embryotoxin (7) neuromuscular dysfunction (8) eye damage (9,10)	immediate toxicity: birds: very high (11) fish: medium (3,12) crustaceans: high to very high (12) bees: very high (14) aquatic insects: very high (13) water: slightly soluble slightly volatile
transformation product(s): sulfoxide analogue of fenthion				oral: high (1)		
sulfone analogue of fenthion				oral: high (1) thirty-six times more toxic than fenthion (2)		
fenuron Beet Kleen (with propham & chlorpropham); Croptex Chrome; Croptex Ruby; Dozer; Dybar; Falisilvan; Fenidim; fenidin (USSR); Fenulon; fenulon (with So. Africa); Gold; Herbon Yellow (with chlorpropham); Karmex FP; Morlex (with chlorpropham & ethofumesate & propham); PDU; PDV; Premalox; Quintex; Urab 1,1-dimethyl-3-phenylurea; 3-phenyl-1,1-dimethylurea CAS # 101-42-8	urea	herbicide	?	oral: low to medium (1) dermal: ? inhalation: ?	suspect mutagen (2)	immediate toxicity: fish: low (3) water: soluble slightly volatile

NAME: Common Trade and Other Chemical CAS Number	Class of Chemical	Chief Pesticide Use; Status	Persistence	Effects on Mammals		Adverse effects on other non-target species Physical properties
				Immediate Toxicity (Acute)	Long-Term Toxicity (Chronic)	
fenvalerate Ectrin; Extrin; Fenkill; fenvalethrin; Mikantop (with dimethoate); Moscade; OMS 2000; Pydrin; Pyrid; S-5602; Sanmarton; Sumibac; Sumicidin; Sumicombi (with fenitrothion); Sumifleece; Sumifly; Sumitomo (with fenitrothion); Sumittick; Tirade; WL 43775 (RS)-α-cyano-3-phenoxybenzyl (RS)-2-(4-chlorophenyl)-3-methylbutyrate CAS # 51630-58-1	pyrethroid	insecticide	mod-pers (1)	oral: ? dermal: ? inhalation: ?	cumulative (2) suspect mutagen (3)	immediate toxicity: birds: low (1) fish: low to very high (1,4) marine invertebrates: high to very high (1) long-term toxicity: fish: adverse effects in gill structure (5) water: insoluble non-volatile to slightly volatile
ferbam Black Fungicide; Carbamate; Coromate; ferbame (France); Ferbert; Fermate; Fermocide; Ferradow; Green Cross kerbam; Hexaferb; Karbam Black; Knockmate; Miller Blue Mold Dust; New Leaf Black Fungicide; Niagar Carbamate; Planters Blue Mold Dust; Sup'r-Flo Ferbam Flowable; Trifungol; Vancide FE 95 iron tris(dimethyldithiocarbamate); ferric dimethyldithiocarbamate CAS # 14484-64-1	thiocar-bamate	fungicide	non-pers (1)	oral: low to medium (2) dermal: ? inhalation: ?	suspect mutagen (3,4) suspect fetotoxin (4,6) kidney damage (5) sperm damage (7)	immediate toxicity: fish: medium to high (1) bees: low to medium (8) long-term toxicity: birds: affects fertility (9) fish: blindness and fin erosion (1); embryotoxic (10) molluscs: "inhibits shell growth" (1) plants: inhibits germination of pollen in some plants (10) water: soluble non-volatile to slightly volatile
transformation product(s): **carbon disulfide** (see carbon disulfide)						
N-nitrosodimethylamine				oral: very high (1)	carcinogen (2,3) mutagen (1,3) liver damage (1)	long-term toxicity: molluscs: may cause reproductive & gastrointestinal damage (4) water: "soluble" oil: "soluble" "volatile" "nonflammable"
ferrous sulfate Aitkens Lawn Sand Plus (with chloroxuron); Ashlade D-Moss (with chloroxuron); copperas; green vitriol; iron protosulfate; iron vitriol; SHL Turf Feed & Weed (with dichlorprop & MCPA) iron(II) sulfate; iron(2+)sulfate CAS # 7720-78-7	metal/ mineral	herbicide wood preservative	?	oral: medium (1) dermal: ? inhalation: ?	?	water: soluble
flamprop-isopropyl Barnon Plus; Commando; Effix; flamprop-M-isopropyl; Lancer; Mataven; Suffix BW; Super Barnon; Supper Suffix; WL 28651; WL 43423; WL 43425 isopropyl N-benzoyl-N-(3-chloro-4-fluorophenyl)-ᴅ-alanine CAS # 63782-90-1	miscel-laneous	herbicide	?	oral: low to medium (1) dermal: ? inhalation: ?	?	immediate toxicity: birds: low (1) fish: medium (1) bees: "nontoxic" (1) water: slightly soluble slightly volatile

NAME: Common Trade and Other Chemical CAS Number	Class of Chemical	Chief Pesticide Use; Status	Persistence	Effects on Mammals		Adverse effects on other non-target species Physical properties
				Immediate Toxicity (Acute)	Long-Term Toxicity (Chronic)	
fluazifop-butyl Fusilade; Fusilade 2000; Fusilade Five; Fusilade Super; Grass-B-Gone; Hache Uno Super; Onecide; PP009; Tornado (with fomesafen); TS-7236 butyl 2-[4-[[5-(trifluoromethyl)-2-pyridinyl]oxy]phenoxy]propananoate; 2-[4-(5-trifluoromethyl-2-pyridyloxy)phenoxy]propionic acid CAS # 69806-50-4	phenoxy	herbicide	non-pers to mod-pers (1,2)	oral: medium (3,4) dermal: low to medium (4) inhalation: ?	?	immediate toxicity: birds: low (4) fish: high (4) water: slightly soluble slightly volatile "nonflammable"
fluchloralin BAS-392-H; Basalin N-(2-chloroethyl)-α,α,α-trifluoro-2,6-dinitro-N-propyl-p-toluidine; N-(2-chloroethyl)-2,6-dinitro-N-propyl-4-(trifluoromethyl)aniline; N-(2-chloroethyl)-2,6-dinitro-N-propyl-4-(trifluoromethyl)benzenamine CAS # 33245-39-5	dinitro-aniline	herbicide	?	oral: low to medium (1,2) dermal: ? inhalation: medium (2)	?	immediate toxicity: birds: low to medium (2) fish: very high (2) water: slightly soluble slightly volatile to volatile flammable
flucythrinate AC 222,705; AI3 29391; Cyanotril (with dimethoate); Cybolt; Cythrin; Flutrin (with dimethoate); Kiedex; Lepister (with chlorpyrifos); OMS 2007; Pay-Off M (with methomyl) (RS)-α-cyano-3-phenoxybenzyl (S)-2-(4-difluoromethoxyphenyl)-3-methylbutyrate CAS # 70124-77-5	pyrethroid	insecticide	non-pers to mod-pers (1,2)	oral: high (3) dermal: high (3) inhalation: very high (4)	?	immediate toxicity: fish: very high (5) crustaceans: very high (1) water: insoluble non-volatile
fluenethyl Flu; Fluenyl; Lambrol; M 2060 2-fluoroethyl (4-biphenyl)acetate; 2-fluoroethyl 4-biphenylacetate; 2-fluoroethyl [1,1'-biphenyl]-4-acetate CAS # 4301-50-2	miscel-laneous	insecticide acaricide not registered, USA	?	oral: high to very high (1,2) dermal: very high (2) inhalation: ?	?	immediate toxicity: bees: medium (3)
transformation product(s): **monofluoroacetic acid** the active principle in sodium fluoroacetate (see chart for details) CAS # 144-49-0						
flufenoxuron Cascade; WL 115110 1-[4-(2-chloro-α,α,α-trifluoro-p-tolyloxy)-2-fluorophenyl]-3-(2,6-difluorobenzoyl)urea; N-[[[4-(2-chloro-4-(trifluoromethyl)phenyl]-2-fluorophenyl]amino]carbonyl]-2,6-difluorobenzamide CAS # 101463-69-8	urea	insecticide	non-pers to mod-pers (1)	oral: low to medium (1) dermal: low to medium (2) inhalation: medium (1)	?	immediate toxicity: birds: low to medium (1) fish: low (1) water: insoluble non-volatile

NAME: Common Trade and Other Chemical CAS Number	Class of Chemical	Chief Pesticide Use; Status	Persistence	Effects on Mammals		Adverse effects on other non-target species Physical properties
				Immediate Toxicity (Acute)	Long-Term Toxicity (Chronic)	
fluometuron C-2059; Ciba 2059; Cotogard (with prometryn); Cotoran multi (with metolachlor); Cotoran Multi 50WP; Cottonex; Croak (with MSMA); Higalcoton; Lanex; Meuturon 4L; Pakhtaran; Zorial 1,1-dimethyl-3-(a,a,a-trifluoro-m-tolyl)urea; 3-(m-trifluoromethylphenyl)-1,1-dimethylurea1 CAS # 2164-17-2	urea	herbicide	non-pers to mod-pers (1,2)	oral: low to medium (1,2) dermal: ? inhalation: ?	spleen damage (3) blood damage (4)	immediate toxicity: fish: low (1,2) bees: low (5) water: slightly soluble to soluble slightly volatile "nonflammable"
fluoroacetamide 1081; Baran; Compound 1081; Fluorakil 100; Fussol (with; Megatox; Yanock 2-fluoroacetamide CAS # 640-19-7	fluoro-acetate	rodenticide cancelled USA, 1989	?	oral: very high (1) dermal: very high (1) inhalation: ? less toxic than fluoroacetate (1)	?	immediate toxicity: birds: high (2) long-term toxicity: secondary poisoning in some species (1) water: "very soluble"
transformation product(s): **fluorocitrate** transformation product of fluoroacetate				L-erythro-fluoro-citrate causes the toxicity (1)	kidney damage (2) inhibits essential enzymes in energy (ATP) production; leads to organ & system failures (3,4)	
fluorodifen C-6989; fluorodiphen; Preforan; Soyex p-nitrophenyl a,a,a-trifluoro-2-nitro-p-tolyl ether; 2-nitro-1-(4-nitrophenoxy)-4-(trifluoromethyl) benzene CAS # 15457-05-3	phenol	herbicide	non-pers to mod-pers (1)	oral: low (2) dermal: low to medium (2) inhalation: ?	?	immediate toxicity: bees: low (3) water: slightly soluble non-volatile
flurecol-butyl Aniten (with MCPA); Florencol; flurecol-n-butylester (USA, Great Britain, So. Africa); flurenol; flurenol-n-butylester; IT-3223 n-butyl 9-hydroxyfluorene-9-carboxylate; 9-hydroxyfluorene-9-carboxylic acid butyl ester CAS # 467-69-6	miscel-laneous	herbicide	?	oral: low (1) dermal: low to medium (2) inhalation: ?	?	immediate toxicity: fish: medium (3) water: slightly soluble oil: soluble slightly volatile
fluridone Brake; Compel; E1-171; Pride; Sonar; Sonar 5P; Sonar A5 1-methyl-3-phenyl-5-(trifluoro-m-tolyl)-4-pyridone CAS # 59756-60-4	miscel-laneous	herbicide	non-pers to mod-pers (1,2)	oral: low (3) dermal: ? inhalation: medium (2)	?	immediate toxicity: birds: medium (2) fish: low to medium (2) crustaceans: medium (2) water: slightly soluble non-volatile
transformation product(s): **monomethylformamide** n-monomethylformamide CAS # 123-39-7					teratogen (1)	

NAME:	Common Trade and Other Chemical CAS Number	Class of Chemical	Chief Pesticide Use; Status	Persistence	Effects on Mammals		Adverse effects on other non-target species Physical properties
					Immediate Toxicity (Acute)	Long-Term Toxicity (Chronic)	
flusilazole DPX-H 6573; Nustar; Olymp; Punch (with carbendazim); Start (with pyrazophos) bis(4-flurophenyl)(methyl)(1H-1,2,4-triazol-]-ylmethyl]silane; 1-[[bis(4-flurophenyl)(methyl) silyl]methyl]-1H-1,2,4-triazole CAS # 85509-19-9		triazole	fungicide	?	oral: medium (1) dermal: low to medium (2) inhalation: low to medium (1)	?	water: slightly soluble to soluble volatile
fluvalinate Klartan; Mavrik; Spur (RS)-α-cyano-3-phenoxybenzyl (R-2-(2-chloro-α,α,α-trifluoro-p-toluidino)-3-methylbutyrate CAS # 69409-94-5		pyrethroid	insecticide acaricide	non-pers to mod-pers (1)	oral: high (2) dermal: low (2) inhalation: ?	?	immediate toxicity: birds: low (1) fish: medium to high (1) freshwater invertebrates: low to medium (1) water: insoluble non-volatile to slightly volatile
τ-fluvalinate Amalux (with quinalphos); Apistan; Klartan; Mavrik 2E; Mavrik 2F; Mavrik HR; Spur; Torero (with clofentezine) (RS)-α-cyano-3-phenoxybenzyl N-(2-chloro-α,α,α-trifluoro-p-tolyl-ᴅ-valinate CAS # 102851-06-9		pyrethroid	insecticide	non-pers (1)	oral: ? dermal: ? inhalation: ?	?	immediate toxicity: fish: very high (1) crustaceans: very high (1)
folpet Folnit; Folpan; Folplan; Folprame (with copper oxychloride); Folsystem; Foltan; Foltapet (with captafol); Foltazip; Foltene; Foltimil (with copper sulphate) trichloromethylthiophthalimide; N-(trichloromethylthio)phthalimide CAS # 133-07-3		phthalate	fungicide	?	oral: low (1) dermal: low to medium (2) inhalation: medium to high 92)	carcinogen (2) suspect mutagen (2) teratogen (3,4)	immediate toxicity: birds: low to medium (5) fish: high to very high (6) crustaceans: medium (6) bees: low to medium (7) phytoplankton: high (8) plants: may be toxic to some (9) long-term toxicity: birds: teratogen (8) water: slightly soluble slightly volatile
fomesafen Dardo; Flex; PP 021; Reflex; Reflex T (with terbutryn); Tornado (with fluazifop-butyl) 5-(2-chloro-α,α,α-trifluro-p-tolyloxy)-N-mesyl-2-nitrobenzamide; 5-(2-chloro-α,α,α-trifluro-p-tolyloxy)-N-methylsulfonyl-2-nitro benzamide; 5-[2-chloro-4-(trifluromethyl) phenoxy]-N-(methylsulfonyl)-2-nitrobenzamide CAS # 72178-02-0		amide	herbicide	non-pers to mod-pers (1)	oral: low to medium (1,2) dermal: ? inhalation: ?	?	immediate toxicity: birds: low (3) fish: low (1) water: slightly soluble slightly volatile "nonflammable"
fonofos Admiral; Capfos; Cudgel; Doubledown (with disulfoton); Dyfonate; Metro; N-2790 O-ethyl S-phenyl (RS)-ethylphosphonodithioate; O-ethyl S-phenyl ethylphosphonodithioate CAS # 944-22-9		organo-phosphate	insecticide	non-pers to mod-pers (1,2)	oral: very high (2,3) dermal: high to very high (1,2) inhalation: high (4)	?	immediate toxicity: birds: high to very high (4) fish: very high (4) water: slightly soluble slightly volatile

NAME: Common Trade and Other Chemical CAS Number	Class of Chemical	Chief Pesticide Use; Status	Persistence	Effects on Mammals		Adverse effects on other non-target species Physical properties
				Immediate Toxicity (Acute)	Long-Term Toxicity (Chronic)	
formaldehyde BFV; D&P 77 Dust; Dyna-Form; Formalin; Formalina; formic aldehyde; Karsan; methanal; methylene oxide; oxomethane; oxymethylene; Victoria methyl aldehyde CAS # 50-00-0	aldehyde	fungicide herbicide	?	oral: medium (1) dermal: low to medium (2) inhalation: high (1)	carcinogen (2,3) suspect mutagen (2,4) liver damage (5)	water: "very soluble" "flammable"
transformation product(s): **formic acid** CAS # 64-18-6					eye damage (1)	
paraformaldehyde (see paraformaldehyde)						
formetanate hydrochloride Carzol; Dicarzol; EP 332; Fundal Forte (with chlordimeform); Schering 36056; SN 36056 N,N-dimethyl-N'-[3-[[(methylamino)carbonyl]oxy]phenyl]methanimidamide; (3-dimethylamino-(methyleneiminophenyl)-N-methylcarbamate hydrochloride CAS # 23422-53-9	carbamate	acaricide insecticide	non-pers (1,2)	oral: very high (1) dermal: low to medium (2) inhalation: medium to high (1)	?	immediate toxicity: birds: very high (1) fish: low to medium (1) crustaceans: very high (1) bees: medium (3) water: very soluble non-volatile
fosamine ammonium DPX 1108; Krenite Brush Control Agent; Krenite S ammonium ethyl carbamoyl phosphonate CAS # 25954-13-6	carbamate	herbicide	non-pers to mod-pers (1,2)	oral: low to medium (3) dermal: ? inhalation: low (1)	?	immediate toxicity: birds: low to medium (1) fish: low (1) long-term toxicity: birds: teratogen (4) water: very soluble slightly volatile combustible
fosetyl-al Aliette; Aliette Extra (with captan & thiabendazole); Mikal (with folpet); Rhodex (with mancozeb) aluminum tris (o-ethylphosphonate) CAS # 39148-24-8	organo-phosphate	fungicide	more persistent on foliage than in soil (1)	oral: low (1) dermal: ? inhalation: ?	suspect carcinogen (1) degenerative effecst on testes (1) delayed fetal development (1) changes in urinary tract development (1)	immediate toxicity: birds: low (2) fish: low (2) bees: low (1) water: soluble non-volatile
GC 6506 GC 6506 O,O-dimethyl O-(4-methylmercaptophenyl)phosphate; dimethyl p-(methylthio)phenyl phosphate CAS # 3254-63-5	organo-phosphate	insecticide acaricide	?	oral: very high (1) dermal: ? inhalation: ?	?	immediate toxicity: bees: "highly toxic to bees" (1)

NAME: Common Trade and Other Chemical CAS Number	Class of Chemical	Chief Pesticide Use; Status	Persistence	Effects on Mammals		Adverse effects on other non-target species Physical properties
				Immediate Toxicity (Acute)	Long-Term Toxicity (Chronic)	

glyphosate
 	
CP67573; Fallow Master (with dicamba); Landmaster (with 2,4-D); Mon 0573
 	
N-(phosphonomethyl)glycine
 	
CAS # 1071-83-6
	miscel-laneous	herbicide	mod-pers (1,2)	oral: medium (3,4)
				
				dermal: medium (3)
				
				inhalation: ?
					suspect carcinogen (4) suspect mutagen (5)	immediate toxicity:
						  birds: low (6)
						  fish: low to medium (6)
						  crustaceans: low to medium (7)
						  bees: low (6)
						
						long-term toxicity:
						  plants: mutagen (8)
						
						water: soluble
						
						oil: insoluble
						
						non-volatile

salt(s): **glyphosate trimesium** Touchdown CAS # 81591-81-3						
isopropylamine salt of glyphosate Pondmaster; Rattler; Rodeo; Roundup; Roundup L&G; Shackle; Shacklet C; Spasor; Sting; Vision N-(phosphomethyl)glycine, isopropylamine salt CAS # 38641-94-0						
sodium salt of glyphosate Polado N-(phophonomethyl)glycine, sodium salt CAS # 70393-85-0						
transformation product(s): **formaldehyde** (see formaldehyde)						
N-nitrosoglyphosate (in contact with nitric acid)			mod-pers (1)		suspect carcinogen (1) suspect mutagen (1)	
surfactant: **polyoxyethyleneamine**				"LD$_{50}$ of POEA is less than 1/3 that of....(glyphosate)" (1)		immediate toxicity: fish: medium to high (2,3)
contaminant of surfactant: **1,4-dioxane** p-dioxane CAS # 123-91-1					carcinogen (1)	

NAME: Common Trade and Other Chemical CAS Number	Class of Chemical	Chief Pesticide Use; Status	Persistence	Effects on Mammals		Adverse effects on other non-target species Physical properties
				Immediate Toxicity (Acute)	Long-Term Toxicity (Chronic)	
heptachlor Drinox H-34; E-3314; heptachlore (France); Heptagran; Heptamul; Termide; Velsicol 104 heptachlorotetrahydro-4,7-methanoindene; 1,4,5,6,7,8,8-heptachloro-3a,4,7,7a-tetrahydro-4,7-methanoindene CAS # 76-44-8	organo-chlorine	insecticide 1984: only use of chlordane/heptachlor permitted in US is fire ant control in power transformers	mod-pers to pers	oral: high dermal: high to very high inhalation: low to high more toxic than chlordane in animals (see chlordane) readily absorbed through skin	carcinogen liver damage may cause cataracts	immediate toxicity: fish: high to very high amphibians: high crustaceans: high to very high bees: very high aquatic insects: very high long-term toxicity: sea urchins: embryotoxin water: insoluble slightly volatile to volatile
transformation product(s):						
heptachlor epoxide Compounds 53-CS-17; Velsicol 2,3,4,5,6,7,7-heptachloro-1a,1b,5,5a,6,6a-hexahydro-2,5-methano-2H-indeno[1,2-b]oxirene CAS # 1024-57-3			?	oral: high dermal: ? inhalation: ? more toxic than heptachlor	cumulative carcinogen	water: insoluble oil: "lipophilic"
hydrogen chloride (see dienochlor)						
carbon monoxide (see methylene chloride)						
hexachlorobenzene Anti-Carie; Anticarie; HCB; No Bunt; perchlorobenzene 1,2,3,4,5,6-hexachlorobenzene CAS # 118-74-1	organo-chlorine	fungicide insecticide cancelled in USA	pers (1,2)	oral: medium (7) dermal: ? inhalation: ?	cumulative (3,4) carcinogen (1,5) teratogen (6) fetotoxin (1) liver damage (3,7) nerve damage (8,9) thyroid damage (10,11) immunotoxin (12,13) porphyria cutanea tarda (7) crosses placenta (6)	immediate toxicity: fish: low to medium (14) long-term toxicity: biomagnification (15) birds: reduced hatchability, reproductive damage, liver damage (2,6) fish: cumulative, liver & kidney damage (2,14) crustaceans: cumulative (14) water: insoluble oil: "lipophilic" slightly volatile to volatile
pentachlorobenzene CAS # 608-93-5				oral: high (1)		
octachlorodibenzo-p-dioxin OCDD octachlorodibenzo-p-dioxin CAS # 3268-87-9	dibenzo-dioxin		non-pers to mod-pers (1)		cumulative (1) acne (2)	water: insoluble non-volatile
transformation product(s): **pentachlorophenol** (see pentachlorophenol)						
pentachlorobenzene CAS # 608-93-5				oral: high (1)		

NAME: Common Trade and Other Chemical CAS Number	Class of Chemical	Chief Pesticide Use; Status	Persistence	Effects on Mammals		Adverse effects on other non-target species
				Immediate Toxicity (Acute)	Long-Term Toxicity (Chronic)	Physical properties
hexachlorophene Hexide; Nabac; Seribak 2,2'-methylene bis(3,4,6-trichlorphenol); dihydroxy hexachlorodiphenyl methane CAS # 70-30-4	organo-chlorine	fungicide antibiotic acaricide cancelled, all products, USA, 1983	?	oral: high (1) dermal: ? inhalation: ?	suspect teratogen (2,3) neurotoxin (4,5)	immediate toxicity: birds: medium (6) water: insoluble oil: "soluble"
contaminant(s): **TCDD** (see chloroneb)						
hexazinone Velpar; Velpar Gridmall; Velpar K (with diuron) 3-cychlohexyl-6-(dimethylamino)-1-methyl-1,3,5-triazine-2,4-(1H,3H)-dione; 3-cyclohexyl-6-(dimethylamino)-1-methyl-s-triazine-2,4(1H,3H)-dione CAS # 51235-04-2	triazine	herbicide	non-pers to mod-pers (1,2)	oral: medium (2) dermal: low to medium (3) inhalation: low to medium (4)	suspect teratogen (4)	immediate toxicity: birds: medium (4) fish: low (4) crustaceans: low (4) water: very soluble slightly volatile "nonflammable"
hydramethylnon Amdro; Arinosu-Korori; Blatex; Combat; Cyaforce; Cyclon; Impact; Matox; Maxforce; pyrimidinone; Wipeout tetrahydro-5,5-dimethyl-2(1H)-pyrimidinone[3-[4-(trifluoromethyl)phenyl]-1-[2-[4-(trifluoromethyl)phenyl]ethenyl]-2-propenylidene]hydrazone; 5,5-dimethylperhydropyrimidin-2-one 4-trifluoromethyl-α-(4-trifluoromethylstyryl) cinnamylidenehydrazone CAS # 67485-29-4	miscel-laneous	insecticide	non-pers (1)	oral: medium (2) dermal: ? inhalation: ?	suspect carcinogen (3)	immediate toxicity: fish: high to very high (1) water: insoluble combustible
hydrogen cyanide Cyclon; HCN; prussic acid hydrocyanic acid CAS # 74-90-8	cyanide	fumigant insecticide restricted use, USA	?	oral: very high (1) dermal: "extremely toxic" (2) inhalation: "intensely poisonous" (3)	?	water: "soluble" flammable
salt(s): **calcium cyanide** A-Dust; Cyanogas; Degesch Calcium Cyanide A-Dust CAS # 592-01-8	organic	fumigant insecticide rodenticide restricted use, USA		oral: very high (1)		
sodium cyanide (see sodium cyanide)						
imazamethabenz Assert; Dagger (+)-2-[4,5-dihydro-4-methyl-4-(1-methylethyl)-5-oxo-1H-imidazol-2-yl]-4-methyl benzoic acid(ii) with (+)-2-2[4,5-dihydro-4-methyl-4-(1-methylethyl)-5-oxo-1H-imidazol-2-yl]-5-methylbenzoic acid (i) CAS # 100728-84-5	miscel-laneous	herbicide	non-pers to mod-pers (1)	oral: low (1) dermal: ? inhalation: low to medium (1)	?	immediate toxicity: birds: medium (1) fish: low (1) crustaceans: low (1) water: slightly soluble non-volatile combustible

NAME: Common Trade and Other Chemical CAS Number	Class of Chemical	Chief Pesticide Use; Status	Persistence	Effects on Mammals		Adverse effects on other non-target species Physical properties
				Immediate Toxicity (Acute)	Long-Term Toxicity (Chronic)	
imazapyr Arsenal; Assault; Chopper; Contain 2-(4-isopropyl-4-methyl-5-oxo-2-imidazolin-2-yl)nicotinic acid CAS # 81334-34-1	miscel-laneous	herbicide	mod-pers to pers (1,2)	oral: low to medium (1,2) dermal: ? inhalation: ?	?	immediate toxicity: birds: low to medium (1) fish: low (1) crustaceans: low (1) water: "soluble" combustible
salt(s): **ammonium salt of imazapyr** 2-(4-isopropyl-4-methyl-5-oxo-2-imidazolin-2-yl)nicotinic acid, ammonium salt CAS # 81510-83-0	imidazo-linone	herbicide	?	oral: low (1) dermal: ? inhalation: ?	?	water: very soluble
ioxynil ACP-63303; Actril; Actril DS (with 2,4-D); Actril S (with bromoxynil & dichlorprop & MCPA); Axall (with bromoxynil & mecoprop); Banlene Solo (with dicamba & dichlorprop); Bantrol; Basagran Ultra (with bentazon & dichlorprop); Bentrol; Bio Lawn Weedkiller (with 2,4-D & dicamba); Brittox (with bromoxynil & mecoprop); Certrol; Dantril (with bromoxynil & dichlorprop); MB-8873; Oxytril 4 (with dichlorprop & bromoxynil & MCPA); Oxytril CM (with bromoxynil); Oxytril P (with dichlorprop & bromoxynil); Totril; toxynil (Europe) 4-hydroxy-3,5-diiodobenzonitrile; 3,5-diiodo-4-hydroxybenzonitrile CAS # 1689-83-4	benzo-nitrile	herbicide not registered for use in U.S.A.	non-pers (1)	oral: high (1) dermal: high (2) inhalation: ?	?	immediate toxicity: birds: high (1) fish: medium (1) water: slightly soluble "nonflammable"
iprodione Chipco-26019; glycophene; Kidan; Rovral 3-(3,5-dichlorophenyl)-N-(1-methylethyl)-2,4-dioxo-1-imidazolidine carboxamide CAS # 36734-19-7	amide	fungicide	mod-pers (1)	oral: medium (1) dermal: low to medium (2) inhalation: low to medium (1)	?	immediate toxicity: birds: low to medium (3) fish: medium (2) crustaceans: low to high (4) water: slightly soluble non-volatile to slightly volatile combustible
isazophos CGA-12223; Miral; Triumph O-(5-chlro-1-methylethyl)-1H-1,2,4-triazol-3-yl O,O-diethyl phosphorothioate CAS # 42509-80-8	organo-phosphate	insecticide nematocide	?	oral: high to very high (1) dermal: ? inhalation: ?	?	immediate toxicity: birds: high (1) water: soluble
isofenphos Amaze; BAY 9214; BAY SRA 12869; Carma (with carbofuran); Disyston O (with disulfoton); Nemacur O (with fenaminophos); Oftanol; Oftanol Combi (with phoxim); Oftanol T (with thiram) 1-methylethyl 2-[ethoxy[(1-methylethyl) amino]phosphinothioyl]oxy]benzoate CAS # 25311-71-1	organo-phosphate	insecticide	mod-pers (1)	oral: high to very high (2) dermal: high to very high (3) inhalation: high (2)	delayed neurotoxicity, irreversible demyelination, paralysis (4,5)	immediate toxicity: birds: very high (6) fish: high (2) long-term toxicity: birds: delayed neurotoxicity (7) water: slightly soluble slightly volatile

NAME: Common Trade and Other Chemical CAS Number	Class of Chemical	Chief Pesticide Use; Status	Persistence	Effects on Mammals		Adverse effects on other non-target species Physical properties
				Immediate Toxicity (Acute)	Long-Term Toxicity (Chronic)	
isopropalin 4-isopropyl-2,6-dinitro-N, N dipropylaniline; 4-(1-methylethyl)-2,6-dinitro-N,N-dipropylbenzanamine CAS # 33820-53-0	dinitro-aniline	herbicide	?	oral: low (1) dermal: low to medium (2) inhalation: ?	?	immediate toxicity: fish: very high (3) water: insoluble slightly volatile flammable
contaminant(s): **dipropylnitrosamine**	·				carcinogen (1)	

isoxaben
 	
AZ 500; Cent-7; Combat; Elset; Flexidor; Gallery; Knock Out; Ratio; Snapshot 80 (with oryzalin); X-Pand
 	
N-[3-(1-ethyl-1-methylpropyl)-5-isoxazolyl]-2,6-dimethoxybenzamide
 	
CAS # 82558-50-7
	amide	herbicide	pers (1)	oral: low (1)
				dermal: ?
				inhalation: ?
					carcinogen (1)	long-term toxicity:
						  birds: reduces hatchability of eggs (1)
						  fish: bioaccumulates (1)
						
						water: slightly soluble
						
						slightly volatile
						
						"nonflammable"

kadethrin AI3-29 117; Kadethrin; RU 15 525; Spray-Tox 5-benzyl-3-furylmethyl (E-(1R)-cis-2,2-dimethyl-3-(2-oxothiolan-3-ylidenemethyl)cyclopropanecarboxylate; [1R-[1a,3a(E)]]-[5-(phenylmethyl)-3-furanyl]methyl 3-[(dihydro-2-oxo-3(2H)-thienylidene)methyl-2,2-dimethyl cyclopropanecarboxylate CAS # 58769-20-3	pyrethroid	insecticide	?	oral: medium (1) dermal: ? inhalation: ?	?	?

leptophos
 	
Abar; Lepton; Phosvel; VCS-506
 	
O-(4-bromo-2,5-dichlorophenyl) O-methylphenylphosphonothioate
 	
CAS # 21609-90-5
	organo-phosphate	insecticide	?	oral: high to very high (1,2)
				
				dermal: high to very high (2)
				
				inhalation: ?
					delayed neurotoxin (1)	immediate toxicity:
						  birds: medium (1)
						  fish: very high (3)
						  bees: high (4)
						
						long-term toxicity:
						  birds: delayed neurotoxin (1)
						
						water: slightly soluble

transformation product(s): **desbromoleptophos**					more neurotoxic than leptophos (1)	
d-limonene Aacess Penetrator; cajeputene; cinene; Dipentene; kautschin 1-methyl-4-(1-methylethenyl)cyclohexene CAS # 5989-27-5	miscel-laneous	insecticide	?	oral: low (5) dermal: ? inhalation: ?	suspect carcinogen (1) suspect teratogen (2,3) immunotoxin (4) kidney damage (3)	water: "practically insoluble"

NAME: Common Trade and Other Chemical CAS Number	Class of Chemical	Chief Pesticide Use; Status	Persistence	Effects on Mammals		Adverse effects on other non-target species Physical properties
				Immediate Toxicity (Acute)	Long-Term Toxicity (Chronic)	

lindane
 	
Aficide; Agrocide; Asparasin; BBH; Bexol; Borer Kill; Detox 25; Forlin; gamma isomer of BHC; gamma-BHC; gamma-HCH (Gr. Britain); Gammaspra; Gammcide; Gexane; Isotox; Jacutin; Kwell; Lindagam; Lindaterra; Lintox
 	
gamma isomer of benzene hexachloride; 1,2,3,4,5,6-hexachlorocyclohexane
 	
CAS # 58-89-9
	organo-chlorine	insecticide
		
		cancelled, most uses, USA, 1983
			pers (1)	oral: high to very high (2)
				
				dermal: high (2)
				
				inhalation: ?
					cumulative (2,3) carcinogen (4,5) suspect mutagen (2) teratogen (2) hormone damage (6) testicular damage (7) immunotoxin (2,8) neurotoxin (9,10) aplastic anemia (11) bone marrow damage (12)	BIOCIDE
						
						immediate toxicity:
						  birds: medium to high (13)
						  fish: very high (14)
						  amphibians: medium (15)
						  crustaceans: very high (14)
						  earthworms: low (15)
						  aquatic worms: high (16)
						  bees: very high (17)
						  toxic to some plants and phytoplankton (15)
						
						long-term toxicity:
						  birds: reduced egg production; eggshell thinning (19)
						  fish: liver damage, behavioral changes (20)
						  amphibians: teratogen (21)
						  plants: mutagen (18)
						
						water: slightly soluble
						oil: "slightly soluble"
						slightly volatile
						combustible

transformation product(s): **hydrogen chloride** hydrochloric acid hydrogen chloride CAS # 7647-01-0				oral: low (1)		
2,4,6-trichlorophenol (see 2,4,6-trichlorophenol)						
benzene (see benzene)						
pentachlorobenzene CAS # 608-93-5				oral: high (1)	cumulative (1)	
pentachlorophenol (see pentachlorophenol)						
phosgene gas (see carbon tetrachloride)						

linuron
 	
Gemini (with chlorimuron); Hoe 2810; Lextra (with trifluralin); Linurex; Lorox; Sarclex
 	
3-(3,4-dichlorophenyl)-1-methoxy-1-methylurea; N-(3,4-dichlorophenyl)-N'-methoxy-N'-methy lurea
 	
CAS # 330-55-2
	urea	herbicide	mod-pers (1)	oral: medium (1)
				
				dermal: ?
				
				inhalation: ?
					suspect carcinogen (2)	immediate toxicity:
						  fish: medium (1)
						  bees: low (3)
						
						long-term toxicity:
						  plants: mutagen (4)
						
						water: slightly soluble
						
						slightly volatile
						
						"nonflammable"

(continued on next page)

NAME: Common Trade and Other Chemical CAS Number	Class of Chemical	Chief Pesticide Use; Status	Persistence	Effects on Mammals		Adverse effects on other non-target species Physical properties
				Immediate Toxicity (Acute)	Long-Term Toxicity (Chronic)	
transformation product(s): **TCAB** TCAB 3,4,3',4'-tetrachloroazobenzene structure is analogous to TCDD (see under 2,4,5-T) CAS # 14047-09-7		also a component of diuron			suspect mutagen (1,2) chloracne & hyperkeratosis (3)	
magnesium phosphide Detiaphos Pellets; Magtoxin Round Tablets; Phostoxin Plates magnesium phosphide CAS # 12057-74-8	metal/ mineral magne- sium	fumigant restricted use, USA	?	oral: ? dermal: ? inhalation: ?	?	water: "sparingly soluble" "spontaneously flammable in oil"
transformation product(s): **phosphine gas** (see aluminum phosphide)						
malathion AC 4049; Carbofos; Cython; Cyuthion; EmmatosAC 4049; For-Mal; Fyfanon; Kop-Thion; Kypfos; Malagram; Malakill; Malamar; Malaphos; Malatal; Malathiozoo; Malaude; Malmed; mercaptothion (South Africa); MLT; Zithiol O,O-dimethyl S-(1,2-dicarbethoxyethyl) dithiophosphate; O,O-dimethyl dithiophosphate of diethyl mercaptosuccinate CAS # 121-75-5	organo- phosphate	insecticide	non-pers (1)	oral: medium to high (2) dermal: medium to high (2) inhalation: medium (2)	suspect mutagen (3) suspect teratogen (4) delayed neurotoxin (5) allergic reactions (6) behavior effects (5) ulcers, gastrointestinal inflammation (7) damage to eyesight (8) abnormal brain waves (9) immunosup- pression (10)	BIOCIDE immediate toxicity: birds: medium to high (11) fish: medium to high (12) bees: very high; nectar of treated plants toxic (13) amphibians: very high (12) crustaceans: medium to very high (12) aquatic worms: medium (13) earthworms: high (12) aquatic insects: very high (12) water: soluble oil: "limited solubility in petroleum oils" slightly volatile to volatile flammable
ester(s): **diethyl fumarate** 2-butenedioc acid, diethyl ester CAS # 623-91-6				oral: medium (1) synergistic with malathion		
transformation product(s): **malaoxon** dimethoxyphosphinylthiodiethyl ester of butandioic acid CAS # 1634-78-2			non-pers (1)		carcinogen (2) ulcers (3)	
O,O,S-trimethyl phosphorothioate				oral: high (1)	delayed toxicity (1)	

NAME: Common Trade and Other Chemical CAS Number	Class of Chemical	Chief Pesticide Use; Status	Persistence	Effects on Mammals		Adverse effects on other non-target species Physical properties
				Immediate Toxicity (Acute)	Long-Term Toxicity (Chronic)	
mancozeb Acarie; Blecar MN; Crittox MZ; Delsene MX 200 (with carbendazim); Dithane M-45; Fore; FT 2M (with Bordeaux mixture); Fubol (with metalaxyl); Furado; Galben (with benalaxyl); Galben M (with benalaxyl); Mancobleu (with copper oxychloride); Manzate 200; manzeb; Manzin; Mycodifol MZ (with folpet); Rhodex (with fosetyl-al); Tedane Extra (with dicofol & dinocap); Turbair Dicamate (with zineb); Vondozeb zinc ion & manganese ethylene bisdithiocarbamate CAS # 8018-01-7	thiocar-bamate	fungicide cancelled, most products, USA	non-pers to mod-pers (1,2)	oral: low (1) dermal: low to medium (3) inhalation: ?	?	immediate toxicity: birds: low to medium (2) bees: low (4) fish: high to very high (2) long-term toxicity: plants: inhibits germination of pollen in some plants (5) water: slightly soluble combustible
transformation product(s): **ethylene thiourea** (see amobam)						
maneb Bolda (with sulfur & carbendazim); Chem-Neb; Clortosip (with chlorothalonil & copper oxychloride); CR-3029; Delsene M (with carbendazim); Dithane M-22; Kypman 80; Lonocol M; M-Diphar; Maneba; manebe (France); Manebgan; Manebza; Manesan; Manex; Manzate; Manzati; Manzin; MEB; MnEBD; Remasan; Rimidine Plus (with carbendazim & fenarimol); Ronilan M (with vinclozolin); Sopranebe; Stannophus; Sup'R Flo; Tersan LSR; Tersane LSR; Trimangol; Tubothane; Turf Fungicide; Unicrop; Vancide; Zyban manganese ethylenebisdithiocarbamate CAS # 12427-38-2	thiocar-bamate	fungicide cancelled, most products, USA	mod-pers (1,2)	oral: low to medium (3,4) dermal: ? inhalation: low to medium (5)	suspect teratogen (6,7)	immediate toxicity: bees: low to medium (8) water: soluble
transformation product(s): **carbon disulfide** (see carbon disulfide)						
ethylene thiourea (see amobam)						
MBT Captax; Dermacid; Mertax; Niocides; Sulfadene; Thiotax; Vancide 51 (with sodium dimethyldithiocarbamate) 2-mercaptobenzothiazole; 2-benzothiazolethiol CAS # 149-30-4	miscel-laneous	fungicide	?	oral: medium (1) dermal: ? inhalation: ?	liver damage (2)	water: "insoluble"
MCPA Agroxone; Bordermaster; Extra; Hormotuho; Kilsem; Legumex; Lontrel Plus; Mephanac; metaxon; Methoxane; Rhomene; Rhonox; Shamrox; Springcorn Extra (with dicamba); Springcorn Plus (with dichlorprop); Tetralex-Plus (with dicamba & mecoprop); Tetroxone M (with dichlorprop & bromoxynil & ioxynil); Weed-Rhap LV-4D; Zelan (4-chloro-2-methylphenoxy)acetic acid; [(4-chloro-o-tolyl-oxy]acetic acid CAS # 94-74-6	phenoxy	herbicide	non-pers to mod pers (1,2)	oral: medium (1,2) dermal: ? inhalation: ?	suspect teratogen (3)	immediate toxicity: birds: low (2) fish: low (2) fresh water invertebrates: low (2) bees: medium (4) water: soluble

NAME: Common Trade and Other Chemical CAS Number	Class of Chemical	Chief Pesticide Use; Status	Persistence	Effects on Mammals		Adverse effects on other non-target species Physical properties
				Immediate Toxicity (Acute)	Long-Term Toxicity (Chronic)	
mecoprop Brittox (with bromoxynil & ioxynil); Chipco Turf Fungicide MCPP; CMPP; Compitox; Cornox-Plus; Fettel (with dicamba & triclopyr); Herrifex DS; Hymec; Iso-Cornox; Kilprop; Liranox; MCPP; Mecomec; Mecoper; Mecopex; Mepro; Propenex-Plus; Ronstar (with 2,4-D); Trimec (with dicamba & 2,4-D); Vipex 2-(2-methyl-4-chlorophenoxy)propionic acid; 2-(4-chloro-2-methylphenoxy)propionic acid CAS # 7085-19-0	phenoxy	herbicide	non-pers (1)	oral: medium (2) dermal: high (1) inhalation: low to medium (3)	mutagen (2) teratogen (2)	immediate toxicity: birds: medium (2) fish: low (2) bees: low to medium (4) water: very soluble "nonflammable"
mercury inorganic forms transpose readily into organic forms, in the body or in the environment, organic forms have high toxicity. CAS # 7439-97-6	metal/mineral mercury	fungicide cancelled, most products, USA	perm	oral: very high (1) dermal: ? inhalation: ? organic forms more toxic than inorganic forms	cumulative (2) brain, kidney, heart colon and lung damage (2) neurotoxin (3) decreased hormone levels (4) acrodynia syndrome (5)	BIOCIDE immediate toxicity: birds: high to very high (1) fish: medium to very high (1) amphibians: very high (1) molluscs: very high (1) crustaceans: very high (1) long-term toxicity: birds: immunotoxin (6) bioaccumulates (1) water: "insoluble" volatile
compound(s):						
mercuric chloride Bical (with mercurous chloride); Calgo-gran (with mercurous chloride); corrosive sublimate; Fungchex (wtih mercurous chloride); mercury chloride; Merfusan; Mersil (with mercurous chloride); Wood Ridge Corrosive Sublimate; Wood Ridge Mixture 21 (with mercurous chloride) mercury (II) chloride CAS # 7487-94-7	inorganic	fungicide		oral: very high (1,2)	carcinogen (3) mutagen (4,5) teratogen (6,7) fetotoxin (7) neurotoxin (8,9) kidney damage (10,11) autoimmune disease (12,13)	immediate toxicity: birds: high (14,15) water: soluble slightly volatile
mercurous chloride Bical (with mercuric chloride); Calgo-gran (with mercuric chloride); Calo-clor (with mercuric chloride); Calogreen; Calomel; Chlorure; Cyclosan; Fungchex (with mercuric chloride); Mersil (with mercuric chloride) mercurous chloride; mercury monochloride subchloride of mercury CAS # 7546-30-7	inorganic	insecticide, fungicide			neurotoxin (1)	long-term toxicity: plants: "phytotoxic" (2) water: slightly soluble

NAME:	Common Trade and Other Chemical CAS Number	Class of Chemical	Chief Pesticide Use; Status	Persistence	Effects on Mammals		Adverse effects on other non-target species
					Immediate Toxicity (Acute)	Long-Term Toxicity (Chronic)	Physical properties
metalaxyl Apron; Fubol (with mancozeb); Proturf; Ridomil; Ridomil Plus (with copper oxychloride); Subdue N-(2,6-dimethylphenyl)-N-(methoxyacetyl)-alanine methyl ester; N-(2-methoxyacetyl)-N -(2,6-xylyl)-DL-alaninate CAS # 57837-19-1			fungicide	mod-pers (1)	oral: medium (2) dermal: low to medium (3) inhalation: ?	?	immediate toxicity: fish: low (2) water: soluble
metaldehyde Antimilace; Ariotox; Bug-geta; Cekumeta; Corry's Slug Death; Halizam; Helarion; Meta; metason; Mifaslug; Namekil; Slug Pellets; Snarol Meal metacetaldehyde; r-2,c-4,c-6, c-8-tetramethyl-1,3,5,7-tetroxocane; polymer of acetaldehyde CAS # 108-62-3	aldehyde	molluscicide restricted USA, 1974	non-pers (1)	oral: medium (2) dermal: ? inhalation: ?	spinal damage leading to paralysis in hindquarters (3)	water: soluble flammable
metam-sodium A7-Vapam; Carbam; Karbation; Maposol; Soil-Prep sodium N-methyldithiocarbamate CAS # 137-42-8	thiocarbamate	fungicide herbicide insecticide nematocide	non-pers (1)	oral: medium to high (2) dermal: high (2) inhalation: ? "may be fatal if swallowed, inhaled, or absorbed through skin" (3) "inhalation must be prevented" (3)	teratogen (4) fetotoxin (5)	immediate toxicity: birds: medium (3) fish: "toxic" (3) bees: "nontoxic" (3) water: very soluble "non-volatile" "nonflammable"
transformation product(s): **carbon disulfide** (see carbon disulfide)							
methyl isocyanate (see bendiocarb)							
methidathion GS 13005; Supracide; Utlracide O,O-diemthylS-[2-methoxy-1,3,4-thiadiazol-5(4H-onyl-(4)-methyl]phosphorodithioate; O,O-dimethyl phosphorodithioate, S-ester with 4-(mercaptomethyl)-2-methoxy Δ²-1,3,4-thiodiazolin-5-one CAS # 950-37-8	organophosphate	insecticide acaricide	non-pers (1)	oral: high to very high (2) dermal: high (3) inhalation: ?	suspect mutagen (4,5)	immediate toxicity: birds: high to very high (1) fish: very high (2) bees: very high (6) water: slightly soluble slightly volatile
methiocarb BAY 37344; BAY H-321; Draza; Grandslam; mercaptodimethur; Mesurol; metmercapturon; Ortho Slug-Geta; Slug-M 4-(methylthio)-3,5-xylyl methylcarbamate; 3,5-dimethyl-4-(methylthio)phenol methylcarbmate CAS # 2032-65-7	carbamate	acaricide insecticide molluscicide bird repellent	non-pers (1)	oral: high to very high (2) dermal: low to high (2) inhalation: low to high (3)	?	immediate toxicity: birds: high to very high (1) fish: medium to high (3) aquatic invertebrates: very high (4) bees: very high (5) water: slightly soluble highly volatile
transformation product(s): **N-hydroxymethyl methiocarb sulfoxide**					oral: low to high (1)		
methiocarb sulfoxide					oral: very high (1)		

NAME: Common Trade and Other Chemical CAS Number	Class of Chemical	Chief Pesticide Use; Status	Persistence	Effects on Mammals		Adverse effects on other non-target species Physical properties
				Immediate Toxicity (Acute)	Long-Term Toxicity (Chronic)	
methomyl S-methyl N-[(methylcarbamoyl)oxy] thioacetimidate CAS # 16752-77-5	carbamate	insecticide nematocide some formulations restricted, USA	non-pers to mod-pers (1)	oral: high to very high (2) dermal: medium (3) inhalation: high (4)	mutagen (5) anemia, respiratory damage, hypersensitivity (6) blood damage (6) liver, kidney, spleen, & bone marrow damage (7)	immediate toxicity: birds: very high (8) fish: medium to high (9) crustaceans: high (9) bees: high (10) aquatic insects: very high (9) water: very soluble slightly volatile
transformation product(s): **acetonitrile**				oral: medium (1)	teratogen (2)	flammable
methoprene Altosand; Altosid; Altosid Briquet; Apex 5E; Diacon; Dianex; Insect & Mite Houseplant Mist; Kabat; Pharoid; Precor; Precor Residual Fogger with Adulticide isopropyl (2E,4E)-11-methoxy- 3,7,11-trimethyl-2,4-dodecadienoate; (E,E-1-methylethyl-11-methoxy-3,7,11- trimethyl-2,4-dodecadienoate; isopropyl (E,E-11-methoxy-3,7,11- trimethyl-2,4-dodecadienoate CAS # 40596-69-8	miscel- laneous	insect growth regulator	non-pers (1)	oral: low (2) dermal: low to medium (3) inhalation: low (3)	?	immediate toxicity: birds: low to medium (4) fish: low to high (5) crustaceans: "very highly toxic" (6) amphibians: very high (7) water: slightly soluble slightly volatile
methoxychlor dianisyltrichloroethane; Dimethoxy-DT; DMDT; Maralate; Marlate; Methoxide; Methoxo; Methoxy-DDT; Moxie; Smo-Cloud Bug Killer 2,2-bis (p-methoxyphenyl)-1,1,1- trichloroethane; 1,1,1-trichloro-2,2-bis (p-methoxyphenyl)ethane CAS # 72-43-5	organo- chlorine	insecticide	pers (1)	oral: medium (2) dermal: medium (3) inhalation: ?	cumulative (3) suspect carcinogen (3,4) fetotoxin (3,5) kidney & liver damage (6) interferes with development of male reproductive system (7,8) reproduction reduced in original & next generation (9) estrogen-like effects (9)	immediate toxicity: birds: low to medium (10) fish: very high (11) amphibians: high (12) crustaceans: very high (11) bees: medium (13) aquatic insects: very high (11) long-term toxicity: fish & molluscs: cumulative (12) water: insoluble
methyl bromide Bedfume; Brom-O-Gas; Brozone; Celfume; Dowfume; Embafume; Fumigant-1; Iscobrome; Kayafume; MeBr; Meth-O-Gas; Pestmaster; Profume; Rotox; Terr-O-Gas (with chloropicrin); Weed Fume bromoethane; monobromomethane CAS # 74-83-9	miscel- laneous	fumigant	non-pers (1,2)	oral: high (3) dermal: ? inhalation: ?	mutagen (4) neurotoxin (5,6) liver & kidney damage (7) brain damage (8)	water: soluble highly volatile nonflammable
methyl isothiocyanate Trapex; Xindex isothiocyanatomethane; methyl mustard oil CAS # 556-61-6		fumigant nematicide	?	oral: high (1,2) dermal: very high (2) inhalation: ?	?	water: "slightly soluble" "flammable"

NAME: Common Trade and Other Chemical CAS Number	Class of Chemical	Chief Pesticide Use; Status	Persistence	Effects on Mammals		Adverse effects on other non-target species Physical properties
				Immediate Toxicity (Acute)	Long-Term Toxicity (Chronic)	
methyl parathion Bladen Extra (with parathion); Cekumethion; Defithion; Folimat Combi (with omethoate); Fostox Metil; Gearphos; Kelthane; Kethane Mixte (with dicofol); metafos; Metaphos; Methyl-bladen; Mixte (with dicofol); Neutrion (with tetradifon); Paralindex (with lindane); Parasoufre Acaricide (with dicofol + sulfur); parathion methyl; Partron M; Sylan Methyl (with endosulfan); Taxylone (with phosalone); Tuver Acaricide (with dicofol & ethion); Verfor; Veromite; Viticarb; Wofatox O,O-dimethyl O-p-nitrophenyl phosphorothioate CAS # 298-00-0	organo-phosphate	insecticide some uses restricted, USA	mod-pers to pers (1,2)	oral: very high (3) dermal: high to very high (3) inhalation: very high (3)	mutagen (4,5) fetotoxin (6) retinal & sciatic nerve damage (3) reduced protein synthesis in fetus (7) immunotoxin (8)	BIOCIDE immediate toxicity: birds: very high (9) fish: medium (3) bees: very high (10) crustaceans: very high (3) long-term toxicity: birds: changes breeding behavior, may reduce reproductive capacity (3,11) fish: reduction in sex hormone, may affect reproduction (12); inhibits feeding behavior (13) plants: chromosome damage (14) water: slightly soluble oil: "slightly [soluble] in petroleum oils" slightly volatile
transformation product(s): **para-nitrophenol** p-nitrophenol CAS # 100-02-7				oral: medium to high (1) dermal: high (2)		water: "moderately soluble"
methylene chloride dichloromethane CAS # 75-09-2	organo-chlorine	fumigant solvent in aerosol pesticides	mod-pers (1)	oral: medium (2) dermal: ? inhalation: low (2)	carcinogen (4,5) mutagen (4) brain damage (6) liver and kidney damage (7) may cause loss of memory, disturbed sleep, hallucinations, & changes in heart beat (8,9) decreased ability to learn (3) alters learning ability (3)	immediate toxicity: fish: low (10) crustaceans: low (10) long-term toxicity: plants: mutagen (5) water: very soluble highly volatile "nonflammable"
transformation product(s): **phosgene gas** (see carbon tetrachloride)						
carbon monoxide				can potentiate cardiovascular stress in diseased heart (1,2)	brain damage (3)	
metiram Carbatene; Ethisul; FMC 9102; NIA 9102; Polyram; Polyram-Combi; Thioneb; Trioneb; Zinc Metiram; zineb-ethylene thiuram disulfide adduct mixture of (ethylenebis(dithiocarbamate)) zinc with ethylenebis(dithiocarbamic acid), bimolecular and trimolecular cyclic anhydrosulfides and disulfides CAS # 9006-42-2	thiocar-bamate	fungicide	?	oral: low (1) dermal: low to medium (2) inhalation: low to medium (1)	suspect mutagen (3)	immediate toxicity: fish: medium (1) bees: low (4) water: "practically insoluble" non-volatile

(continued on next page)

NAME:	Common Trade and Other Chemical CAS Number	Class of Chemical	Chief Pesticide Use; Status	Persistence	Effects on Mammals		Adverse effects on other non-target species Physical properties
					Immediate Toxicity (Acute)	Long-Term Toxicity (Chronic)	
transformation product(s): **ethylene thiourea** (see amobam)							
metolachlor Bicep (with atrazine); Codal (with prometryn); Cotodon (with dipropetryn); Cotoran multi (with fluometuron); Dual; Milocep (with propazine); Pennant; Primagram & Primextra (with atrazine); Primagram (with atrazine); Primextra (with atrazine); Turbo (with metribuzine) 2-chloro-6'-ethyl-N-(2-methoxy-1-methylethyl)acet-o-toluidide CAS # 51218-45-2	amide	herbicide	non-pers (1)	oral: medium (2) dermal: low to medium (2) inhalation: low to high (2)	suspect carcinogen (4) testicular atrophy (4)	immediate toxicity: birds: low to medium (3) fish: medium (2) water: soluble slightly volatile
transformation product(s): **nitrosamines** (see dinoseb)							
metribuzin BAY 6159H; BAY 94337; Canopy (with chlorimuron); DIC-1468; Lexone Sencor; Salute (with trifluralin); Sencoral (France); Sencorer; Sencorex (Gr. Britain); Turbo (with metolachlor) 4-amino-6-tert-butyl-3-(methylthio)-as-triazin-5(4H)-one CAS # 21087-64-9	triazine	herbicide	non-pers to mod-pers (1)	oral: medium (1) dermal: low (2) inhalation: ?	liver & kidney damage (3)	immediate toxicity: birds: high (4) fish: low (4) bees: low to medium (5) water: soluble slightly volatile "nonflammable"
mevinphos Apavinfos; Duraphos; Menite; Mevinox; OS-2046; Phosdrin; Phosfene methyl 3-(dimethoxyphosphinoloxy)but-2-enoate; 2-methoxycarbonyl-1-methylvinyl dimethyl phosphate CAS # 26718-65-0	organo-phosphate	insecticide acaricide restricted use, USA	non-pers (1)	oral: high to very high (2,3) dermal: very high (2) inhalation: medium (2)	eye damage (4)	immediate toxicity: birds: very high (5) fish: very high (6) crustaceans: very high (6) bees: very high (7) aquatic insects: medium (6) water: "soluble" oil: slightly soluble slightly volatile flammable
mexacarbate Dowco 139; Zectran 4-(dimethylamino)-3,5-xylyl methylcarbamate CAS # 315-17-4	carbamate	insecticide acaricide molluscicide	non-pers (1)	oral: very high (2) dermal: high (1) inhalation: ?	?	immediate toxicity: birds: very high (3) fish: medium (4) amphibians: medium (3) crustaceans: very high (4) aquatic insects: very high (4) bees: very high (5) water: soluble oil: soluble slightly volatile
MGK R11 R 11 2,3,4,5-bis(2-butylene) tetrahydro-2-furaldehyde CAS # 126-15-8	aldehyde	insect repellent voluntary cancellation, USA 1990	?	oral: medium (1) dermal: ? inhalation: ?	suspect carcinogen (2) teratogen (2) ovarian atrophy (2)	immediate toxicity: fish: low to medium (1) water: "practically insoluble"

NAME: Common Trade and Other Chemical CAS Number	Class of Chemical	Chief Pesticide Use; Status	Persistence	Effects on Mammals		Adverse effects on other non-target species Physical properties
				Immediate Toxicity (Acute)	Long-Term Toxicity (Chronic)	
milky spore disease Doom; Grub Attack; Japidemic; Japonex; Milky Spore Powder; milky white disease *Bacillus popillae* or *Bacillus lentimorbus* (former most frequently used)	biological	insecticide	"pers" (1)	oral: low (2) dermal: low (2) inhalation: low (2)	?	immediate toxicity: toxic only to Japanese beetle grubs (larvae) and to other soil-feeding beetle grubs (2) "nonflammable"
mirex Dechlorane; ferriamide; GC 1293 dodecachlorooctahydro-1,3,4-metheno-2*H*-cyclobuta[c,d]pentalene CAS # 2385-85-5	organo-chlorine	insecticide cancelled USA, 1976	pers (1)	oral: medium to high (2,3) dermal: high (3) inhalation: ?	cumulative (4,5) carcinogen (11) heart damage (6)	immediate toxicity: birds: low to medium (7) fish: low (8) bees: high (9) long-term toxicity: bioaccumulates (10) water: "insoluble" oil: "lipophilic"
transformation product(s): **chlordecone** (see chlordecone)						
monocrotophos Azodrin; Bilobran; C 1414; Crisodrin; Monocron; Nuvacron; Plantdrin; SD or Shell SD 9129; Susvin; Ulvair dimethyl phosphate of 3-hydroxy-*N*-methyl-*cis*-crotonamide CAS # 919-44-8	organo-phosphate	insecticide cancelled USA, 1988	non-pers (1)	oral: very high (2) dermal: high (1) inhalation: very high (2)	mutagen (3,4)	immediate toxicity: birds: very high (5) fish: low to medium (2) bees: very high (6) long-term toxicity: crustaceans: reproductive damage (7) water: very soluble oil: "slightly soluble" slightly volatile
monolinuron Afesin; Aresin; Arresin; Gramonol (with paraquat dichloride); Hoe 2727; Premalin 3-(*p*-chlorophenyl)-1-methoxy-1-methylurea; 3-(4-chlorophenyl)-1-methyoxy-1-methylurea CAS # 1746-18-2	urea	herbicide	non-pers (1)	oral: medium (1) dermal: ? inhalation: ?	?	immediate toxicity: birds: low to medium (1) fish: low to medium (1) slightly volatile combustible
monuron Chlorfenidim; CMU; Karmex Monuron Herbicide; Monurex; Monurox; Monuruon; Rosuran; Telvar; Telvar Monuron Weedkiller 3-(*p*-chlorophenyl)-1,1-dimethylurea CAS # 150-68-5	urea	herbicide cancelled, most uses, USA, 1977	mod-pers (1)	oral: medium (2) dermal: ? inhalation: ?	carcinogen (3) liver, kidney and spleen damage (4)	immediate toxicity: fish: low to medium (5) bees: low (6) phytoplankton: very high (5) long-term toxicity: inhibits soil nitrification causing accumulation of nitrite in soil (4) water: soluble oil: soluble slightly soluble "nonflammable"
salt(s): **monuron TCA** Telvar; Urox 3-(4-chlorophenyl)-1,dimethyluronium trichloroacetate (salt of monuron) CAS # 140-41-0						

NAME: Common Trade and Other Chemical CAS Number	Class of Chemical	Chief Pesticide Use; Status	Persistence	Effects on Mammals		Adverse effects on other non-target species Physical properties
				Immediate Toxicity (Acute)	Long-Term Toxicity (Chronic)	
nabam Chem Bam; Dithane-40 or D-14; DSE; Kaybam; nabame (France); Parzate; Spring-Bak disodium ethylene-1,2-bisdithiocarbamate CAS # 142-59-6	thiocarbamate	fungicide cancelled, most products, USA	non-pers (1)	oral: high (2,3) dermal: ? inhalation: high (4)	?	immediate toxicity: birds: medium (5) amphibians: high (5) long-term toxicity: amphibians: teratogen (6) water: soluble
transformation product(s): **carbon disulfide** (see carbon disulfide)						
ethylene thiourea (see amobam)						
naled bromchlophos (So. Africa); Bromex (with chlorbromuron); Dibrom; Hibrom; Ortho 4355; RE 4355 1,2-dibromo-2,2-dichloroethyl dimethylphosphate CAS # 300-76-5	organophosphate	insecticide acaricide	?	oral: high (1) dermal: high (2) inhalation: ?	suspect mutagen (3)	immediate toxicity: birds: high to very high (4) fish: medium to very high (5) amphibians: medium (6) crustaceans: very high (5) bees: very high (7) aquatic insects: very high (5) water: "practically insoluble" slightly volatile
transformation product(s): **dichlorvos** (see dichlorvos)						
naphthalene camphor tar; moth balls; moth flakes; naphthalin; tar camphor; white tar naphthalene the main constituent of coal tar CAS # 91-20-3	phenol	insecticide fumigant repellent	non-pers (1)	oral: medium (2) dermal: low to medium (3) inhalation: "toxic" (4) dermal exposure most dangerous to newborns (4)	cataracts (3) corneal damage (4) transported across the placenta blood damage (1) liver, kidney damage (1)	immediate toxicity: crustaceans: medium (5) long-term toxicity: fish: cumulative (6) crustaceans: decreased oxygen uptake (5) water: slightly soluble oil: "very soluble" volatile combustible
transformation product(s): **1-naphthol** (see carbaryl)						
naphthaleneacetic acid Fruitone-N; Keriroot (with captan); NAA-800; Phymone; Planofix; Rhizopon B; Tipoff; Tre-Hold Sprout Inhibitor; Tre-Hold Wound Dressing 2-(1-naphthyl)acetic acid CAS # 86-87-3	miscellaneous	plant growth regulator	?	oral: medium (1) dermal: medium (2) inhalation: ?	liver damage (3)	immediate toxicity: birds: low to medium (3) water: soluble oil: "slightly soluble"
ester(s): **ethyl ester**				oral: medium (1) dermal: low to medium (1)		

NAME: Common Trade and Other Chemical CAS Number	Class of Chemical	Chief Pesticide Use; Status	Persistence	Effects on Mammals		Adverse effects on other non-target species Physical properties
				Immediate Toxicity (Acute)	Long-Term Toxicity (Chronic)	
naptalam Alanap-L; Dyanap (with dinoseb); Rescue (with dinoseb) N-1-naphtylphthalamic acid (I) CAS # 132-66-1	miscel-laneous	herbicide	non-pers (1)	oral: low (1) dermal: ? inhalation: low to medium (2)	eye damage (3)	immediate toxicity: bees: low (4) long-term toxicity: plants: mutagen (5) water: soluble
salt(s):						
sodium salt				oral: medium (1)		water: very soluble
niclosamide Aten; Atenase; Bayluscid; Bayluscide; Bayluscit; Cestacide; Clonitralid; Clonitralide; Copharten; Fenasal; Grandal; Helmiantin; Iometan; Kontal; Lintex; Manosil; Mato; Molluscicide Bayer 73; Molutox; Yomesan 2',5-dichloro-4'-nitrosalicylanilide 2-aminoethanol salt; 2',5-dichloro-4'-nitrosalicylanilide, ethanolamine salt; 5-chloro-N-(2-chloro-4-nitrophenyl)-2-hydroxybenazmide, 2-aminoethanol salt CAS # 1420-04-8	amide	molluscicide	non-pers (1)	oral: low to medium (1) dermal: ? inhalation: low (1)	suspect mutagen (2) sperm damage (3)	water: soluble oil: "lipophilic" slightly volatile
nicotine Black Leaf 40 (nicotine sulfate); Destruxol Orchid Spray; Emo-Nik; Fumetobac; Mach-Nic; Niagara P.A. Dust; Nic-Dust; Nic-Sal; Nico-Fume; Nicocide; Ortho N-4 & N-5 Dusts; Tendust 1,3-(1-methyl-2-pyrrolidinyl) pyridine CAS # 54-11-5	botanical	insecticide	?	oral: high (1) dermal: high (2) inhalation: ?	?	immediate toxicity: birds: medium (3) long-term toxicity: birds: teratogen (4) water: "soluble" highly volatile
transformation product(s): **N-nitrosonornicotine**					carcinogen (1)	
nitrapyrin N-Serve; nitrapyrine 2-chloro-6-trichloromethylpyridine CAS # 1929-82-4	miscel-laneous	antibiotic	non-pers (1)	oral: medium (2) dermal: medium (2) inhalation: ?	fetotoxin (3)	immediate toxicity: birds: high (2) fish: medium (2) long-term toxicity: fish: cumulative (1) water: slightly soluble slightly volatile
nitrofen FW-925; niclofen; NIP; nitrofene (France); nitrophen; TOK; TOK E-25; TOK-2 2,4-dichlorophenyl p-nitrophenyl ether; 2,4-dichloro-1-(4-nitrophenoxy)benzene CAS # 1836-75-5	miscel-laneous	herbicide not made or sold in USA	non-pers to mod-pers (1)	oral: medium (2) dermal: medium (3) inhalation: low (3)	cumulative (4) carcinogen (5) suspect mutagen (5) teratogen (6) liver damage (3) kidney damage (7) blood damage (4)	water: slightly soluble slightly volatile

NAME: Common Trade and Other Chemical CAS Number	Class of Chemical	Chief Pesticide Use; Status	Persistence	Effects on Mammals		Adverse effects on other non-target species Physical properties
				Immediate Toxicity (Acute)	Long-Term Toxicity (Chronic)	

norflurazon
 	
Evital; Solicam; Telok (with simazine); Zorial
 	
4-chloro-5-methylamino-2-(α,α,α-trifluro-m-tolyl)pyridazin-3(2H)-one
 	
CAS # 27314-13-2
	miscel-laneous	herbicide	non-pers to mod-pers (1)	oral: low (2)
				
				dermal: low (1)
				
				inhalation: ?
					liver damage (3) thyroid damage (3)	immediate toxicity:
						  birds: low to medium (2)
						  fish: low (2)
						
						water: slightly soluble
						
						non-volatile
						
						"nonflammable"

nuclear polyhedrosis virus Biotrol VHZ; Elcar; Heliothis NPV, (Elcar); Lymantria dispar NVP (Gypcheck); Orgyia pseudotsugata NPV (TM Biocontrol-1) CAS # 240194-80-1	biological	insecticide		"no adverse effects were observed in any acute oral, dermal, inhalation, and intravenous test" (1)		immediate toxicity: "NPV poses a minimal to non-existent risk to nontarget wildlife" (1)

O-benzyl-p-chlorophenol
 	
Clorophene; Santophen 1; Septiphene
 	
4-chloro-2-(phenylmethyl)phenol;
4-chloro-α-phenyl-o-cresol;
  2-benzen-4-chlorophenol;
  5-chloro-2-hydroxydiphenylmethane;
o-benzyl-p-chlorophenol
 	
CAS # 120-32-1
	phenol	antibiotic	?	oral: medium (1)
				
				dermal: ?
				
				inhalation: ?
					?	?

omethoate
 	
BAY 45432; Baythroid F (with cyfluthrin); dimethoate-met; dimethoxon; Folimat Combi (with methyl parathion); Folimat T (with tetradifon)
 	
dimethyl S-(N-methylcarbamoylmethyl) phosphorothioate;
O,O-dimethyl S-[2-(methylamine)-2-oxoethyl]phosphorothioate
 	
CAS # 1113-02-6
	organo-phosphate	insecticide	non-pers (1)	oral: high (1)
				
				dermal: high (2)
				
				inhalation: high (1)
					suspect mutagen (3)	immediate toxicity:
						  fish: low to medium (1)
						  bees: "toxic" (2)
						
						water: very soluble
						
						oil: "almost insoluble"
						
						non-volatile

oryzalin
 	
Dirimal; Dirimal Extra (with diuron); EL-119; Rycelan; Ryzelan; Snapshot 80 (with isoxaben); Surflan
 	
3,5-dinitro-N⁴N⁴-dipropylsulfaniliamide;
4-(dipropylamino)-3,5-dinitro-benzenesulfonamide
 	
CAS # 19044-88-3
	dinitro-aniline	herbicide	mod-pers (1)	oral: low to medium (2,3)
				
				dermal: low to medium (2)
				
				inhalation: low to high (2)
					suspect carcinogen (4)	immediate toxicity:
						  birds: low to medium (2)
						  fish: medium (3)
						
						water: slightly soluble
						
						non-volatile
						
						"flammable"

oxadiazon
 	
Ronstar; RP 17623
 	
2-tert-butyl-4-(2,4-dichloro-5-isopropoxyphenyl)-Δ² 1,3,4-oxadiazoline-5-one;
3-[2,4-dichloro-5-(1-methylethoxy)phenyl]-5-(1,1-dimethylethyl-1,3,4-oxadizol-2(3H)-one
 	
CAS # 19666-30-9
	miscel-laneous	herbicide	mod-pers (1)	oral: low to medium (2,3)
				
				dermal: medium (3)
				
				inhalation: low to medium (3,4)
					suspect carcinogen (4,5)	immediate toxicity:
						  birds: low to medium (3)
						  fish: low to medium (3)
						  bees: low to medium (6)
						  crustaceans: medium to high (7)
						
						water: insoluble
						
						slightly volatile
						combustible

NAME: Common Trade and Other Chemical CAS Number	Class of Chemical	Chief Pesticide Use; Status	Persistence	Effects on Mammals		Adverse effects on other non-target species Physical properties
				Immediate Toxicity (Acute)	Long-Term Toxicity (Chronic)	
oxamyl Vydate-G; Vydate-L N,N-dimethyl-2-methylcaaramoyloxyimino-2-(methylthio) acetamide; S-methyl N',N'-dimethyl-N-(methylcarbamoyloxy)-1-thio-oxamimidate (I) CAS # 23135-22-0	carbamate	insecticide some uses restricted, USA	non-pers (1)	oral: very high (1) dermal: medium (2) inhalation: very high (1)	eye damage (3)	immediate toxicity: birds: high to very high (1) fish: medium (1) crustaceans: medium to high (4) bees: medium (5) water: very soluble slightly volatile
oxydemeton-methyl Croneton MR; Dipterex MR; Ecombi (with parathion); Metasystox R S-2-ethylsulfinylethyl O,O-dimethylphosphorothiote CAS # 301-12-2	organo-phosphate	insecticide	non-pers (1)	oral: very high (2) dermal: very high (3) inhalation: high (2)	mutagen (4) suspect teratogen (5,6) sperm damage (6)	immediate toxicity: birds: high to very high (7) fish: low to medium (8) crustaceans: high (8) bees: high (9) aquatic insects: high (10) water: miscible oil: sparingly soluble slightly volatile
oxyfluorfen Goal; Koltar; RH-2915 2-chloro-α,α,α-trifluro-p-tolyl 3-ethoxy-4-nitrophenyl ether CAS # 42874-03-3 contaminant(s): **perchloroethylene** CAS # 127-84-4	miscel-laneous	herbicide cancelled, most products, USA 1982	non-pers (1)	oral: low (1) dermal: low to medium (2) inhalation: ?	suspect mutagen (4) blood, kidney, liver, and thyroid damage (4) carcinogen (1) suspect fetotoxin (1)	immediate toxicity: birds: low to medium (3) fish: high (3) water: insoluble slightly volatile
oxytetracycline hydrochloride Biosolomycin; Hydrocyclin; Liquamycin; Otetryn; Oxlopar; terramicin; terramitsin; Terramycin Hydrochloride 2-Naphthacene carboxamide, 4(dimethylamino)-1,4,4a,5,5a,6, 11,12a-octahydroxy-6-methyl-1, 11-dioxo-monohydrochloride CAS # 2058-46-0	biological	antibiotic	?	oral: low (1) dermal: ? inhalation: ?	carcinogen (2)	water: soluble oil: "insoluble"
paradichlorobenzene Di-chloride; Para Crystals; Para Nuggets; Paracide; Paradow; Paramoth 1,4-dichlorobenzene; p-dichlorobenzene CAS # 106-46-7	organo-chlorine	fumigant	?	oral: medium (1) dermal: ? inhalation: ?	suspect carcinogen (2) suspect mutagen (3) liver & kidney damage (2,3) lung damage (2) anemia (2)	immediate toxicity: fish: medium (4) long-term toxicity: plants: mutagen (3) water: slightly soluble oil: "lipophilic" highly volatile
paraformaldehyde Formagene; Paraform polyoxymethylene; polymerized formaldehyde CAS # 30525-89-4 transformation product(s): **formaldehyde** (see formaldehyde)	aldehyde	fungicide	?	oral: low to medium (1) dermal: low to medium (1) inhalation: medium (1)	?	?

NAME: Common Trade and Other Chemical CAS Number	Class of Chemical	Chief Pesticide Use; Status	Persistence	Effects on Mammals		Adverse effects on other non-target species Physical properties
				Immediate Toxicity (Acute)	Long-Term Toxicity (Chronic)	
paraquat Actor (with diquat); Cyclone (with diquat); Dexuron (with diuron); Farmon PDQ (with diquat); Gramazine (with simazine); Gramixel (with diuron); Gramuron (with diuron); Groundhog (with amitrole & diquat & simazine); methyl viologen; Pardi-Weedol (with diquat); Pathclear (with amitrole & diquat & simazine); Preglone (with diquat); Reglox (with diquat); Soltair (with diquat & simazine); Spraygrow (with diquat); Sprayseed (with diquat); Spraytop (with diquat); Surefire (with diuron); Talent (with asulam); Totacol (with diuron) 1,1'-dimethyl-4,4'-bipyridinium CAS # 4685-14-7	bypiridyl	herbicide voluntary withdrawal from market, Norway, 1981 banned in Sweden, 1983 Banned in Netherlands, 1989	non-pers (1,2)	oral: high to very high (3) dermal: very high (3) inhalation: high (3)	suspect mutagen (4) suspect teratogen (5) suspect neurotoxin (6) fingernail loss (7) liver, kidney, pancreas, gastrointestinal tract, adrenal, nerve, brain, heart, muscle and eye damage (6,8) irreversible lung injury within hours of ingestion (7) implicated in Parkinson's disease (9,10)	immediate toxicity: birds: high fish: low to medium (2,11) amphibians: low (2) crustaceans: low to medium (2) aquatic insects: low (11) microorganisms: toxic to some (12) long-term toxicity: birds: embryotoxin (13); growth rate, adverse effects; blood damage (14,15) amphibians: clinical syndrome of Parkinson's disease (9) blue-green algae: mutagen (16) water: "soluble"
related compound(s): emetic (formulated into technical paraquat in case of ingestion)					heart damage (1)	
transformation product(s): formaldehyde (see formaldehyde)						
paraquat dichloride Dextrone (with diquat); Galgo-quat; Gramonol (with monolinuron); Gramoxone (with MCPA); Gramoxone Special; Liro-paraquat; Longlife Plus; Ortho paraquat; Prelude; Protex; R-Bix; Radex; Scythe; Sipquat; Speedway; Sweep; Terraklene (with simazine); Violan 1,1'-dimethyl-4,4'-bypiridinium dichloride CAS # 1910-42-5	bypiridyl	herbicide dessicant defoliant plant growth regulator	?	oral: high to very high (1,2) dermal: high (3) inhalation: very high (4) many long-term symptoms may begin immediately from single or minimal exposure (see long-term toxicity)	suspect carcinogen (4) suspect mutagen (4) brain damage (5)	immediate toxicity: birds: medium to high (6) fish: low to medium (1) crustaceans: medium to high (4) long-term toxicity: birds: egg hatchability, adverse effects (3) amphibians: teratogen (7) water: "soluble" "non-volatile"
transformation product(s): QINA						
parathion Alkron; Alleron; Bladen Extra (with methyl parathion); Corothion; Folidol E-65; Genthion; Geofos; Lethalaire; Malatox (with malathion); Niran; Orthophos; Paradusto; Paraspra; Penncap-E; Sopragram (with lindane); Stathion; Tamaron (with acephate-met); thiophos (USSR); Vapophos O,O-diethyl O-p-nitrophenyl phosphorothioate CAS # 56-38-2	organo-phosphate	insecticide restricted use: USA uses withdrawn 1991, USA	mod-pers (1)	oral: very high (2) dermal: very high (2) inhalation: very high (2)	suspect carcinogen (2) suspect mutagen (2) suspect teratogen (3) retinal degeneration (2) neurotoxin (2) limited immunotoxin (4)	BIOCIDE immediate toxicity: birds: very high (5) fish: medium to very high (6) bees: very high (8) amphibians: high (7) aquatic worms: very high (9) crustaceans: very high (6) aquatic insects: very high (6)

NAME: Common Trade and Other Chemical CAS Number	Class of Chemical	Chief Pesticide Use; Status	Persistence	Effects on Mammals		Adverse effects on other non-target species Physical properties
				Immediate Toxicity (Acute)	Long-Term Toxicity (Chronic)	
parathion (*cont.*)						long-term toxicity: birds: disrupts incubation behavior (10); reduction in sex hormones, possibly impairing reproduction (11); embryotoxin (12) fish: reduced sex hormone and ovarian activity (13) bioconcentrates in fish, molluscs and amphibians (14) water: slightly soluble slightly volatile
transformation product(s): **para-nitrophenol** p-nitrophenol CAS # 100-02-7				oral: medium to high (1) dermal: high (2)		water: "moderately soluble"
paraoxon diethyl p-nitrophenol ester of phosphoric acid CAS # 311-45-5				oral: very high (1) dermal: very high (1) inhalation: ?		water: very soluble
PCBs Arochlor; Chlophen; Chlorextrol; Clophen; Dykanol; Fenchlor; Inerteen; Kanechlor; Noflamol; Phenochlor; Pyralene; Pyranol; Sentotherm; Therminol polychlorinated biphenyls chlorinated biphenyls chlorinated diphenyls The total number of PCB's is over 200 half of which are commonly found among the commericial preparations. CAS # 1336-36-3	organo-chlorine	formerly used to reduce vapor pressure and prolong residual activity of pesticides cancelled, all products, USA, 1970	pers (1)	oral: low to medium (2)	cumulative (3) carcinogen (4) immunotoxin (5) liver damage (2) neurotoxin (6) decreased homoglobin (1) chloracne (2) reduced fertility (7)	BIOCIDE immediate toxicity: fish: low to high (8) crustaceans: medium to very high (8) bees: low to medium (9) long-term toxicity: birds: behavioral defecits; reduced egg shell thickness. (10,11) biomagnification (1) water: insoluble oil: soluble slightly volatile to volatile "nonflammable"
contaminant(s): **pentachlorodibenzofuran**	dibenzo-furan					
chlorinated naphthalenes					chloracne (1) liver damage (1) hyperkeratosis (1)	
tetrachlorodibenzofuran	dibenzo-furan					

NAME: Common Trade and Other Chemical CAS Number	Class of Chemical	Chief Pesticide Use; Status	Persistence	Effects on Mammals		Adverse effects on other non-target species Physical properties
				Immediate Toxicity (Acute)	Long-Term Toxicity (Chronic)	
PCNB Avicol; Botrilex; Brassicol; Earthcide; Folosan; Kobu; Kobutol; Pentagen; PKhNB (USSR); quintozene; Terrachlor; terraclor (Turkey); Tilcarex; Tri-PCNB; Tritisan pentachloronitrobenzene CAS # 82-68-8	organo-chlorine	fungicide cancelled, most products, USA, 1982	mod-pers (1)	oral: medium (1) dermal: medium to high (2) inhalation: ?	suspect mutagen (4)	immediate toxicity: birds: medium (5) fish: very high (1) water: insoluble volatile
contaminant(s): **pentachlorobenzene** CAS # 608-93-5				oral: high (1)	cumulative (1)	
hexachlorobenzene (see hexachlorobenzene)						
transformation product(s): **pentachloroaniline** PCA					cumulative (1)	
pentachlorophenylmethylsulfide PCMS					cumulative (1)	
pendimethalin AC 92553; Accotab; Go-Go-San; Herbadox; Horbadox; Prowl; Sipaxol; Squadron (with imazaquin); Stomp; Way Up 2,6-dinitro-3,4-xylidine CAS # 40487-42-1	dinitro-aniline	herbicide	non-pers to mod-pers (1)	oral: medium (2,3) dermal: low to medium (3) inhalation: low (3)	?	immediate toxicity: fish: high (4) birds: low (2) crustaceans: high (4) water: insoluble oil: very soluble slightly volatile flammable
transformation product(s): **N-nitrosopendimethalin** N-(1-ethylpropyl)-N-nitroso-3,4-dimethyl-2,6-dinitrobenzamine			mod-pers (1)	oral: ? dermal: ? inhalation: ?		
pentachlorophenol Acutox; Chem-Penta; Chemtrol; chlorophen; Cryptogil ol; Dow Pentachlorophenol DP-2 Antimicrobial; Dowicide 7; Durotox; EP 30; Fungifen; Fungol; Grundier Arbezol; Lauxtol A; Moosuran; Ontrack WE-1; PCP; Penta Plus 40; Santobri; Term-i-Trol; Weed-Beads 2,3,4,5,6-pentachlorophenol CAS # 87-86-5	organo-chlorine	insecticide fungicide herbicide defoliant wood preservative molluscicide cancelled, USA, 1984	non-pers to pers (1,2)	oral: high to very high (3) dermal: high to very high (4) inhalation: ? toxic by all portals of entry (5)	carcinogen (6) teratogen (7) fetotoxin (7) embryotoxin (8) immunotoxin (1,9) porphyria (1) aplastic anemia (1) chloracne (1) nerve & liver damage (1) most chronic toxic effects associated with PCP are due to the presence of contaminants (see below) (1)	BIOCIDE immediate toxicity: birds: medium to high (11) fish: medium to very high (3) amphibians: high to very high (12) crustaceans: medium to very high (11,13) bees: low to medium (14) aquatic insects: high to very high (11) aquatic worms: high to very high (11) technical grade PCP significantly more toxic than pure PCP to aquatic organisms (11) long-term toxicity: biomagnification (1) birds: suspect immunotoxin (15,16); fish: inhibited growth

NAME: Common Trade and Other Chemical CAS Number	Class of Chemical	Chief Pesticide Use; Status	Persistence	Effects on Mammals		Adverse effects on other non-target species Physical properties
				Immediate Toxicity (Acute)	Long-Term Toxicity (Chronic)	
pentachlorophenol (*cont.*)						(12); increases vulnerability to predation (17); delayed egg hatchability (18) water: slightly soluble volatile "flammable"
contaminant(s):						
2,4,5-trichlorophenol (see 2,4,5-trichlorophenol)						
hexachlorobenzene (see hexachlorobenzene)						
heptachlorodibenzo-p-dioxin HCDD; HpCDD heptachlorodibenzo-*p*-dioxin CAS # 35822-46-9	dibenzo-dioxin				suspect immunotoxin (1)	
hexachlorodibenzofuran	dibenzo-furan		mod-pers (1)	oral: very high (1)	suspect carcinogen (2) teratogen (3) thymus, liver, and spleen damage (2)	
octachlorodibenzo-p-dioxin OCDD octachlorodibenzo-*p*-dioxin CAS # 3268-87-9	dibenzo-dioxin		non-pers to mod-pers (1)		cumulative (1) acne (2)	water: insoluble non-volatile
hexachlorodibenzo-p-dioxin HCDD; HxCDD hexachlorodibenzo-*p*-dioxin CAS # 34465-46-8	dibenzo-dioxin			oral: very high (1,2)	carcinogen (3) suspect teratogen (2,3) suspect fetotoxin (2,3) immunotoxin (4) acnegenic (2) thymus, liver, kidney, & spleen damage (2)	water: insoluble
tetrachlorophenol CAS # 25167-83-3	phenol	fungicide		oral: high (1)		
TCDD (see chloroneb)						
transformation product(s):						
tetrachlorocatechol tetrachloropyrocatechol; tetrachloro-1,2-benzenediol; 3,4,5,6-tetrachloro-1,2-benzenediol CAS # 1198-55-6				oral: high	suspect mutagen	
tetrachlorohydroquinone CAS # 87-87-6				oral: high (1)	suspect mutagen (2)	water: slightly soluble
tetrachlorophenol CAS # 25167-83-3	phenol	fungicide		oral: high (1)		

NAME: Common Trade and Other Chemical CAS Number	Class of Chemical	Chief Pesticide Use; Status	Persistence	Effects on Mammals		Adverse effects on other non-target species / Physical properties
				Immediate Toxicity (Acute)	Long-Term Toxicity (Chronic)	
perfluidone Destun; MBR-825; perfluoridone 1,1,1-trifluoro-N-[2-methyl-4-(phenylsulfonyl)phenyl]methane sulfonamide CAS # 37924-13-3	amide	herbicide	non-pers to mod-pers (1,2)	oral: medium (3) dermal: low to medium (4) inhalation: medium (5)	liver damage (6)	immediate toxicity: fish: low (3) water: slightly soluble non-volatile to slightly volatile "nonflammable"
permethrin Ambush; Atroban; Bio Flydown; Corsair; Dragon; Ectiban; Expar; Gard-Star; Hard-Hitter; Insectiban; Jureong; Kafil; Nix; Over-Time; Permectrin; Pounce; Quamlin; Rondo; Stockade; Tornade; Torpedo 3-phenoxybenzyl (1RS)-cis,trans-3-(2,2-dichlorovinyl)-2,2-dimethyl-cyclopropanecarboxylate CAS # 52645-53-1	pyrethroid	insecticide acaricide	non-pers (1)	oral: low to high (2) dermal: ? inhalation: low to high (3)	blood damage (4)	immediate toxicity: fish: very high (5) birds: "practically non-toxic" (2) marine invertebrates: very high (6) bees: "toxic" (7) water: insoluble to slightly slightly soluble non-volatile combustible
petroleum oils arranged from most volatile to least volatile-each product may have a range of components and volatility compound(s):	oils					
fuel oil diesel oil						long-term toxicity: birds: embryotoxin & teratogen when are exposed to eggs (1)
kerosene CAS # 8008-20-6				oral: high (1) inhalation: low (1)	suspect mutagen (2) lung damage(3) neurotoxin (2)	
mineral oil					carcinogen (1,2)	
mineral spirits (overlaps naphtha) CAS # 8032-32-4				"jeopardize survival of test animals"		immediate toxicity: flash point varies within range
naphtha petroleum naphtha CAS # 64741-64-6 8030-30-6 8002-05-9		solvent		oral: medium (6)	carcinogen (1) suspect mutagen (2) anemia (3) liver damage (3) heart, lung damage, from more volatile reactions (5)	immediate toxicity: fish: medium (4) crustaceans: medium (4) water: "insoluble" flammable
stoddard solvents CAS # 8052-41-3				oral: low (1)		
phenmedipham Betaflow; Betamix (with desmedipham); Betanal; Betanal E; Betanal Perfekt (with ethofumesate); Betanal Plus; Betanal Progress (with ethofumesate & desmedipham); Betanal Tandem (with ethofumesate); Betaron (with ethofumesate); Campaign (with clopyralid); Goliath; Gusto; Pistol-400; Protrum K; Spin-aid; Vangard	carbamate	herbicide	non-pers to mod-pers (1)	oral: "low" (1) dermal: medium (1) inhalation: ?	?	immediate toxicity: birds: "very low" (1) fish: medium to high (1) bees: "very low" (1) water: slightly soluble

NAME: Common Trade and Other Chemical CAS Number	Class of Chemical	Chief Pesticide Use; Status	Persistence	Effects on Mammals		Adverse effects on other non-target species Physical properties
				Immediate Toxicity (Acute)	Long-Term Toxicity (Chronic)	
phenmedipham (*cont.*) methyl 3-(3-methylcarbaniloyloxy)carbanilate; 3-methoxycarbonylaminophenyl 3'-methylcarbanilate CAS # 13684-63-4						
phenol carbolic acid hydroxybenzene; monohydroxybenzene; phenic acid; phenyl alcohol; phenyl hydroxide; phenylphenol CAS # 108-95-2; CAS # 90-43-7	phenol	fungicide antibiotic	?	oral: medium (1) dermal: ? inhalation: ?	suspect carcinogen (2)	immediate toxicity: fish: medium (3) water: very soluble oil: "almost insoluble" highly volatile combustible
phenothrin d-phenothrin; fenothrin; Multicide Concentrate F-2271; Neo-Pynamin 5/1/30 (with allethrin & piperonyl butoxide); Neo-Pynamin Forte Aerosol (with tetramethrin); OMS 1810; Pesguard (with allethrin & piperonyl butoxide); Pesguard FS (with tetramethrin); Pesguard Insect Killer (with tetramethrin); Pesguard NX; Pesguard Plant Spray (with tetramethrin); phenothrine; S-2539; Sumithrin; Sumithrin A Plus (with tetramethrin); Sumithrin B Plus (with tetramethrin); Sumithrin Plus (with allethrin) 3-phenoxybenzyl (±)-*cis,trans*-chrysanthemate CAS # 26002-80-2	pyrethroid	insecticide	"more stable to photolysis than other pyrethoids" (1)	oral: low (2) dermal: ? inhalation: ?	?	immediate toxicity: birds: "very low" (3) fish: very high (3) crustaceans: low (3) water: slightly soluble slightly volatile
phorate AAstar (with flucythrinate); Corliss (with terbufos); Cygard (with terbufos); Geomet; Granutox; Thimet O,O-diethyl S-ethylthiomethyl phosphorodithiote CAS # 298-02-2	organo-phosphate	insecticide restricted use, USA	non-pers (1)	oral: very high (2) dermal: very high (2) inhalation: very high (1)	mutagen (3)	immediate toxicity: birds: very high (4) fish: very high (5) amphibians: high (4) crustaceans: very high (5) bees: medium (6) aquatic insects: very high (5) water: slightly soluble volatile
phosalone Azofene; Benzphos; Fozalon; NIA-9241; NPH-1091; phosalon; Ranbeck (with dichlorvos); Ransbeck (with dichlorvos); RP 11974; Rubitox; Taxylone (with methyl parathion); Zolone; Zolone Flo; Zolone Liquid S-6-chloro-2,3-dihydro-2-oxo-1,3-benzoxazol-3-ylmethyl O,O-diethyl phosphorodithioate CAS # 2310-17-0	organo-phosphate	insecticide acaricide	non-pers (1)	oral: high (2) dermal: high (2) inhalation: ?	?	immediate toxicity: birds: high (3) fish: high to very high (3) crustaceans: very high (4) bees: high (5) oil: "insoluble" non-volatile
phosmet Imidan; Prolate N-(mercaptomethyl)phthalimide-S-(O,O-dimethylphosphorodithioate) CAS # 732-11-6	organo-phosphate	acaricide insecticide	non-pers (1)	oral: high to very high (1) dermal: medium to high (1) inhalation: very high (1)	suspect carcinogen (1) suspect teratogen (2) suspect fetotoxin (2)	immediate toxicity: birds: medium to high (1) fish: very high (1) bees: very high (1) crustaceans: very high (1) aquatic insects: very high (1) water: slightly soluble volatile

NAME: Common Trade and Other Chemical CAS Number	Class of Chemical	Chief Pesticide Use; Status	Persistence	Effects on Mammals		Adverse effects on other non-target species Physical properties
				Immediate Toxicity (Acute)	Long-Term Toxicity (Chronic)	
phosphamidon Apamidon; Dimecron 2-chloro-2-diethylcarbamoyl-1-methylvinyl dimethyl phosphate; 2-chloro-3-dimethoxyphosphinoyloxy-N,N-diethylbut-2-enamide CAS # 13171-21-6	organo-phosphate	insecticide, acaricide restricted use, USA	?	oral: very high (1) dermal: very high (1) inhalation: high (2)	mutagen (3) embryotoxin (4) testicular damage (5) liver damage (6)	BIOCIDE immediate toxicity: birds: very high (7) fish: medium to high (8) crustaceans: very high (8) aquatic insects: very high (8) long-term toxicity: birds: "teratogenic" (9) water: very soluble slightly volatile
phoxim Baython; Muscatox (with cyfluthrin); Oftanol Combi (with isofenphos); Omnicur (with carbofuran); Volaton diethoxyphosphinothioyloxyimino (phenyl)acetonitrile; 2-(diethoxyphosphinothioloxyimino)-2-phenylacetonitrile CAS # 14816-18-3	organo-phosphate	insecticide	?	oral: high (1) dermal: high (1) inhalation: ?	suspect mutagen (2)	immediate toxicity: birds: high to very high (3) fish: medium to high (4) bees: very high (5) water: slightly soluble
transformation product(s): **TEPP** (see TEPP)						
picloram Amdon; Borolin; Debroussaillant 4323 DP (with dichlorprop); Hydon (with bromacil); K Pin; piclorame; Printazol N (with 2,4-D & MCPA); Printazol Total (with 2,4-D & MCPA & mecoprop); Spica 66; Tordon; Tordon 101 Mixture (with 2,4-D); Tordon 10K & 22K; Tordon RTU; Torgal 4-amino-3,5,6-trichloropyridine-2-carboxylic acid CAS # 1918-02-1	miscel-laneous	herbicide restricted use, USA	non-pers to mod-pers (1,2)	oral: low to medium (1) dermal: medium (3) inhalation: high (2)	suspect carcinogen (4) teratogen (5) liver damage (6) kidney damage (4) testicular atrophy (4)	immediate toxicity: birds: low to medium (7) fish: low to high (8) crustaceans: very high (8) aquatic insects: low to very high (8) bees: low to medium (9) water: soluble oil: slightly soluble slightly volatile flammable
contaminant(s): **nitrosamines**					suspect carcinogen (1)	
hexachlorobenzene (see hexachlorobenzene)						
piperonyl butoxide Butacide; Derringer (with resmethrin); Detrans (with esbiothrin); Duracide 15 (with tetramethrin); FMC 5273; K-O (with deltamethrin & esbiothrin); K-Obiol (with deltamethrin); Kefil (with permethrin & bioallethrin); Neo-Pynamin 5/1/30 (with phenothrin & allethrin); Pesguard (with allethrin & phenothrin); Pesguard NSB (with tetramethrin & fenitrothion); Pyrenone; Vapona Flykiller (with permethrin & bioallethrin) (butylcarbityl) (6-propylpiperonyl) ether; α-[2-(2-butoxyethyoxy)ethoxy] 4,5-(methylenedioxy)-2-propyltoluene derived from safrole via dihydrosafrole CAS # 51-03-6	methylene dioxy phenyl	synergist with insecticides	?	oral: low (1) dermal: high (2) inhalation: ?	can reduce ability of body to detoxify other toxins (neurotoxins, carcinogens, etc.) (3,4) kidney, liver, & adrenal damage (5) anorexia (5)	immediate toxicity: fish: very high (6) amphibians: high (7) crustaceans: very high (6) water: "slightly soluble" oil: "soluble" slightly volatile combustible

NAME: Common Trade and Other Chemical CAS Number	Class of Chemical	Chief Pesticide Use; Status	Persistence	Effects on Mammals		Adverse effects on other non-target species Physical properties
				Immediate Toxicity (Acute)	Long-Term Toxicity (Chronic)	
pirimiphos-ethyl Fernex; Primicid (with drazoxolon) O-2-diethylamino-6-methylpyrimidin-4-yl O,O-diethyl phosphorothiote CAS # 23505-41-1	organo-phosphate	insecticide, acaricide	non pers to mod pers (1)	oral: high to very high (1) dermal: high (1) inhalation: very high (2)	suspect mutagen (3)	immediate toxicity: birds: very high (2) fish: very high (2) bees: very high (4) long-term toxicity: birds: teratogen (5) water: insoluble volatile
pirimiphos-methyl Actellic; Actellifog; Attack (with permethrin); Blex; Cyperallic (with cypermethrin); Giustiziere; Pirigrain; PP-511; Silo-San; Silosan; Singsing (with cypermethrin); Sybol 2 O-[2-(diethylamino)-6-methyl-4-pyrimidinyl] O,O-dimethylphosphorothioate; 2-dimethylamino-6-methylpyrimidin-4-yl dimethyl phosphorothioate CAS # 29232-93-7	organo-phosphate	insecticide, acaricide	non-pers (1)	oral: medium (2) dermal: medium to high (3) inhalation: ?	mutagen (4)	immediate toxicity: birds: high to very high (5) fish: "toxic" (5) water: slightly soluble volatile
potassium azide Kazoe 10G potassium azide CAS # 20762-60-1	azide	herbicide fungicide nematocide insecticide soil fumigant	non-pers (1)	oral: high to very high (1) dermal: very high (1) inhalation: "avoid breathing dust or vapor" (1)	?	immediate toxicity: birds: very high (1) fish: high (2) aquatic insects: medium (2) crustaceans: medium (2) long-term toxicity: plants & bacteria: mutagen (1) water: very soluble
transformation product(s): **hydrazoic acid**						"highly volatile" (1)
potassium bromide potassium bromide CAS # 7758-02-3	miscel-laneous	antibiotic algacide	?	oral: low (1) dermal: low (1) inhalation: ?	?	immediate toxicity: birds: "practically non-toxic" fish: "highly toxic" (1) crustaceans: "highly toxic" (1)
potassium permanganate potassium permanganate CAS # 7722-64-7	miscel-laneous	antibiotic, algacide, fungicide	?	oral: medium (1) dermal: "corrosive to eyes and skin" inhalation: ?	?	immediate toxicity: fish: medium (3) water: soluble
procymidone Sialex; Sumiboto; Sumilex; Sumisclex 3-(3,5-dichlorophenyl)-1,5-dimethyl-3-azabicyclo[3.1.0]hexane-2,4-dione; N-(3,5-dichlorophenyl)-1,2-dimethyl-1,2-cyclopropane dicarboximide CAS # 32809-16-8	amide	fungicide	non-pers (1)	oral: low (2) dermal: ? inhalation: ?	carcinogen (3)	water: slightly soluble slightly volatile
prometon Atratol (with atrazine); Conquer Liquid Vegetation Killer; G-31435; Gesafram;	triazine	herbicide	mod-pers to pers (1)	oral: medium (2) dermal: medium to high (3)	?	immediate toxicity: birds: "very low" (4) fish: low to medium (5) bees: "nontoxic" (6)

(continued on next page)

NAME: Common Trade and Other Chemical CAS Number	Class of Chemical	Chief Pesticide Use; Status	Persistence	Effects on Mammals		Adverse effects on other non-target species Physical properties
				Immediate Toxicity (Acute)	Long-Term Toxicity (Chronic)	
prometon (*cont.*) Gesafram 50; Ontracic 800; Ontrack; Parch (with pentachlorophenol); Pramitol; Pramitol 25E; prometone; Triox Vegetation Killer (with pentachlorophenol) 2,4-bis(isopropylamino)-6-methoxy-triazine; 6-methoxy-*N,N*-bis(1-methylethyl)-1,3,5-triazine-2,4-diamine CAS # 1610-18-0				inhalation: medium (4)		long-term toxicity: birds: suspect embryotoxin (7) water: soluble slightly volatile
prometryn Codal (with metolachlor); Gesagard; Gesagard 500; Peaweed (with terbutryn); prometryne *N²,N⁴*- di-isopropyl-6-methylthio-1,3, 5-triazine-2,4-diamine CAS # 7287-19-6	triazine	herbicide	mod-pers (1)	oral: medium (1) dermal: medium (2) inhalation: medium to high (2)	kidney, liver, and bone marrow damage (3) testicular damage (3)	immediate toxicity: birds: low (3) fish: medium (3) crustaceans: medium (3) molluscs: decreased shell growth (5) bees: low to medium (6) water: slightly soluble "nonflammable"
pronamide Clanex; Kerb; Kerb Mix Matrikerb (with clopyralid); Kerb-50-W; propyzamide; RH-315; Siden (with simazine) *N*-(1,1-dimethyl-2-propynyl) 3,5-dichlorobenzamide; 3,5-dichloro-*N*-(1,1-dimethyl-2-propynyl) benzamide CAS # 23950-58-5	amide	herbicide cancelled, most uses, USA, 1979	mod-pers (1)	oral: low (1) dermal: low (1) inhalation: low to medium (2)	suspect carcinogen (3) liver damage (4)	immediate toxicity: birds: low (5) fish: low (5) bees: low (6) water: slightly soluble oil: ? slightly volatile "nonflammable"
transformation product(s): **N-nitrosamide**						
propachlor Bexton; CP 31393; Decimate (with DCPA); propachlore (France); Ramrod 2-chloro-*N*-isopropylacetanilide; vinyl 2-chloroethyl CAS # 1918-16-7	amide	herbicide	mod-pers (1)	oral: medium (1) dermal: low (1) inhalation: ?	cataracts (2)	immediate toxicity: birds: high (1) fish: very high (1) crustaceans: medium (1) water: soluble volatile "nonflammable"
component(s): **aniline**				oral: high (1)	skin photosen-sitization (2)	combustible
propanil Trio (with bromoxynil & 2,4-D) *N*-(3,4-dichlorophenyl) propanamide CAS # 709-98-8	amide	herbicide restricted use, USA	non-pers (1)	oral: medium (2) dermal: medium (1) "should not be breathed or allowed to get in eyes or on skin" (1)	methemo-globinemia (2) hemolytic anemia (3)	immediate toxicity: birds: high (4) fish: medium (5) crustaceans: medium (4) long-term toxicity: birds: suspect teratogen (6) bees: low to medium (7)
components: **tetrachloroazoxy benzene**						water: soluble slightly volatile

NAME: Common Trade and Other Chemical CAS Number	Class of Chemical	Chief Pesticide Use; Status	Persistence	Effects on Mammals		Adverse effects on other non-target species Physical properties
				Immediate Toxicity (Acute)	Long-Term Toxicity (Chronic)	
3,3,4,4 tetrachloroazobenzene					chloracne and hyperkeratosis (1) liver damage (2)	
propargite Comite; Omite; Omite TD (with tertadifon); Uniroyal DO14 2-(p-tert-butylphenoxy)cyclohexyl 2-propynyl sulfite CAS # 2312-35-8	metal/ mineral sulfur	acaricide	mod-pers (1)	oral: medium (1) dermal: "severe eye and skin irritant" inhalation: ?	fetotoxin (1) eye damage (1)	immediate toxicity: birds: low (1) fish: "highly toxic" (1) crustaceans: very high (2) bees: low to medium (3) water: insoluble
propazine Gesamil; Milocep (with metalochlor); Milogard 6-chloro-N^2,N^4-di-isopropyl-1,3,5-triazine-2,4-diamine CAS # 139-40-2	triazine	herbicide	mod-pers (1)	oral: low (1) dermal: ? inhalation: ?	carcinogen (1) suspect mutagen (1) fetotoxin (1)	immediate toxicity: birds: low (1) fish: medium (1) bees: low (2) water: slightly soluble non-volatile
propham Gold (with chlorpropham & fenuron); Morlex (with fenuron & chlorpropham & ethofumesate); Pink C (with chlorpropham & diuron); Premalox (with fenuron & chlorpropham); Quintex (with fenuron & chlorpropham); Tuberite (with chlorpropham) isopropyl carbanilate; isopropyl phenylcarbamate CAS # 122-42-9	carbamate	herbicide	non-pers (1)	oral: low to medium (2) dermal: medium (3) inhalation: low (2)	suspect mutagen (4) fetotoxin (1)	immediate toxicity: fish: low (5) crustaceans: medium (5) bees: low (6) water: slightly soluble to soluble volatile "nonflammable"
propionic acid ethylformic aicd; Luprosil; methylacetic acid; propanoic acid; Propionic Acid Grain Preserver; Sentry Grain Preserver CAS # 79-09-4	miscel-laneous	fungicide	?	oral: medium (1) dermal: high (2) inhalation: ?	?	water: "soluble" flammable
propoxur aprocarb; arprocarb; BAY 39007; Baygon Spray (with dichlorvos & cyfluthrin); Blattanex Residual Spray (with dichlorvos); Boygon; Brygou; Chemagro 9010; Isocarb; o-IMPC; PHC; Raid Ant & Roach Killer; Raid Wasp & Hornet Killer; Sendran; Suncide; Tat Ant Trap; Unden o-isopropoxyphenyl-N-methyl carbamate; 2-(1-methylethoxy)phenol methyl carbamate CAS # 114-26-1	carbamate	insecticide	mod-pers (1)	oral: very high (2) dermal: medium to high (2) inhalation: ?	carcinogen (3,4) suspect mutagen (5) learning disability (6)	BIOCIDE immediate toxicity: birds: very high (7) fish: high (8) bees: high (11) amphibians: medium (9) crustaceans: very high (9) aquatic insects: very high (9) aquatic worms: very high (10) long-term toxicity: toxic to some plants (12) water: soluble slightly volatile
transformation product(s):						
n-nitroso propoxur					mutagen (1)	long-term toxicity: plants: mutagen (2)

NAME: Common Trade and Other Chemical CAS Number	Class of Chemical	Chief Pesticide Use; Status	Persistence	Effects on Mammals		Adverse effects on other non-target species Physical properties
				Immediate Toxicity (Acute)	Long-Term Toxicity (Chronic)	
pyrethrum "Insect powder" *Chrysanthemum cinaeraraefolum;* mixture of pyrethrin I & II, cinerin I & II, jasmolin I & II	botanical	insecticide	non-pers (1)	oral: medium to high (1,2) dermal: ? inhalation: ?	liver damage, especially with synergists and Freon propellant (3) allergic reactions (4) neurotoxin (4)	immediate toxicity: birds: low (5) fish: very high (6) crustaceans: very high (6) water: "not soluble in water" oil: "100% in petroleum distillate" combustible
red squill Bonide; Bonide Topzol Rat Baits & Killing Syrup; Dethdiet; Rat Nots; Rat Snax; Rat's End; Rat-O-Cide Rat Bait; Rat-Pak; Rats Squill; Ratspax; Rodene; Rodine; Rough & Ready Rat Bait & Rat Paste; Squill; Topzol extract of bulbs of *Urginea maritima* (sea onion) containing cardiac glycosides scilliroside, and scillarens A & B (small amounts of the latter two) CAS # 507-60-8	botanical	rodenticide	?	oral: very high (1) dermal: ? inhalation: ?	?	water: "slightly soluble" oil: "practically insoluble"
resmethrin Benzofurolin; Benzyfuroline; Chryson; Chrysron; FMC 17370; For-Synm; Isathrine; NIA 17370; NRDC 104; OMS 1206; Premgard; Pynosect; Pyretherm; resmethrine; Respond; RU 48440; SBP-1382; Scourge; Synthin; Vectrin 5-benzyl-3-furylmethyl[1RS,cis,trans]-2,2-dimethyl-3-(2,2-dimethylvinyl)cyclopropane carboxylate CAS # 10453-86-8	pyrethroid	insecticide	?	oral: low to high (1,2) dermal: medium (1) inhalation: low to medium (1)	suspect neurotoxin (4) suspect immunotoxin (4) decrease in hormone release from the brain (5)	immediate toxicity: birds: low to medium (6) fish: low to high (1,4) crustaceans: low to medium (4) water: insoluble non-volatile combustible
isomer(s): **bioresmethrin** Biobenzyfuroline; NRDC 107; Resbuthrin; RU 11484 5-benzyl-3-furylmethyl[1R,trans]-chrysanthemate; *d-trans*-resmethrin; (±)*trans*-resmethrin CAS # 28434-01-7						immediate toxicity: birds: low (1) water: insoluble slightly volatile
cismethrin 5-benzyl-3-furylmethyl[1R,cis]-chrysanthemate; (±)*cis*-resmethrin CAS # 35764-59-1				oral: high (1)		
ronnel Ectoral; Etrolene; fenchlorphos (Gr. Britain, France, USSR); Korlan; Nankor; Trolene; Viozene *O,O*-dimethyl *O*-(2,4,5-trichlorophenyl) phosphorothioate CAS # 299-84-3	organo-phosphate	insecticide not manufac-tured since 1979	?	oral: low to medium (1,2) dermal: high (2) inhalation: ?	suspect mutagen (6)	immediate toxicity: birds: low to medium (4) fish: high (5) crustaceans: medium (5) bees: high (3) water: slightly soluble oil: "soluble" volatile
contaminant(s): **TCDD** (see chloroneb)						

NAME: Common Trade and Other Chemical CAS Number	Class of Chemical	Chief Pesticide Use; Status	Persistence	Effects on Mammals		Adverse effects on other non-target species Physical properties
				Immediate Toxicity (Acute)	Long-Term Toxicity (Chronic)	
rotenone Cenol Flea Powder; Cenol Garden Dust; Chem-Mite; Cibe Extract; Curex Flea Duster; Derrin; Green Cross Warble Powder; Nicouline; Noxfish; Powder & Root; Rotefive; Rotefour; Rotessenol; Rotocide; Tubatoxin; Warbicide extracts from *Derris* or *Lonchocarpus* (barbasco; cube; haiari; nekos; timbo) plants: the principle active ingredient is 1,2,12,12a-tetrahydro-2-*a*-isopropyl-8,9-dimethoxy(1)benzopyrano(2,4-b)-furo(2,3-h) (1)benzopyrano-6(6aH)-one CAS # 83-79-4	botanical	insecticide acaricide piscicide	non-pers (1)	oral: medium to very high (2,3) dermal: very high (2) inhalation: ?	suspect carcinogen (4,5,8) suspect teratogen (6) suspect fetotoxin (7) liver and kidney damage (2)	immediate toxicity: birds: low to medium (9) crustaceans: medium (10) aquatic insects: medium to high (11) water: slightly soluble oil: slightly soluble
contaminant(s):						
tetrachloroethylene (see tetrachloroethylene)						
xylene (see xylene)						
trichloroethene CAS # 79-01-6					carcinogen (1,2) suspect mutagen (1,3) neurotoxin (4)	long-term toxicity: plants: mutagen (1)
ethylbenzene CAS # 100-41-4					teratogen (1)	
ryania Bonide Ryatox; Ryanex; Ryanexcel; Ryanicide ryanodine, the active insecticidal principle from shrub *Ryania speciosa* CAS # 8047-13-0 (ryania) 15662-33-6 (ryanodine)	botanical	insecticide	?	oral: medium (1) dermal: ? inhalation: ?	?	immediate toxicity: fish: medium (2) bees: low (3) oil: "insoluble"
sabadilla caustic barley; Cavdilla; Sabacide; Sabane Dust; sevadilla; Shirlan; Veratrine from lily plant *Schoenocaulon officinale* (*Asagraea officinalis, Sabadilla officinalum*); active ingredients are of a complex of alkaloids known collectively as verarin; two of these are cevadine & veratridine CAS # 8028-57-7	botanical	insecticide	"lost activity rapidly on exposure to sunlight" (1)	oral: medium (1) dermal: ? inhalation: ?	?	immediate toxicity: bees: medium (2) water: "slightly soluble" oil: "slightly soluble"
safrole AA Outdoor Dog Repellent; Rhyuno oil; Shikimol; Shikimole; yellow camphor oil 4-allyl-1,2-methylene dioxybenzene; oil of camphor sassafrass CAS # 94-59-7	methylene dioxy phenyl	repellent voluntary cancellation USA, 1977	?	oral: medium (1) dermal: ? inhalation: ?	carcinogen (2,3) crosses placenta (3) liver damage (4)	water: "insoluble" "combustible"
transformation product(s):						
dihydrosafrole 1,2-methylenedioxy-4-propylbenzene CAS # 94-58-6	methylene dioxy phenyl	intermediate in the synthesis of piperonyl butoxide		oral: medium (1)	carcinogen (1,2)	

NAME: Common Trade and Other Chemical CAS Number	Class of Chemical	Chief Pesticide Use; Status	Persistence	Effects on Mammals		Adverse effects on other non-target species Physical properties
				Immediate Toxicity (Acute)	Long-Term Toxicity (Chronic)	
siduron Supersan in Trey Triple Action Lawn Aid; Trey; Tupersan 1-(2-methylcyclohexyl)-3-phenylurea CAS # 1982-49-6	urea	herbicide	non-pers to mod-pers (1,2)	oral: low (1) dermal: ? inhalation: ?	?	immediate toxicity: bees: low (3) water: slightly soluble volatile
silica aerogel Cab-O-Sil; Dri-Die; Drianone; Drione (with pyrethrins); Santocel C; SG-67; silica gel; Silikil; Silox; Sprotive Dust SG-67; Warpath (with pyrethrins) amorphous (fumed) silica dust: 3 micron particiles of silicon dioxide; many formulation contain less than 5% ammonium fluosilicate CAS # 7631-86-9	metal/ mineral silica	insecticide dessicant	pers	oral: low (1) dermal: low (1) inhalation: ?	lung damage (2)	immediate toxicity: injures mushrooms (3)

silvex
 	
2,4,5-TP; Aqua-Vex; Ded-Weed; Double Strength; Esteron; Fenoprop; Fruitone T; Garlon; Kuron; Kurosal; Scott's O-X-D; Silvi-Rhap; Weedone 2,4,5-TP; Weedone TP
 	
2-(2,4,5-trichlorophenoxy)propionic acid
 	
CAS # 93-72-1
	phenoxy	herbicide
		
		cancelled USA, 1985
			mod-pers (1)	oral: medium (2)
				
				dermal: ?
				
				inhalation: ?
					teratogen (3) liver and kidney damage (4)	immediate toxicity:
						  birds: low to medium (5)
						  fish: medium to high (1)
						  amphibians: low to medium (1)
						  crustaceans: high (1)
						  bees: low (4)
						
						long-term toxicity:
						  fish: reproductive and liver damage; cumulative (1)
						  molluscs: decrease in shell growth (1)
						  water: soluble
						
						non-volatile to slightly volatile

contaminant(s): **TCDD** (see chloroneb)						

simazine
 	
Amizine (with amitrole); Aquazine; Batazina; CDT; CET; Framed; Gesatop; Herbazin; Herbex; Herbox; Herboxy; Hungazin DT; Premazine; Primatol S; Princep; Printop; Radocon; Simadex; Zeapur
 	
2-chloro-4,6-bis(ethylamino)-s-triazine
 	
CAS # 122-34-9
	triazine	herbicide soil sterilant	mod-pers to pers (1,2)	oral: low to medium (3,4)
				
				dermal: low to medium (4)
				
				inhalation: low to medium (4)
					testes, kidneys, liver & thyroid damage (5) disturbances in sperm production (5)	immediate toxicity:
						  birds: low (6)
						  fish: low to medium (4)
						  crustaceans: low (4)
						  molluscs: low to high (4)
						  bees: low (7)
						  aquatic insects: medium to high (6)
						
						water: slightly soluble
						
						oil: slightly soluble
						
						non-volatile
						
						"nonflammable"

soap
 	
Cryptocidal Soap; De-Moss; Safer's Fungicidal Soap; Safer's Herbicidal Soap; Safer's Insecticidal Soap
 	
potassium salts of selected fatty acids; cocoa fatty acids
	miscel- laneous	miticide insecticide algacide herbicide for moss, lichen & liverwort fungicide	?	oral: low (1)
				
				dermal: ?
				
				inhalation: ?
					?	immediate toxicity:
						  birds: low (1)
						  fish: low (1)
						  bees: "very low" (1)
						
						long-term toxicity:
						  plants: toxic to some in high concentrations (2)
						
						water: "highly soluble"

NAME: Common Trade and Other Chemical CAS Number	Class of Chemical	Chief Pesticide Use; Status	Persistence	Effects on Mammals		Adverse effects on other non-target species Physical properties
				Immediate Toxicity (Acute)	Long-Term Toxicity (Chronic)	
sodium azide Benzide; Noxide; Smite 15G sodium azide CAS # 26628-22-8	azide	herbicide fungicide nematocide insecticide soil fumigant	non-pers (1)	oral: very high (2) dermal: very high (2) Inhalation: "avoid breathing dust of vapor" (1)	suspect mutagen (3)	BIOCIDE immediate toxicity: fish: high (4) crustaceans: medium (4) aquatic insects: medium (4) long-term toxicity: plants & bacteria: mutagen (1) water: very soluble
transformation product(s): **hydrazoic acid**						"highly volatile" (1)
sodium chlorate Altacide Extra (with atrazine); Altavar (with atrazine & 2,4-D); Chlorax; De-Fol-Ate; Drop-Leaf; Fall; Helena "Clean Up" (with bromacil); Klorex; Kusatol; MBC*; Polybor chlorate; Rasikal; Shed-A-Leaf; Tumbleaf; Ureabor; Weed-Killer #50 (with bromacil & sodium metaborate) *also used for carbendazim CAS # 7775-09-9	miscel-laneous	herbicide	?	oral: low (1) dermal: ? inhalation: ?	methemoglo-binemia (2) kidney damage (2)	immediate toxicity: fish: low (3) long-term toxicity: birds: reduction in egg production and fertility (3) water: very soluble "dangerously flammable"
sodium cyanide Cymag; M-44 devices hydrocyanic acid, sodium salt CAS # 143-33-9	cyanide	fumigant insecticide rodenticide cancelled most uses, USA, 1988	?	oral: very high dermal: ? inhalation: ?	?	?
transformation product(s): **hydrogen cyanide** (see hydrogen cyanide)						
sodium fluoroacetate Compound 1080; Fratol; Yasoknock sodium fluoroacetate CAS # 62-74-8	fluoro-acetate	rodenticide suspended, 1972 cancelled, most uses, USA, 1985	?	oral: very high (1,2) dermal: very high (8) inhalation: very high (8) can cause secondary poisoning (3) no antidote (8)	neurotoxin, heart damage (4,5) kidney, liver, nerve, and thyroid damage (6) delayed convulsant (fluoroacetate) (7)	immediate toxicity: birds: very high (2) amphibians: high (2) insects: "toxic" (8) water: "very soluble" oil: "low solubility" "non-volatile"
transformation product(s): **fluorocitrate** (see fluoroacetamide)						
sodium hypochlorite CAS # 7681-52-9	miscel-laneous	antibiotics, algacide, fungicide	?	oral: high (1) dermal: "corrosive" (1)	?	

NAME: Common Trade and Other Chemical CAS Number	Class of Chemical	Chief Pesticide Use;[a] Status	Persistence	Effects on Mammals		Adverse effects on other non-target species Physical properties
				Immediate Toxicity (Acute)	Long-Term Toxicity (Chronic)	
sodium omadine Omadine sodium 1-hydroxy-2(1H)-pyridinethione ,sodium salt (CA) CAS # 15922-78-8	miscel-laneous	antibiotic	?	oral: medium (1) dermal: ? inhalation: ?	suspect fetotoxin (2) hind limb paralysis (2)	immediate toxicity: birds: "slightly toxic" (2) fish: "slightly to very toxic" aquatic invertebrates: "very toxic" (2)
streptomycin O-2-deoxy-2-methylamino-α-L-glucopyranosyl-(1→2)-O-5-deoxy-3-C-formyl-α-L-lyxofuranosyl-(1→4)-N³,N³-diamidino-ᴅ-strepamine CAS # 57-92-1	antibiotic	antibiotic, fungicide		oral: low (1) dermal: ? inhalation: ?	ear damage (2)	water: very soluble
strychnine strychinidin-10-one CAS # 57-24-9	botanical	rodenticide, avicide most uses cancelled, USA by 1988	"stable in the enviornment" (1)	oral: very high (2) dermal: ? inhalation: "do not inhale" (3)		immediate toxicity: birds: very high (1) amphibians: very high (2) long-term toxicity: secondary poisoning water: slightly soluble "nonflammable"
sulfoTEPP ASP-47; Baldafume; BAY-E-393; Bladafum; dithio; dithione; dithioTEPP; Formula 40; Formula GH-200; Lethalaire G-57 Aerosol Insecticide; Sulfatep; sulfotep; sulfotepp; TEDP; thiotepp O,O,O,O-tetraethyldithiopyrophosphate; tetraethyl thiodiphosphate CAS # 3689-24-5	organo-phosphate	insecticide acaricide	?	oral: very high (1) dermal: very high (2) inhalation: very high (1)	?	immediate toxicity: birds: high (3) water: slightly soluble volatile
sulfur Bolda (with maneb & carbendazim); brimstone; Colsul; Corosul D and S; Cosan; flour sulfur; flowers of sulfur; Hexasul; Kolo-100 (with dichlone); Kolofog; Kolospray; Kumulus S; Magnetic 70, 90 and 95; precipitated sulfur; Sofril; Spersul; Sulforon; Sulkol; Thiolux; Thiovit sulfur CAS # 7704-34-9	metal/ mineral sulfur	fungicide acaricide	perm	oral: low (1) dermal: high (1) inhalation: low to medium (1)	?	immediate toxicity: birds: low (1) fish: low (1) estuarine/marine organisms: low (1) aquatic invertebrates: low (1) water: "insoluble"
compound(s):						
sulfuryl fluoride sultropene (France); Vikane CAS # 2699-79-8	metal/ mineral sulfur	fumigant restricted use, USA, all formulations		oral: high (1) inhalation: medium (1)	liver and kidney damage (2) mottled teeth (2) osteosclerosis (3)	immediate toxicity: plants: toxic to some (1) water: soluble highly volatile
sulprofos Bolstar; Helothion O-ethyl O-4-(methylthio)phenyl S-propylphosphorodithioate CAS # 35400-43-2	organo-phosphate	insecticide restricted use, USA	non-pers to mod-pers (1)	oral: high (1) dermal: high (1) inhalation: ?	?	immediate toxicity: birds: high (1) fish: "highly toxic" (2) water: soluble

NAME: Common Trade and Other Chemical CAS Number	Class of Chemical	Chief Pesticide Use; Status	Persistence	Effects on Mammals		Adverse effects on other non-target species Physical properties
				Immediate Toxicity (Acute)	Long-Term Toxicity (Chronic)	
2,4,5-T Bandock (with dicamba & mecoprop); Brushtox; Dacamine; Ded-Weed; Fence Rider; Forron; Fruitone A; Inverton 245; Line Rider; Nettle-Ban (with 2,4-D & dicamba); Reddon; Spontox (with 2,4-D); Tormona; Transamine; Tributon; Trinoxol; Veon 245; Verton 2T; Visko Rhap; Weedar; Weedone 2,4,5-trichlorophenoxyacetic acid CAS # 93-76-5	phenoxy	herbicide cancelled USA, 1985	non-pers to mod-pers (1,2)	oral: medium to high (16) dermal: high (7,8) inhalation: ?	suspect carcinogen (16) teratogen (7,8) fetotoxin (7) liver & kidney damage (9)	immediate toxicity: birds (chicks): high (9) fish: low to high (4) bees: low to medium (10) long-term toxicity: birds: may cause behavioral effects (11) suspect embryotoxin (12) water: "insoluble" esters non-volatile to slightly volatile
contaminant(s): **2,4,5-trichlorophenol** (see 2,4,5-trichlorophenol) transformation product(s): **2,4,5-trichlorophenol** (see 2,4,5-trichlorophenol)						
TCDD (see chloroneb)						
tebuthiuron Brulan; Brush Bullet; Bushwacker; Combine; E-103; Perfmid; Preflan; Prefmid; Reclaim; Scrubmaster; Spike; Tebulan; Tiurolan 1-(5-*tert*-butyl-1,3,4-thiadiazol-2-yl)-1,3-dimethylurea CAS # 34014-18-1	urea	herbicide	mod-pers to pers (1)	oral: medium to high (2) dermal: ? inhalation: ?	?	immediate toxicity: birds: low to medium (3) fish: low (3) water: soluble slightly volatile "nonflammable"
tefluthrin Elancolan K (with napropamide); Forca; Force; Forza; PP 993; tefluthrine 2,3,5,6-tetrafluoro-4-methylbenzyl (Z)-(1RS)-cis-3-(2-chloro-3,3,3-trifluroprop-1-enyl]-2,2-dimethylcyclopropanecarboxylate CAS # 79538-32-2	pyrethroid	insecticide	non-pers to mod-pers (1,2)	oral: very high (1) dermal: high (1) inhalation: high (1)	?	immediate toxicity: birds: low to medium (1,2) fish: very high (1,2) crustaceans: very high (2) molluscs: high (2) water: insoluble slightly volatile combustible
temephos Abat; Abate; Abathion; Abazan (with trichlofon); AC 52160; Biothion; Difenthos; Nimitex; Nimitox; Swebate; temophos O,O,O',O'-tetramethyl O,O'-thiodi-p-phenylene phosphorothioate; O,O'-(thio-4,1-phenylene)bis[O,O-dimethyl phosphorothioate] CAS # 3383-96-8	organo-phosphate	insecticide	non-pers (1)	oral: low to medium (2,3) dermal: medium (4) inhalation: ?	liver damage (3)	immediate toxicity: birds: high to very high (5) fish: low to high (6) amphibians: low to medium (5) crustaceans: very high (7) aquatic insects: very high (6) bees: high (8) water: insoluble
TEPP Fosuex; G-52 and 56; Gy-TET 40; HETP; Killex; Kilmite; Lethalaire; Lico-40; Nifos T; Pyfos; Pyro-Phos; Teep; TEP; Tetradusto 100; Tetraspra; Tetron; Tetron-100; Vaptone hexaethyl tetraphosphate; tetraethyl prophosphate CAS # 107-49-3	organo-phosphate	insecticide acaricide cancelled USA, 1984	?	oral: very high (1) dermal: very high (1) inhalation: very high (1)	?	BIOCIDE immediate toxicity: birds: very high (2) fish: high (2) amphibians: high (2) crustaceans: very high (2) bees: very high (3)

NAME: Common Trade and Other Chemical CAS Number	Class of Chemical	Chief Pesticide Use; Status	Persistence	Effects on Mammals		Adverse effects on other non-target species Physical properties
				Immediate Toxicity (Acute)	Long-Term Toxicity (Chronic)	
terbacil DuPont Herbicide 732; Geonter; Sinbar 3-*tert*-butyl-5-chloro-6-methyl uracil; 5-chloro-3-(1,1-dimethylethyl)-6-methyl-2,4-(1*H*,3*H*)-pyrimidinedione CAS # 5902-51-2	uracil	herbicide	mod-pers (1)	oral: low (1) dermal: ? inhalation: ?	?	immediate toxicity: birds: low to medium (2) fish: low (2) crustaceans: low (2) bees: low (3) molluscs: low to medium (2) water: soluble slightly volatile "nonflammable"
terbucarb Azac; Azak; Azar; Hercules 9573; terbutol (discontinued) 2-6-di-*tert*-butyl-*p*-tolyl methylcarbamate CAS # 1918-11-2	carbamate	herbicide	mod-pers (1)	oral: low (2) dermal: medium (3) inhalation: ?	?	immediate toxicity: bees: low (4) water: slightly soluble oil: insoluble
terbufos AC 92100; Aragran; Contraven; Corliss (with phorate); Counter; Counter 15G Soil Insecticide; Counter Plus; Cygard (with phorate); Dispell; ST100 *S*-[*tert*-butylthiomethyl] *O*,*O*-diethylphosphorodithioate CAS # 13071-79-9	organo-phosphate	insecticide nematocide restricted use, USA	non-pers (1)	oral: very high (2) dermal: very high (3) inhalation: very high (1)	eye & stomach damage (4) "disturbance to fetal development" (4)	immediate toxicity: birds: very high (3) fish: very high (2) crustaceans: very high (3) water: slightly soluble volatile flammable
transformation product(s): **formaldehyde** (see formaldehyde)						
terbutryn Clarosan; Gesaprim combi (with atrazine); GS 14260; Igran; Plantonit; Preban; Prebane (Gr. Britain); Reflex (with fomesafen); Short Stop E; terburyne; Terbutrex 2-(*tert*-butylamino)-4-(ethylamino)-6-(methylthio)-*s*-triazine; *N*-(1,1-dimethylethyl)-*N*′-ethyl-6-(methylthio)-1,3,5-triazine-2,4-diamine CAS # 886-50-0	triazine	herbicide	non-pers (1)	oral: medium (1,2) dermal: low (2) inhalation: low to medium (3)	suspect carcinogen (2)	immediate toxicity: birds: low to medium (4) fish: medium to high (2) crustaceans: medium (2) bees: low (5) water: slightly soluble slightly volatile "nonflammable"
tetrachloroethylene Ankilostin; carbon bichloride; carbon dichloride; Didakene; ethylene tetrachloride; Nema; per; perc; perchloethylene; perchlor; perchloroethylene; Perclene; perk; Tetracap; tetrachloroethene; Tetropil 1,1,2,2-tetrachloroethylene CAS # 127-18-4	organo-chlorine	fumigant	pers (1)	oral: low to medium (2) dermal: ? inhalation: low (2)	carcinogen (3,4) suspect mutagen (3,5) suspect teratogen (3,4) kidney, liver and respiratory damage (3,6)	immediate toxicity: fish: low to medium (1) long-term toxicity: crustaceans: mutagen (7) plants: mutagen (5) water: soluble highly volatile "nonflammable"
transformation product(s): **hexachlorobenzene** (see hexachlorobenzene)						

NAME: Common Trade and Other Chemical CAS Number	Class of Chemical	Chief Pesticide Use; Status	Persistence	Effects on Mammals		Adverse effects on other non-target species Physical properties
				Immediate Toxicity (Acute)	Long-Term Toxicity (Chronic)	

tetrachlorvinphos
 	
Z-2-chloro-1-(2,4,5-trichlorophenyl)vinyl dimethylphosphate
 	
CAS # 22248-79-9
	organo-phosphate	insecticide	non-pers (1)	oral: medium (1)
				
				dermal: low to medium (2)
				
				inhalation: ?
					carcinogen (3)	immediate toxicity:
						  birds: low to medium (4)
						  fish: medium to high (1)
						  bees: high (5)
						
						water: slightly soluble
						
						non-volatile

tetradifon
 	
Childion; Dorvert; Folimat T (with omethoate); Kelthion; Neutron (with methyl parathion); Tedane (with dicofol); Tedane Combi PB (with dicofol & dinocap); Tedane Extra (with dincap & dicofol); Tedion V-18; Tedov; Turbair Acaricide (with dicofol)
 	
4-chlorophenyl 2,4,5-trichlorophenyl sulfone; 2,4,4',5-tetrachlorodiphenyl sulfone
 	
CAS # 116-29-0
	organo-chlorine	acaricide
		
		suspended USA
			pers (1)	oral: low (2)
				dermal: medium (3)
				
				inhalation: ?
					suspect mutagen (7)
					suspect teratogen (1)
					liver and kidney damage (1)
						immediate toxicity:
						  birds: low to medium (4)
						  fish: medium to high (5)
						  crustaceans: very high (5)
						  bees: low (6)
						
						water: insoluble
						
						oil: soluble
						
						non-volatile
						
						"combustible"

contaminant(s):						
2,4,5-T						
chlorinated dibenzodioxins (see class description)	dibenzo-dioxin					
tetramethrin Butamin; Doom; Duracide 15 (with piperonyl butoxide); Ecothrin; FMC 9260; Multicide; Neo-Pynamin Forte Aerosol (with phenothrin); OMS 1011; Pesguard ANS (with fenitrothion); Pesguard FS (with phenothrin); Pesguard Insect Killer (with phenothrin); Pesguard NS (with fenitrothion); Pesguard NSB (with fenitrothion & piperonyl butoxide); phthalthrin; Py-Kill; Residrin; SP 1103; Sprigone; Spritex; tetramethrine 3,4,5,6-tetrahydrophthalimidomethyl (±)-cis,trans-chrysanthemate CAS # 7696-12-0	pyrethroid	insecticide	?	oral: low to medium (1,2) dermal: low to medium (3) inhalation: ?	?	water: slightly soluble non-volatile to slightly volatile
isomer(s):						
tetramethrin (1R)-isomers	pyrethroid	insecticide		oral: low (1)		immediate toxicity: fish: very high (1) water: slightly soluble slightly volatile

thiabendazole
 	
Aliette Extra (with fosetyl-al & captan); Apl-Luster; Arbotect; Bioguard; Elmpro; Mertect; RPH; TBZ; Tecto; Thibenzole; Tobaz
 	
2-(4-thiozolyl)-benzimidazole
 	
CAS # 148-79-8
	benzimi-dazole	fungicide	mod-pers (1)	oral: medium (2)
				dermal: ?
				
				inhalation: "low" (3)
					mutagen (4)
					suspect teratogen (5)
					blood and bone marrow damage (6)
						immediate toxicity:
						  earthworm: very high (7)
						  fish: low (7)
						
						water: slightly soluble
						
						"volatile at 310 degrees"

NAME: Common Trade and Other Chemical CAS Number	Class of Chemical	Chief Pesticide Use; Status	Persistence	Effects on Mammals		Adverse effects on other non-target species Physical properties
				Immediate Toxicity (Acute)	Long-Term Toxicity (Chronic)	

thiophanate ethyl
 	
Cercobin; Nemafax; Topsin; Verdamax
 	
diethyl 4,4'-(o-phenylene)bis(3-thioallophanate)
 	
CAS # 23564-06-9
	carbamate	fungicide
		
		cancelled all products, USA
			?	oral: low (1)
				
				dermal: low to medium (2)
				
				inhalation: medium (2)
					thyroid damage (3)	immediate toxicity:
						  birds: low to medium (2)
						  fish: medium (2)
						  aquatic invertebrates: medium

thiophanate methyl
 	
Cercobin M; Ditek; Fungo; Hitrun (with vinclozolin); Labilite; Mildothane; NF-44; TD-1881; Thiophanate M; Topsin M; Zyban
 	
dimethyl[1,2-phenylene)bis (iminocarbonothioyl)] bis[carbamate]; dimethyl 4,4'-o-phenylenebis (3-thioallophanate); bis[3-(methoxycarbonyl)-2-thioureido]benzene
 	
CAS # 23564-05-8
	carbamate	fungicide	non-pers (1)	oral: low (2)
				
				dermal: low to medium (3)
				
				inhalation: low to medium (3)
					decreased body weight of newborn (4)	immediate toxicity:
						  birds: low (3)
						  fish: medium to high (3)
						  earthworms: very high (5)
						
						water: "insoluble"

transformation product(s): **carbendazim** (see carbendazim)						

thiram
 	
Accelerator Thiuram; Aules Chipco Thiram 75; Cyuram DS; Deksan; Ekagom TV; Fernasan; Hexthir; Mercuram; Nobencutan; Panoram; Pomarsol Forte; Royal TMTD; Spotrete; Thioknock; Thirasan; Trameton; Tripomol; Tuads; Tues; Tulisan; Vulkacit MTIC
 	
bis(dimethylthiocarbamoyl)disulfide; tetramethylthiuram disulfide; tetramethylthiperoxydicarbonic diamide
 	
CAS # 137-26-8
	thiocar-bamate	fungicide animal repellent	non-pers (1)	oral: medium (2)
				
				dermal: ?
				
				inhalation: low to high (3)
				
				avoid alcohol ingestion before or after exposure (see disulfiram) (14)
					cumulative (4) suspect mutagen (5,6) teratogen (7,8) liver damage (9,10)	immediate toxicity:
						  birds: medium to high (3)
						  fish: medium to high (3)
						  bees: medium (11)
						
						long-term toxicity:
						  birds: excess build-up of cartilage in legs (12); reproductive damage (13)
						  plants: suspect mutagen (8)
						
						water: slightly soluble

related compound(s): **disulfiram** Antabuse; deters alcohol ingestion; ethyl analogue of thiram tetraethylthiuram disulfide				neurological effects; 10 times less toxic than thiram (1)	suspect teratogen (1,2) liver damage (1)	
transformation product(s): **N-nitrosodimethylamine** (see ferbam)						
tin CAS # 7440-31-5	metal/ mineral tin		perm	oral: ? dermal: ? inhalation: ?	?	

NAME:	Common Trade and Other Chemical CAS Number	Class of Chemical	Chief Pesticide Use; Status	Persistence	Effects on Mammals		Adverse effects on other non-target species Physical properties
					Immediate Toxicity (Acute)	Long-Term Toxicity (Chronic)	
tin (cont.) compound(s):							
cyhexatin cyhexan; Dowco 213; Plictran tricyclohexyl hydroxytin; tricyclohexyhydroxystannane; tricyclohexyltin hydroxide CAS # 13121-70-5		organic	acaricide cancelled USA, 1987		oral: medium (1) dermal: low to high (1,2) inhalation: ?		immediate toxicity: birds: medium (2) bees: medium (3) water: "insoluble"
fenbutatin-oxide Bendex; Neostanox; SD-14114; Torque; Vendex hexakis(2-methy-2-phenylpropyl)distannoxane; bis[tris(2-methyl-2-phenylprop yl)tin]oxide CAS # 13356-08-6		organic	acaricide		oral: low to medium (1)		immediate toxicity: birds: low to high (2) fish: high (1) bees: low (3) water: "insoluble"
fentin hydroxide Dowco 186; Du-Ter; Duter; Super Tin; Suzu H; TPTH; TPTOH; triphenyltin hydroxide (USA, So. Africa); Tubotin; Vancide KS triphenyltin hydroxide; hydroxytriphenylstannane CAS # 76-87-9		organic	fungicide some formulations restricted use, USA	non-pers to mod-pers (1)	oral: high (1) dermal: high to very high (2) inhalation: very high (1)	teratogen (1) immunotoxin (1)	immediate toxicity: birds: high (3) fish: very high (4) bees: low (5) water: slightly soluble non-volatile combustible
tributyltin chloride complex Tin San tributyltin chloride complex CAS # 1461-22-9		organic	restricted use, USA				immediate toxicity: fish: very high (1) long-term toxicity: fish: liver damage, reduces growth (1)
tributyltin fluoride fluorobutylstannane; tributyl stannane fluoride CAS # 1983-10-4		organic	restricted use, USA				immediate toxicity: amphibians: very high (1)
tributyltin oxide bioMET; Butinox; C-Sn-9; TBTO bis(tributyltin) oxide; bis(tri-n-butyltin) oxide; hexabutyldistannoxane CAS # 56-35-9		organic	fungicide (wood preservative) insect repellent restricted use, USA		oral: high (1)	suspect mutagen (2) teratogen (3)	immediate toxicity: birds: high (4) amphibians: very high (5)
triphenyltin acetate acetoxy-triphenylstannane; Batasan; Brestan; fentin acetate; phentinoacetate; Suzu; Tinamte; TPTA triphenyltin acetate CAS # 900-95-8		organic	fungicide algacide mollusicide		oral: high (1) dermal: high (2) inhalation: very high (1)		immediate toxicity: birds: high (3) fish: very high (4) water: slightly soluble
toluene Methacide; methylbenzene; phenylmethane; toluol toluene CAS # 108-88-3		aromatic hydro-carbon	solvent	?	oral: low (1) dermal: ? inhalation: ?	neurotoxin (2) addictive (3) brain damage (3) hearing loss (4) behavioral effects (5)	immediate toxicity: fish: low to medium (6) long-term toxicity: fish: embryotoxin (6) crustaceans: teratogen (7) water: "slightly soluble" flammable

NAME: Common Trade and Other Chemical CAS Number	Class of Chemical	Chief Pesticide Use; Status	Persistence	Effects on Mammals		Adverse effects on other non-target species
				Immediate Toxicity (Acute)	Long-Term Toxicity (Chronic)	Physical properties
toxaphene Agricide Maggot Killer; Alltox; camphechlor (So. Africa); Camphoclor; Camphofene Huileux; Chem-phene; Clor Chem T-590; Cristoxo; Estanox; Fasco-Terpene; Geniphene; Motox; Octachlorocamphene; Penphene; Phenacide; Phenatox; polychlorocamphene (USSR); Toxadusto-10; Toxakil; Toxaspra-8 mixture of various chlorinated camphenes CAS # 8001-35-2	organo-chlorine	insecticide acaricide cancelled, all products, USA, 1982	pers (1)	oral: high to very high (2) dermal: ? inhalation: ?	carcinogen (1,3) suspect mutagen (4) suspect fetotoxin (3,5) liver & kidney damage (1,3) neurotoxin (8) adrenal damage (6) blood damage (1)	immediate toxicity: birds: medium to very high (9) fish: very high (10) amphibians: very high (2) crustaceans: high to very high (2) bees: medium (11) molluscs: high to very high (2) aquatic insects: very high (2) long-term toxicity: bioaccumulates in aquatic plants, invertebrates, fish, & birds (2) birds: thyroid damage; affects egg production; decrease in backbone collagen (2) fish: liver & kidney damage; decrease in backbone collagen; spinal deformities; reduced egg viability (1) crustaceans: reduced offspring; reduced growth (1) molluscs: effects on shell growth & reproduction (12) phytoplankton: decreased photosynthesis (12) water: slightly soluble oil: "soluble" highly volatile
tralomethrin HAG 107; NU 831; OMS 3048; RU 25474; Scout; Scout X-Tra; Tracker; Tralate; tralomethrin (S)-α-cyano-3-phenoxybenzyl(1R,3S)-2,2,-dimethyl-3[(RS)-1,2,2,2-tetrabromoethyl]cyclopropanecarboxylyate CAS # 66841-25-6	pyrethroid	insecticide	?	oral: medium to high (1) dermal: low to medium (1) inhalation: hign (2)	?	immediate toxicity: birds: medium (2) fish: very high (1) crustaceans: very high (1) water: slightly soluble non-volatile
triadimefon Amiral; BAY ME B6447; Bayleton; Bayleton Total (with carbendazim); Bayleton Triple (with captafol & carbendazim) 1-(4-chlorophenoxy)-3,3-dimethyl-1-(1H-1,2,4-triazol-1-yl)-2-butanone CAS # 43121-43-3	triazole	fungicide	mod-pers to pers (1)	oral: low to medium (1) dermal: low to medium (2) inhalation: low to high (1)	?	immediate toxicity: birds: medium (1) fish: low to medium (1) water: slightly soluble slightly volatile flammable
triallate Buckle (with trifluralin); CP 23426; DATC-BW; Far-Go; tri-allate	thiocar-bamate	herbicide	mod-pers (1)	oral: medium (2) dermal: low to medium (1) inhalation: ?	suspect carcinogen (3) suspect mutagen (4) suspect neurotoxin (3)	immediate toxicity: birds: low to medium (1) fish: high (2) crustaceans: high (2) bees: low to medium (6)

NAME: Common Trade and Other Chemical CAS Number	Class of Chemical	Chief Pesticide Use; Status	Persistence	Effects on Mammals		Adverse effects on other non-target species Physical properties
				Immediate Toxicity (Acute)	Long-Term Toxicity (Chronic)	
triallate (*cont.*) S-2,2,3-trichloroallyl di-isopropylthiolcarbamate CAS # 2303-17-5					brain, liver, and spleen damage (5)	long-term toxicity: birds: liver, kidney, and reproductive damage; (7) delayed neurotoxicity (3) water: slightly soluble volatile "nonflammable"
triazophos Deltaphos (with deltamethrin); Hostathion O,O-diethyl O-1-phenyl-1H-1,2,4-triazol-3-yl phosphorothiote CAS # 24017-47-8	organo-phosphate	insecticide	mod-pers (1)	oral: high (2) dermal: low to medium (2) inhalation: ?	suspect mutagen (3)	immediate toxicity: birds: very high (4) fish: medium (4) water: slightly soluble slightly volatile
trichlorfon Anthon; BAY L 13/59; Bovinox; Briten; Cekufon; chlorofos (USSR); Ciclosam; Danex; Denkaphon; Dipterex; Diptetes; Ditrifon; Equino-Acid; Leivasom; metrifonate; Neguvon; Proxol; trichlorofon; Trinex; Tugon dimethyl (2,2,2-trichloro-1-hydroxyethyl)-phosphonate CAS # 52-68-6	organo-phosphate	insecdticide	mod-pers (1)	oral: high (2) dermal: low to medium (2) inhalation: low to high (3)	suspect carcinogen (4,5) suspect mutagen (4,7) suspect teratogen (4) fetotoxin (8) bone marrow & liver damage(4,9) immunotoxin (10)	BIOCIDE immediate toxicity: birds: high to very high (11) fish: medium to high (12) aquatic insects: very high (12) crustaceans: medium to very high (12) aquatic worms: very high (13) bees: medium (14) long-term toxicity: plants: suspect mutagen (15) water: very soluble oil: "insoluble" slightly volatile
transformation product(s): **dichlorvos** (see dichlorvos)						
2,4,5-trichlorophenol 2,4,5-TCP; Collunosol; Dowicide 2; Preventol 2,4,5-trichlorophenol CAS # 95-95-4	organo-chlorine	antibiotic fungicide cancelled USA, 1987	non-pers to mod-pers (1)	oral: medium (2) dermal: ? inhalation: ?	kidney & liver damage (3)	long-term toxicity: plants: suspect mutagen (3) water: soluble highly volatile
contaminant(s): **pentachlorodibenzo-p-dioxin** (see dibenzodioxin class)	dibenzo-dioxin					
1,3,6,8-tetrachlorodibenzo-p-dioxin (see dibenzodioxin class)	dibenzo-dioxin					water: insoluble
2,7-dichlorodibenzo-p-dioxin (see 2,4-D)						
TCDD (see chloroneb)						

NAME: Common Trade and Other Chemical CAS Number	Class of Chemical	Chief Pesticide Use; Status	Persistence	Effects on Mammals		Adverse effects on other non-target species Physical properties
				Immediate Toxicity (Acute)	Long-Term Toxicity (Chronic)	
2,4,6-trichlorophenol 2,4,6-TCP; Dowicide 2S; Omal 2,4,6-trichlorophenol CAS # 88-06-2	organo-chlorine	bactericide insecticide fungicide wood preservative	?	oral: medium (1,2) dermal: ? inhalation: ?	carcinogen (1) may cause central nervous system damage (3)	?
contaminant(s): **TCDD** (see chloroneb)						
pentachlorodibenzofuran (see PCB's)						
hexachlorodibenzofuran (see pentachlorophenol)						
triclopyr Broadshot (with dicamba & 2,4-D); Crossbow; Dowco 233; Fettel (with dicamba & mecoprop); Garlon (also a name for silvex); Garlon 3A; Garlon 4; Herbaron B (with dicamba & mecoprop); Turflon 3,5,6-trichloro-2-pyridinyloxyacetic acid CAS # 55335-06-3	organo-chlorine	herbicide	mod-pers (1)	oral: medium to high (2,3) dermal: ? inhalation: ?	suspect carcinogen (4,5) suspect mutagen (4)	immediate toxicity: birds: medium (2) fish: low, (Garlon 4: high) (3) crustaceans: low (6) water: soluble slightly volatile combustible
trifluralin Buckle (with triallate); Cannon (with alachlor); Carpidor; Commence (with clomazone); Ipersan; Janus; Laurel; Lextra (with linuron); Mudekan; Salute (with metribuzin); Su Seguro Cardidor; Trefanocide; Treficon; Treflan; trifluraline (France) α,α,α-trifluoro-2,6-dinitro-N,N-dipropyl-p-toluidine CAS # 1582-09-8	dinitro-aniline	herbicide cancelled, most uses, USA, 1982	mod-pers (1,2)	oral: low (3) dermal: low to medium (4) inhalation: ?	suspect carcinogen (5,6) suspect mutagen (7,8) suspect teratogen (9) fetotoxin (5)	immediate toxicity: birds: low (5) fish: high to very high (10) amphibians: very high (11) crustaceans: high to very high (10) bees: low to medium (12) aquatic insects: medium (10) water: insoluble volatile
contaminant(s): **N-nitroso-di-n-propylamine**				oral: high (1)	carcinogen (2,3) mutagen (4,5)	
trimethacarb Broot; Landrin; OMS 597; UC 27867 3,4,5-trimethylphenyl methylcarbamate CAS # 2686-99-9	carbamate	insecticide	non-pers to mod-pers (1)	oral: high (2) dermal: low to medium (2) inhalation: ?	?	immediate toxicity: "toxic to fish and wildlife" (3) water: slightly soluble slightly volatile
vamidothion Kilval; Kilvar; Trucidor; Vamidoate O,O-dimethyl S-2-(1-methylcarbamoylethylthio) ethylphosphorothioate CAS # 2275-23-2	organo-phosphate	insecticide not registered for use in USA	non-pers (1)	oral: high to very high (2) dermal: ? inhalation: ?	?	immediate toxicity: birds: very high (1) fish: low (1) water: very soluble oil: "soluble" "negligible vapor pressure"

NAME: Common Trade and Other Chemical CAS Number	Class of Chemical	Chief Pesticide Use; Status	Persistence	Effects on Mammals		Adverse effects on other non-target species Physical properties
				Immediate Toxicity (Acute)	Long-Term Toxicity (Chronic)	

vernolate
 	
Reward (with atrazine); Surpass (with dichlormid); Vernam (with atrazine)
 	
S-propyl dipropylthiocarbamate
 	
CAS # 1929-77-7
	thiocar-bamate	herbicide	non-pers (1)	oral: medium (2)	?	immediate toxicity:
						  fish: medium (3)
						  crustaceans: medium (3)
						  bees: low to medium (4)
						
						water: slightly soluble
						
						oil: "miscible"
						
						volatile
						
						combustible

vinclozolin
 	
BAS-3520; Hitrun (with thiophanate-methyl); Konker (with carbendazim); Ornalin; Ronilan; Ronilan M (with maneb); Ronilan S Combi (with sulfur); Ronilan Spezial (with chlorothalonil); Silbos (with thiram); Vorlan
 	
(R,S)-3-(3,5-dichlorophenyl)-5-methyl-5-vinyl-1,3-oxazolidinedione
 	
CAS # 50471-44-8
	amide	fungicide	?	oral: low (1)
				
				dermal: low to medium (2)
				
				inhalation: low (1)
					?	immediate toxicity:
						  birds: low to medium (2)
						  fish: low (1)
						
						water: slightly soluble
						
						oil: soluble
						
						slightly volatile

warfarin Arab Rat Deth; coumafene (France); Dethmor; Eastern States Duocide; Fasco Fascrat Powder; Fatal; Kypfarin; Martin's Mar-Frin; Rat & Mice Bait; Rat Gard; Rat-Death; Rat-Kill; Rat-Mix; Rat-Nix; Rat-O-Cide; Rat-Ola; Rataway; Twin Light Rat Away; Warfarat; zoocoumarin (USSR & Netherlands) 3-(α-acetonylbenzyl)-4-hydroxycoumarin CAS # 81-81-2	coumarin	rodenticide	?	oral: high (1) dermal: ? inhalation: ?	teratogen (2,3)	water: slightly soluble slightly volatile

xylene
 	
ksylene (Poland); xiloli (Italian); xylenen (Dutch); xylol (German)
 	
dimethylbenzene
 	
CAS # 1330-20-7
  108-38-3 (meta isomer)
  95-47-6 (ortho isomer)
  106-42-3 (para isomer)
 	
(commercial xylene is a mixture of isomers)
 	
contaminant(s):
benzene
(see benzene)
	aromatic hydro-carbon	solvent herbicide	non-pers (1)	oral: medium (2)
				
				dermal: ?
				
				inhalation: low (3)
					suspect teratogen (3)
					suspect neurotoxin (3)
					skin damage (4)
						water: "insoluble"
						
						highly volatile
						
						flammable

zinc phosphide
 	
Gopha-Rid; idall-Zinc; Kilrat; Mole and Gopher Bait; Mouse-con; Phosvin; Rodent Pellets; Rumetan
 	
zinc phosphide
 	
CAS # 1314-84-7
 	
transformation product(s):
phosphine gas
(see aluminum phosphide)
	metal/mineral
	
	zinc
		insecticide rodenticide
		
		restricted use, USA
		
		"confined, in many countries, to trained personnel"
			non-pers
			"under exposed acid-free conditions will remain active for long periods of time" (1)
			
			"stable when dry" (2)
				oral: very high (1)
				
				dermal: ?
				
				inhalation: ?
				
				"do not inhale, avoid skin contact" (3)
					?	immediate toxicity:
						  birds: very high (4)
						  fish: "negligible" (5)
						  crustaceans: "in stream killed many" (6)
						
						water: insoluble

CHEMICAL CLASSES OF PESTICIDES

NAME: Common Trade and Other Chemical CAS Number	Class of Chemical	Chief Pesticide Use; Status	Persistence	Effects on Mammals		Adverse effects on other non-target species Physical properties
				Immediate Toxicity (Acute)	Long-Term Toxicity (Chronic)	
zineb Asporum; Blightox; Blizene; Cineb; Crittox; Crystal Zineb; Hexathane; Kupratsin; Kypzin; Lonacol; Micide; Miltox; Novozin N 50; Parzate; Pomarsol S Forte; Thiodow; Tritofterol; Zebtox; Zidan; Zinosan zinc ethylene bisdithiocarbamate CAS # 12122-67-7	thiocar-bamate	fungicide cancelled, most products, USA	non-pers to mod-pers (1)	oral: low (2) dermal: low to medium (3) inhalation: ?	anemia (3)	immediate toxicity: birds: low to medium (4) long-term toxicity: fish: embryotoxin (5) amphibians: teratogen (6) plants: toxic to some; inhibits germination of pollen in some plants (7) water: slightly soluble combustible
transformation product(s): **ethylene thiourea** (see amobam)						
ziram Corozate; Cuman; Fuclasin Ultra; Fuklasin; Hexazir; Karbam; Methasen; Mezene; Milbam; Niagara Z-C Spray; Opalate; Pomarsol Z Fote; Prodaram; Tricarbamix Z; Triscabol; Vancide MZ-96; Z-C Spray; Zerlate; Zincmate; Ziram Technical bis(dimethyldicarbamato)zinc; zinc dimethyl dithiocarbamate CAS # 136-30-4	thiocar-bamate	fungicide	non-pers (1)	oral: medium (2) dermal: ? inhalation: ?	suspect carcinogen (3,4) suspect mutagen (3,5) suspect teratogen (6,7) bone damage (8)	immediate toxicity: fish: high (1) bees: "nontoxic" (3) long-term toxicity: birds: reproductive damage (9,10)
transformation product(s): **carbon disulfide** (see carbon disulfide)						
dimethylamine (see dimethylamine)						
N-nitrosodimethylamine (see ferbam)						

Chapter Five

Chemical Classes of Pesticides

This section has two chief purposes: to give more information on immediate toxicity beyond the rating of degree on the charts, with the kinds of reactions that may occur, and to indicate the characteristics of families of pesticides. If we have little data on a member of a group, some general idea of its character may be inferred from the qualities typical of its fellow pesticides, allowing for differences in intensity and type of reaction.

Immediate toxicity reactions are more difficult to find in the literature than are LD_{50}s or data on some long-term effects. The importance of recognizing symptoms that indicate poisoning is clearly great, but the clues listed are the best we could find.

Many pesticide active ingredients have elements of more than one chemical group, so placing them in families that share toxicological effects can be difficult. Classification can be arbitrary. Our intention is to assemble classes with consistent toxicology. In some cases it might be advisable to check the various components in a pesticide. Classes and subunits are presented in alphabetical order.

The mode of action given is that for mammals; it may be the same for non-mammal pests in some cases. The effects listed are typical of the group, while some may apply most seriously to certain members. These are the reactions that can occur, in various degrees of severity. Not all would be apt to occur in any one case. The manner and amount of exposure, the vulnerability of the victim, and the medical treatment provided can alter cases. Immediate effects can also include death in severe cases; we have not always listed this because it can be implied from degree of toxicity and symptoms.

References for each class include those for the separate pesticides in each group. Additional references that apply to the group as a whole are given for each class. For most of these, we also consulted those references listed here, referred to by numbers 1, 2, and 3 in the reference sections. (Note: in chapter 6, these references are numbered 9, 12, and 24, respectively).

(1) Gosselin, R. S., R. P. Smith, and H. C. Hodge. 1984. *Clinical toxicology of commercial products*. Baltimore, MD: Williams and Wilkins.

(2) Hayes, W. J., and E. R. Laws. 1991. *Handbook of pesticide toxicology*, vols. 1, 2, and 3. San Diego, CA: Academic Press.

(3) Morgan, D. P. 1989. *Recognition and management of pesticide poisoning*. U.S. EPA doc. no. 540/9-88-001.

ALCOHOL

allyl alcohol   ethyl hexanediol

Mode of action: Depression of central nervous system; mucous membrane irritation.

Immediate effects: Very irritating to skin and mucous membranes; eye damage; first or second degree burns; readily absorbed through skin; pulmonary edema; central nervous system depression.

Long-term effects: Suspect mutagen; liver damage.

REFERENCES

(1)
Brenner, D. E., et al. 1987. *Cancer Res.* 47:3259–3265.
Budavari, S. M., et al. 1989. *The Merck index*, 11th ed. Rahway, NJ: Merck and Co.
Lutz, D., et al. 1982. *Mutat. Res.* 93:305–315.

Esters

ethyl formate

Mode of action: An ester of ethyl alcohol, twice as toxic as other esters, perhaps by its hydrolysis to formic acid. Acts by depressing the central nervous system, leading to stupor and coma.

Immediate effects: Abdominal pain; central nervous system depression; convulsions; turning blue; dermatitis; dizziness; eye irritation leading to conjunctivitis; double vision; mucous membrane irritation including respiratory tract; leading to edema; bronchospasm; cough; respiratory depression and failure; narcosis; gastrointestinal tract: nausea, vomiting.

REFERENCES

(1)
Budavari, S. M., et al. 1989. *The Merck index*, 11th ed. Rahway, NJ: Merck and Co.

ALDEHYDE

acrolein   paraformaldehyde
formaldehyde   MGK R11
metaldehyde

Mode of action: Degeneration of mucosal lining in respiratory tract and liver and kidneys.

Immediate effects: Eye and nasal irritation; excessive fluid in the lungs; lung damage; skin irritation and burns; central nervous system depression if combined with alcohol intoxication; abdominal pain; nausea; vomiting; diarrhea; fever; convulsions; labored breathing; bronchitis; difficult or no urination.

Long-term effects: Suspect mutagens; teratogens; embryotoxins; lung, liver, and kidney damage; carcinogens; spinal damage leading to paralysis; dermatitis; asthma; sleeplessness; ovarian atrophy.

REFERENCE

(1)

AMIDE

acetochlor	diphenamid
alachlor	fomesafen
benzadox	iprodione
butachlor	isoxaben
butam	metalachlor
carboxim	niclosamide
CDAA	perfluidone
chlordimeform	procymidone
cycloheximide	pronamide
DEET	propachlor
dichlofluanid	propanil
dichlormid	vinclozolin
diethatyl	

Mode of action: Not fully understood.

Immediate effects: Skin irritant and sensitizer; irritating to eyes and respiratory tract; nausea; headache; uncoordination; stiffness of movement; salivation; tremors; muscle weakness; sensitivity to light.

Long-term effects: Chloracne via dioxin contaminants (propanil); carcinogens; mutagens, irreversible eye damage; kidney and liver damage; suspect teratogens; immunotoxins; cardiovascular effects; embryotoxins; sperm damage.

Environmental effects: Groundwater contaminants; N-nitroso contaminants.

REFERENCES

(2), (3)

ANTIBIOTIC

antimycin A	blasticidin S
antimycin A3	streptomycin

Mode of action: May affect nervous system.
Immediate effects: Eye irritation; headache; nau-

sea; abdominal pains; fever; coughing; labored breathing.

Long-term effects: Eye damage.

REFERENCES

(1); (2)

Farm chemicals handbook. 1991. Willoughby, OH: Meister Printing Co.

AROMATIC HYDROCARBON

benzene	toluene
diphenyl	xylene

Mode of action: Damage to central nervous system and other tissues, such as heart.

Immediate effects: Irritation of respiratory tract: excessive fluid in lungs; headache; vertigo; impotence; kidney dysfunction; lung damage; central nervous system excitement or depression; uncoordination; slurred speech; emotional swings; involuntary rapid eye movement; tremors; coma; fatigue.

Long-term effects: Dermatitis; brain damage; carcinogens; suspect mutagens; teratogens; blood damage; bone damage; paralysis; convulsions; central and peripheral nervous system damage; liver and kidney damage; hearing loss.

REFERENCE

(1)

AZIDE

potassium azide	sodium azide

Mode of action: Inhibits enzymes.

Immediate toxicity: Irritates mucous membranes, leading to bronchitis; pulmonary edema; low blood pressure; fast breathing and heartbeat; acidosis; headache; nausea; vomiting; diarrhea; EKG changes; high levels of white blood cells. Swelling of brain and lungs; liver damage.

REFERENCE

(1)

BENZIMIDAZOLE

benomyl	fenazaflor
carbendazim	thiabendazole

Mode of action: Interferes with cellular respiration.

Immediate effects: Dizziness; nausea; vomiting;

tremors; convulsions; decreased respiratory rate; lethargy; pupil dilation, eye irritation.

Long-term effects: Defective or incomplete development of bone marrow; suspect carcinogens; suspect mutagens; testicular damage; mutagens; anemia; teratogens; liver damage; reduced sperm; blood damage.

REFERENCE

(1)

BENZONITRILE

bromoxynil   dichlobenil
chlorothalonil   ioxynil

Mode of action: May be due to uncoupling of oxydative phosphorylation and inhibiting of electron transport, with inhibition of some enzymes.

Immediate effects: Irritation of skin, mucous membranes; dermatitis. Ioxynil: excess blood in all organs; edema of lungs and brain. Bromoxynil: dizziness; elevation of some enzymes; headache; hyperthermia; muscle pain; thirst; vomiting; weakness; weight loss; anorexia. Chlorothalonil: hyperexcitability.

Long-term effects: Carcinogens; teratogen; skin, eye, kidney damage. Suspected—dichlobenil: anorexia; blood in urine; kidney damage; liver damage; reproductive changes with postnatal damage. Chlorothalonil: growth suppression; pre- and postnatal damage; kidney destruction.

REFERENCES

(3)
Hallenbeck, W. H., and K. M. Burns. 1985. *Pesticides and human health.* New York: Springer-Verlag.

BIOLOGICALS

abamectin
Bacillus thuringiensis
milky spore disease
nuclear polyhedrosis virus

These compounds are collected under the classes title of *biologicals* because they are either organisms in their own right or compounds extracted from organisms other than plants (see Botanicals). Biological pesticides such as bacteria, viruses, and fungi are being used more and more as new strains are found with better ways to prolong their viability. Most of these biological pesticides are species-specific or specific to a group of organisms. Though adverse effects are unlikely, these pesticides should be used with utmost care. Inert ingredients

such as solvents and carriers can be anything, and, as mentioned in the inerts explanation, can be more potent than the pesticide itself.

BOTANICALS

camphor   rotenone
endod   ryania
nicotine   sabadilla
pyrethrum   strychnine
red squill

This class of pesticides is composed of a variety of compounds from many plant species. These pesticides are not grouped by chemical relationships or mode of action as are organophosphates or organochlorines. Each botanical pesticide is treated individually and as extensively as possible.

Camphor

Mode of action: Stimulates the central nervous system.

Immediate effects: Nausea; vomiting; feeling of warmth; headache; confusion; vertigo; excitement; restlessness; delirium; hallucinations; tremors; convulsions; coma; death due to respiratory failure.

Long-term effects: Liver damage; gastric distress.

REFERENCE

(1)

Endod

Environmental effects: Used as a mollusicide against the snails that carry schistosomiasis, endod is toxic to fish.

REFERENCE

Stabens, J. K., et al. 1990. *Vet. Hum. Toxicol.* 2(3):212–216.

Nicotine

Mode of action: Stimulates and eventually depresses nervous system.

Immediate effects: Readily absorbed through skin; agitation; headache; sweating; dizziness; confusion; weakness; uncoordination; rapid breathing; high blood pressure; slow pulse; constricted pupils; irregular heartbeat; tremors; convulsions; depression leading to dilated pupils, low blood pressure, and rapid pulse; faintness prostration; death due to respiratory failure.

Long-term effects: Transformation product N-nitrosonornicotine is a carcinogen.

REFERENCE

(1)

Pyrethrum

Mode of action: Blocks nerve impulse transmission.

Immediate effects: Skin irritation; asthmatic reactions; (those with asthma problems should avoid pyrethrum use, high doses yield tremors, ataxia, labored breathing, and salivation); numbness of lips and tongue; vomiting; diarrhea; headache; uncoordination; stupor. Allergic reactions as from other *Compositae* such as ragweed and chrysanthemum.

Long-term effects: Piperonyl butoxide, carbamates, and organophosphates may be combined with pyrethrum in various formulations. These added ingredients may result in symptoms listed under the appropriate class description.

Environmental effects: Highly toxic to fish.

REFERENCES

(1); (3)
Hallenbeck, W. H., and K. M. Burns. 1985. *Pesticides and human health.* New York: Springer-Verlag.

Red Squill

Mode of action: Cardiac glycoside, which decreases potassium in cells and increases sodium cells leading to irregular heart action.

Immediate effects: Similar to that of digitalis. Nausea; vomiting; irregular heartbeat; convulsions.

Long-term effects: Cardiac effects may last for several weeks.

REFERENCES

(1); (3)
Hallenbeck, W. H., and K. M. Burns. 1985. *Pesticides and human health.* New York: Springer-Verlag.

Rotenone

Mode of action: Inhibits cell respiration and blocks conduction of nerve impulses.

Immediate effects: Numbness of mouth and tongue; nausea; vomiting; gastric pain; muscle tremors; uncoordination; irritation of skin and respiratory tract; respiratory stimulation followed by depression and death.

Long-term effects: May be mixed with piperonyl butoxide in various formulations resulting in symptoms of that compound; suspect carcinogen; suspect teratogen; suspect fetotoxin; liver and kidney damage.

REFERENCES

(1)
Hallenbeck, W. H., and K. M. Burns. 1985. *Pesticides and human health.* New York: Springer-Verlag.

Ryania

Immediate toxicity: Retraction of eyes into socket; vomiting; weakness; diarrhea; slow deep breathing; salivation; central nervous system depression; coma; death due to respiratory failure.

REFERENCE

(1)

Sabadilla

Mode of action: Similar to that of digitalis.

Immediate effects: Irritating to upper respiratory tract and skin; vomiting; headache; giddiness; weakness; twitching; convulsions; hypothermia; death due to respiratory or cardiovascular failure.

REFERENCES

(1); (3)

Strychnine

Mode of action: Renders the spinal nerves open for excess stimulation; may inhibit cholinesterase.

Immediate effects: Convulsion; restlessness; apprehension; heightened perception; rarely vomiting; dehydration; death due to respiratory failure.

Long-term effects: Kidney damage.

Environmental effects: Highly toxic to wildlife.

REFERENCES

(1)
Hallenbeck, W. H., and K. M. Burns. 1985. *Pesticides and human health.* New York: Springer-Verlag.

BIPIRIDYL

chlormequat	diquat
chlormequat chloride	paraquat
difenzoquat	paraquat dichloride

Mode of action: May interfere with cell respiration and damage cell membranes.

Immediate effects: Skin damage; loss of fingernails; nosebleed; eye damage; excess fluid in lungs; irritation of mucosal linings in mouth, throat, chest, and upper abdomen; giddiness; headache; muscle pain; diarrhea; blood and or pus in urine; decreased urine secretion; jaundice; kidney failure; vomiting.

Long-term effects: Irreversible lung damage preventing exchange of oxygen (lung damage usually manifests itself 2–14 days after exposure); prolonged dermal exposure, damaging skin, can result in more rapid absorption through the skin; kidney damage; suspect mutagen, suspect teratogen; suspect fetotoxin; suspect embryotoxin; liver damage; cataracts. Protracted dermal exposure can lead to death.

REFERENCES

(1); (3)

CARBAMATE

aldicarb	ethiofencarb
aminocarb	formetanate hydrochloride
asulam	fosamine ammonium
barban	methiocarb
bendiocarb	methomyl
benthiocarb	mexacarbate
bufencarb	oxamyl
butoxycarboxim	phenmedipham
carbanolate	propham
carbaryl	propoxur
carbofuran	terbucarb
chlorpropham	thiophanate ethyl
desmedipham	thiophanate methyl
dioxacarb	trimethacarb
diram	

Mode of action: Inhibits acetochlolinesterase and so damages nerve function.

Immediate effects: Sensory and behavioral disturbances; uncoordination; depressed motor functions; malaise; muscle weakness; dizziness; sweating; headache; salivation; nausea; vomiting; abdominal pain; slurred speech; difficult breathing; blurred vision; muscle twitching spasms; convulsions; diarrhea; depression of cholinesterases even more prominently in fetus; skin sensitization.

Long-term effects: Memory loss; behavioral defects; suspect mutagens; mutagens, carcinogens; cataracts; suspect carcinogens; teratogens; spleen, bone marrow, liver, and testes damage; reduced sperm levels; fetotoxins; suspect viral enhancers; increased organ weights; decreased body weights; anemia; decreased hemoglobin; decreased fertility from ovary and testis damage; may convert to N-nitroso compounds in soil and in vivo with saliva.

Environmental effects: Can disrupt schooling behavior of fish; teratogens in fish; toxic to earthworms (carbendazim, thiophonate methyl), reduction in earthworm populations and invertebrate populations (WHO 1986, 56–57); groundwater contaminants.

REFERENCES

(3)

Cambon, C. et al. 1979. Effect of the insecticidal carbamate derivatives (carbofuran, primicarb, aldicarb) on the activity of a cetylcholinesterase in tissues from pregnant rats and fetuses. *Toxicol. Appl. Pharmacol.* 49:203–208.

Desi, I. 1974. Neurotoxicologic studies of two carbamate pesticides in subacute animal experiments. *Toxicol. Appl. Pharmacol.* 27:465–476.

Suspected carbamate; intoxications—Nebraska. 1979. *Morbidity/Mortality Weekly Report* 28(12).

World Health Organization. 1986. *Carbamate pesticides: A general introduction.* Environmental Health Criterial 64. Geneva: WHO.

COUMARIN

brodifacoum	dicumarol
bromadiolone	difenacoum
coumafuryl	warfarin

Mode of action: Blocks vitamin K–dependent synthesis of blood clotting substance prothrombin; predisposes animal to widespread internal bleeding.

Immediate toxicity: Nosebleed; bleeding gums; blood in the urine and feces; bruises due to ruptured blood vessels; skin damage.

Long-term toxicity: Teratogen; paralysis due to cerebral hemorrhage.

Environmental effects: Toxicity to susceptible nontarget mammals.

REFERENCE

(3)

CYANIDE

acrylonitrile	hydrogen cyanamide
ammonium thiocyanate	hydrogen cyanide
calcium cyanamid	sodium cyanide
calcium cyanide	trichloroisocyanuric acid

Mode of action: A chemical asphyxiant, it prevents the tissues from using oxygen, rapidly affecting the organs most sensitive to oxygen loss, the brain and heart; extremely fast-acting.

Immediate effects: Caustic skin and respiratory system irritant causing respiratory paralysis, numbness of throat, stiff jaw, salivation, nausea, vomiting, dizziness, convulsions, paralysis, incontinence, pink skin, and brain damage.

Long-term effects: Carcinogen; mutagen; teratogen; thyroid, blood, and respiratory damage; neurotoxin—sometimes psychiatric aftereffects.

REFERENCES

(1); (3)

DIBENZODIOXIN : DIBENZOFURAN

These groups of chemicals occur in pesticides as contaminants, generated in the production process, or later when products are heated. Some are among the most toxic synthetic chemicals. The degree of toxicity depends on the number and location of the halogen (chlorine) atoms. There are 75 chlorodibenzo-p-dioxins, and 135 chlorodibenzo-furans, ranging from mono- to octo- categories. The most studied is 2,3,7,8-tetradibenzo-p-dioxin (2,3,7,8-TCDD), the toxicity of which is closely matched by 2,3,7,8-tetrachlorodibenzofuran. Some confusion in terminology has come from the practice of calling 2,3,7,8-TCDD by the name *dioxin*. There are many chemicals that can be so called, with a wide range of toxicity.

Those listed specifically as contaminants of the pesticides covered here are:

Chlorodibenzo-p-dioxins
  trichlorodibenzo-p-dioxin
  1,3,7-trichlorodibenzo-p-dioxin
  tetrachlorodibenzo-p-dioxin (TCDD)
  1,3,6,8-tetrachlorodibenzo-p-dioxin
  2,3,7,8-tetrachlorodibenzo-p-dioxin*
  hexachlorodibenzo-p-dioxin
  octachlorodibenzo-p-dioxin
  pentachlorodibenzo-p-dioxin
Chlorodibenzofurans
  tetrachlorodibenzofuran
  2,3,7,8-TCDF
  hexachlorodibenzofuran
  heptachlorodibenzofuran
  octachlorodibenzofuran
  pentachlorodibenzofuran

Mode of action: Attack several organ systems with toxicity enhanced by their being cumulative in the body, persistent, and so bioaccumulative in food chains. High toxicity to the thymus and immune system is a key reason for their broad effects.

Immediate effects: Weight loss; edema; chloracne; loss of fingernails and toenails.

Long-term effects: Anorexia to starvation; carcinogen; teratogen; suspect mutagen; embryotoxin; fetotoxin; neurotoxin; anemia (aplastic); immune system damage: atrophy of thymus, lymphatic system, and lympocytes; damage to liver, spleen, bone marrow, blood, thyroid, adrenal cortex, gastrointestinal tract (hemorrhage and necrosis), urinary tract, skin and sebaceous glands; brain hemorrhage; abnormal

*Highly toxic in a few parts per trillion.

eye movement; enzyme imbalance leading to hyperpigmentation and hirsutism; impaired sight, hearing, smell, taste.

Environmental effects: Similar effects on many animals. Especially noted in birds is reduced egg production and low viability of young; 2,3,7,8-TCDF fatal to fish embryos in parts per trillion.

REFERENCES

Huff, J. E., et al. 1980. Long-term hazards of polychlorinated dibenzodioxins and polychlorinated dibenzofurans. In *Environmental health perspectives*, vol. 36, 221–240.
Kimbrough, R. D., ed. 1980. *Halogenated biphenyls, terphenyls, naphthalenes, dibenzodioxins, and related products*. New York: Elsevier.

DINITROANILINE

benefin	fluchloralin
butralin	isopropalin
dichloran	oryzalin
dinitramine	pendimethalin
ethafluralin	trifluraline

Mode of action: Interfere with cell respiration.

Immediate effects: Skin and eye irritation.

Long-term effects: Cataracts; suspect mutagen; liver and kidney damage; carcinogens, teratogens; fetotoxins.

REFERENCE

Hallenback, W. H., and K. M. Burns. 1985. *Pesticides and human health*. New York: Springer-Verlag.

FLUOROACETATE

sodium fluoroacetate (1080)
fluoroacetamide (1081)
fluenethyl

Mode of action: In the body of the victim, these compounds change to fluorocitrate, which blocks the citrate and succinate metabolism, a lethal synthesis. Cardiac and central nervous systems are damaged, other organ damage is found in poisoned animals. Symptoms are somewhat delayed, over a quarter of an hour, so the victim has time to absorb a lethal dose. Death occurs from an hour to a day later. No effective antidote is known.

Immediate effects: Central nervous system effects include hyperactivity, outcries, convulsions leading to respiratory paralysis. Cardiac effects are blanching of the retina, muscular weakness, spasmodic convulsions, and ventricular fibrillation. A third syndrome, the depressive, causes decreased activity, respiratory depression, and very slow pulse rate. Even in the short time before death, damage to heart, liver, aorta, brain, and testes occur. Gastrointestinal and lung hemorrhages are common.

Long-term effects: With sublethal doses, the organ damage listed above occurs, and the poison may accumulate with repeated exposures.

Environmental effects: Secondary poisoning is the greatest hazard, since a poisoned animal may contain a lethal dose for any predator that eats it. The sequence can pass from one victim to another. Animal populations of an area can be greatly altered; special hazard to some endangered species noted.

REFERENCES

(1); (2); (3)

INDANDIONE

chlorphacinone   diphacinone

Mode of action: Blocks vitamin K-dependent synthesis of blood clotting substance prothrombin; predisposes animal to widespread internal bleeding.

Immediate effects: Nosebleed; bleeding gums; blood in urine and feces; bruises due to ruptured blood vessels; skin damage.

Long-term effects: Nerve and heart damage, unlike coumarins.

Environmental effects: Dichacinone has been known to be toxic to vampire bats when low level doses were injected into cattle in Mexico.

REFERENCES

(2); (3)

METAL/MINERAL

aluminum	magnesium
arsenic	mercury
cadmium	silica
calcium	sulfur
copper	tin
lead	zinc

Toxic effects of these metals and minerals vary, but they share the attribute of being elements: essential, irreducible substances that cannot be decomposed by chemical means. They are therefore permanent in themselves, though combinations with other substances can degrade. They may react differently depending on whether they are in organic (containing carbon) or inorganic form.

Aluminum

aluminum phosphide

Mode of action: Reaction with water, in air, or in the stomach if ingested, produces phosphine gas.

Immediate effects: Fatigue; nausea; headache; dizziness; thirst; cough; tremor; shortness of breath; paresthesis (abnormal sensations); jaundice; pulmonary edema leading to death.

Long-term effects: Liver, heart, kidney, gastrointestinal, lung damage.

REFERENCES

(1); (3)

World Health Organization. 1988. *Phosphine and selected metal phosphides.* Environmental Health Criteria 73. Geneva: WHO.

Phosphine, Produced also by Calcium, Magnesium, and Zinc Phosphides

Mode of action: Not clearly understood, but it affects cellular oxygen intake, and changes in some enzymes. Loss of cell viability and membrane integrity may account for liver enzyme changes, kidney cell swelling, bronchial damage, and myocardial bleeding. It reaches the liver and nervous system, and blood, initially, reacts with some haemo- and copper-containing proteins in vitro, and with mammalian haemoglobin in presence of oxygen.

Immediate effects: Abdominal pain; acidosis; nausea; diarrhea; vomiting; garlic smell on breath; imbalance; dizziness; convulsions; coma; turn blue; eye, skin, mucous membrane irritation with double vision; headache; hallucinations; hemorrhage; blood damage; palpitations; EKG abnormalities; liver damage including elevated enzymes; jaundice; fatty degeneration; necrosis; hypothermia; high blood pressure; kidney damage including uremia, reduction of urine, blood in urine, respiratory bronchospasm, cough, congestion, difficulty breathing, rales, and edema of larynx and lungs; nosebleed; tremor. Death may be due to pulmonary edema, cardiac arrest, circulatory collapse, or respiratory failure, and may be delayed from four days to two weeks after exposure.

REFERENCES

(1); (3)

Hallenback, W. H., and K. M. Burns. 1985. *Pesticides and human health.* New York: Springer-Verlag.

World Health Organization. 1988. *Phosphine and selected metal phosphides.* Environmental Health Criteria 73. Geneva: WHO.

Arsenic

ammonium arsenate (inorganic)
ammonium methanearsenate (organic)
arsenic acid (inorganic)
arsenic pentoxide (inorganic)
arsenic trioxide (inorganic)
cacodylic acid (organic)
calcium methanearsenic (organic)
calcium arsenate (inorganic)
calcium arsenite (inorganic)
calcium propanearsenate (organic)
chromated copper arsenate (CCA) (inorganic)
copper acetoarsenite (organic)
cupric arsenite (inorganic)
DSMA (organic)
lead arsenate (inorganic)
methyl arsonic acid (organic)
MSMA (organic)
OBPA (organic)
sodium arsenate (inorganic)
sodium arsenite (inorganic)

Arsenic is rarely found in its pure form, but is usually combined with other elements to make inorganic forms, or with carbon and hydrogen to make organic forms. The organic forms are generally less toxic than the inorganic forms.

Mode of action: Damages cells in the nervous system, blood vessels, liver, kidneys, and other tissues by combining with thiol groups and substituting arsenic anions for phosphate in many reactions.

Immediate effects: Severe inflammation of mouth through gastrointestinal tract; thirst; vomiting; diarrhea; headache, dizziness; muscle weakness and spasms; hypothermia; lethargy; delirium, convulsions; coma; severe inflammation of mucous membranes of nose; larynx, bronchi, peripheral nervous system disturbances.

Long-term effects: Dermal symptoms include hyperkeratosis; hyperpigmentation; peeling of skin; edema of face, eyelids, and ankles; white striations on nails; loss of nails or hair; anorexia; weight loss; peripheral neuropathy with abnormal sensations; pain; anesthesia; muscular uncoordination; liver injury and jaundice; cirrhosis; high blood pressure of the portal veins; kidney damage; EKG abnormalities; anemia; reduced leucocytes and platelets in blood; skin cancer; lung cancer; degenerative disease of brain.

REFERENCES

(1); (2); (3)
Environmental Protection Agency. 1984. Health assessment document for inorganic arsenic. Final Report, EPA-600/8-83-021F.

Cadmium

cadmium carbonate (organic)
cadmium chloride (inorganic)
cadmium oxide (inorganic)
cadmium sebacate (organic)
cadmium succinate (organic)
cadmium sulfate (inorganic)

Mode of action: Absorbed more freely through inhalation than swallowing, in part because it induces vomiting. Accumulates in body.

Immediate effects: Irritation of respiratory and gastrointestinal tracts, headache, persistent cough, labored breathing, chest pain, fever; ingesting causes nausea, vomiting, diarrhea.

Long-term effects: Liver and kidney damage, defective bone structure. Some products are teratogens, mutagens, carcinogens.

REFERENCES

(1); (2); (3)

Calcium

calcium chlorate (inorganic)
calcium cyanamide (organic) (see cyanide)
calcium cyanide (organic) (see cyanide)
calcium hypochlorite (inorganic)
calcium phosphide (inorganic) (see aluminum)
calcium proprionate (organic)
calcium sulfate (inorganic) (see sulfur)
calcium polysulfide (inorganic) (see sulfur)

Mode of action: Most of the toxicity in calcium compounds is related to the other components. Calcium at high levels in the body can cause heart and kidney damage. Calcium phosphide with water transforms to phosphine (see aluminum phosphide).

Immediate effects: Calcium chlorate: gastrointestinal irritation; calcium polysulfide: irritates skin and mucous membranes; calcium itself if heated can create fumes highly irritating to skin, eyes, mucous membranes.

Long-term effects: Include kidney, heart damage, gastrointestinal damage; calcium phosphide releases phosphine gas (see aluminum phosphide); the cyanide compounds may be carcinogenic.

REFERENCE

Budavari, S. M., et al. 1989. *The Merck index*, 11th ed. Rahway, NJ: Merck and Co.

Copper

basic copper carbonate (organic)
basic copper chloride (inorganic)
basic cupric acetate (organic)
Bordeaux mixture (inorganic)
copper ammonium carbonate (organic)
copper ammonium sulfate (inorganic)
copper bis(3-phenylsalicylate) (organic)
copper carbonate (organic)
copper chelate (organic)
copper citrate (organic)
copper ethylenediamine complex (organic)
copper hydroxide (inorganic)
copper linoleate (organic)
copper naphthenate (organic)
copper nitrate (inorganic)
copper oxalate (organic)
copper ozide (inorganic)
copper oxychloride (inorganic)
copper oxychloride sulfate (inorganic)
copper sulfate (inorganic)
copper sulfate monohydrate (inorganic)
copper tea complex (organic)
copper zinc chromate (inorganic)
cupric hydrazinium sulfate (organic)
oxine copper (organic)
zinc coposil (inorganic)

Mode of action: Interferes with enzymes, including lipase, intracellular diastase, and some that metabolize glucose. Copper sulfate is most toxic to animals that cannot vomit, otherwise it is an emetic. Soluble salts are more toxic than insoluble.

Immediate effects: Skin irritation; anorexia; anemia; capillary damage; muscle spasms; lung and eye irritation; gastrointestinal irritation; diarrhea; headache; sweating; weakness; shock; other blood damage.

Long-term effects: Liver; kidney; blood damage; childhood cirrhosis; and hereditary Wilson's disease, caused by high accumulation of copper due to genetic inability to deal with intake.

Environmental effects: Soil levels that build up to levels toxic to plants, different for particular species, can kill. At 400 ppm, several crops have symptoms, in citrus, 15–30 ppm in soil had effects. Among crops tested, clover and alfalfa were most sensitive, spinach and gladiolas were affected at 98–130 ppm. The copper builds up in the top inches of the soil. In water, very toxic to one-celled animals.

REFERENCES

(1); (2); (3)
Yearbook of Agriculture: Soils. 1957. USDA.

Magnesium

magnesium phosphide

Immediate effects: Transforms to phosphine with water (see aluminum phosphide).

REFERENCE

(1)

Mercury

mercuric chloride (inorganic)
mercurous chloride (inorganic)

Mode of action: Both of these compounds are inorganic mercury, less easily absorbed than the organic forms. Inorganic salts are corrosive. Toxicity of the metallic form, quicksilver, or organic compounds have additional toxicity.

Immediate effects: When ingested, necrosis of mouth, throat, esophagus, and stomach follow immediately, then pain, vomiting, purging, to death. If one survives, severe hemolytic colitis may follow days after. Slower reaction can include mouth inflammation, colitis, kidney damage, kidney failure leading to death.

Long-term effects: Carcinogen; mutagen; teratogen; fetotoxin; neurotoxin; kidney; gastrointestinal damage; autoimmune disease.

Environmental effects: Mercury can move through the environment freely, being water soluble, and transform readily into organic and inorganic forms, with effects on a wide range of organisms, plant and animal. Any form can transform into highly toxic methylmercury or bimethyl mercury, which can then be retransformed into metallic mercury, which can evaporate. Transformation occurs in fresh- or saltwater and in the body as metabolite. Bioaccumulates.

REFERENCES

(1); (2); (3)
For other forms of mercury, check toxicology sources in reference list and National Research Council, *An assessment of mercury in the environment.* Washington, DC: National Academy Press, 1978.

Lead

lead arsenate (see arsenic)

Adds to the toxicity of arsenic.
Mode of action: Lead is a general protoplasmic poison, changing many organs. Nervous system and kidneys are affected, and it is very cumulative.

Immediate effects: Gastrointestinal pain; diarrhea; irritability; headache; drowsiness; confusion; muscular weakness; toxic psychosis; vomiting; convulsions; coma; all reflecting damage to nervous system especially the brain.

Long-term effects: Lead is very cumulative, so effects increase with prolonged exposure. Pain in abdomen, legs, and arms; muscle weakness; sterility; kidney damage; anemia; thyroid damage; in children, lowers mental capacity.

Environmental effects: Because both arsenic and lead are permanent elements, areas that were repeatedly sprayed with lead arsenate, apple orchards, for example, still retain high levels of lead arsenate in the soil, to the point of inhibiting young tree growth. This may be mainly from arsenic, but contaminated soil can have various effects on the area ecology.

REFERENCES

(1); (2); (3)

Silica

diatomaceous earth   silica aerogel

Mode of action: Very drying effect, by absorbing moisture from skin and mucous surfaces in mammals.

Immediate effects: Irritates by desiccating surfaces it touches; the diatomaceous earth has a scratching effect as well.

Long-term effects: The aerogel has very small particles, 3 microns, which may accumulate in the lungs, as do the same size asbestos particles, beyond the capacity of lung cilia to move them up and out. Possible damaging effect is not known. Not implicated in silicosis, but avoid inhaling.

REFERENCES

(1); (3)

Sulfur

AMS (ammonium amidosulfate) (inorganic)
calcium polysulfide
calcium sulfate
carbon disulfide (organic)
dimethyl sulfide
propargite (organic)
sulfur (inorganic)
sulfuryl fluoride*

*The effects of the fluoride component in sulfuryl fluoride add toxic aspects in the liver and kidney damage, mottled teeth, and osteosclerosis (unnatural hardening and density of bone).

Mode of action: Transforms to hydrogen sulfide in the stomach, causing the principal toxic reactions except for contact irritation. Nerve damage may be axonal.

Immediate effects: Eye, skin, and respiratory tract irritation, the latter from both dermal and inhalation exposure. Gastrointestinal irritation; nausea; diarrhea; catharsis; dehydration and electrolyte depletion; speech problems. More severe poisoning: convulsions; unconsciousness; fall in blood pressure; respiratory arrest.

Long-term effects: With severe hydrogen sulfide poisoning survivors may show neurological effects such as amnesia, neurasthenia, disturbance of equilibrium, and more serious brain and cortical damage. Repeated lower exposures might cause comparable damage. Also heart damage, teratogen, fetotoxin (carbon disulfide).

REFERENCES

(1); (2); (3)

Tin

cyhexatin (organic)
fenbutatin-oxide (organic)
fentin hydroxide (organic)
tributyltin chloride (complex organic)
tributyltin fluoride (note added effects from
    fluoride) (organic)
tributyltin oxide (organic)
triphenyltin acetate (organic)

Immediate effects: Irritating to eyes, respiratory tract, skin. Damages nervous system, with effects on the brain causing nausea, headaches, vomiting, dizziness, convulsions, paralysis, loss of consciousness. Aversion to light, mental disturbances, and midabdominal pain occur. Blood sugar can be reduced to very low levels. Eye and skin burns, slow healing. Toxicity varies widely between different species of animals.

Long-term effects: Suspect mutagen; teratogen; kidney, liver, brain, adrenal, and eye damage, which can lead to blindness, and paraplegic paralysis. Also reduced fertility; atrophy of testes; reduced size of ovary.

Environmental effects: Use of tin compounds in antifouling paints on ships was restricted because of widespread toxicity to a wide range of aquatic organisms, at low levels from the dispersal of paints.

REFERENCES

(1); (2); (3)
Hallenbeck, W. H., and K. M. Cunningham. 1985. *Pesticides and human health.* New York: Springer-Verlag.

Zinc

zinc phosphide
zinc coposil (see copper)
zineb, ziram (see thiocarbamate)

Mode of action: Transforms to phosphine in contact with acid (see aluminum phosphide).

REFERENCE

(2)

MISCELLANEOUS

Under this heading we have put pesticides that do not fit well into other classes, or that are one of a kind, or for which we have insufficient information to fit them into a group. Since the long-term effects for each pesticide are on the regular chart, this list is limited to immediate effects, as well as mode of action if known, and whatever environmental effects are known.

If we have found no specific data on immediate effects, but the label of the product being used specifies protective clothing and thorough washing of eyes, skin, and clothing after use, perhaps with the precaution of washing clothing separately from other laundry, this is a general warning of immediate effects.

Amitraz

Immediate effects: General debilitation; hypothermia; hyperglycemia; slow heartbeat; hypotension; dilation of pupils; sedation; excessive urination; vomiting; loss of balance; depression of central nervous system.

REFERENCES

(1); (2)

Antu

Immediate effects: Breathing difficulty; turning blue; eczema; pulmonary rales; vomiting; hyperglycemia; urinary failure.

REFERENCES

(1); (2); (3)

Arosurf

Mode of action: Reduces surface tension on water, so air-breathing insects drown.
Immediate effects: Skin irritation.

REFERENCE

U.S. EPA. 1984. *Prosurf MSF Chemical Information Fact Sheet.* Office of Pesticide Programs.

Assert

Immediate effects: Skin irritation; corrosive to eyes.

REFERENCE

American Cyanamid Material Safety Datasheet, February 12, 1990.

Auramine

Immediate effects: Severe skin irritation and destruction; vomiting; fever; headache; yellow vision.

REFERENCE

Arena, J. M., and R. H. Drew, eds. 1986. *Poisoning: Toxicology, symptoms, treatments,* 5th ed. Springfield, IL: Charles C. Thomas.

Avitrol 100 and 200

Immediate effects: Hyperexcitability; salivation; tremors; muscular uncoordination; convulsions; heart or respiratory failure.
Environmental effects: Hazard to nontarget birds, domestic animals.

REFERENCES

(1); (2); (3)

Azacosterol

Immediate effects: ?

Azobenzene

Mode of action: Spleen damage.
Immediate effects: Jaundice; loss of consciousness; cardiac insufficiency; related to the transformation product, azomyte.

REFERENCES

(1); (2)

Benalaxyl

Immediate effects: ?

Benazolin

Immediate effects: Irritates skin and eyes.

REFERENCE

(2)

Bentazon

Immediate effects: Irritates eyes and mucous membranes; vomiting; diarrhea; weight loss; apathy; anorexia; loss of balance; prostration. Irritates the prostrate gland.

REFERENCES

(2); (3)
EPA health advisory summary, bentazon. 1989. Washington, DC: EPA.

6-Benzyladenine

Immediate effects: ?

Bromethalin

Mode of action: Uncouples oxidative phosphorylation, largely through the transformation product, desmethylbromethalin. Edema of brain and spinal column, nerves, leading to loss of myelin nerve sheath, reduction of nerve impulses.
Immediate effects: Skin and eye irritation; hind leg weakness; loss of tactile sensation; death by respiratory arrest.

REFERENCE

van Lier, R. B. L., and L. Cherry. 1988. The toxicity and mechanism of action of bromethalin: a new single-feeding rodenticide. *Fundamental and Applied Toxicology* 11:664–672.

Bromopropylate

Immediate effects: Skin irritation.

REFERENCE

(1)

Bronopol

Immediate effects: Allergic contact dermatitis.

REFERENCE

Robertson, M. H., et al. 1982. *Archives of dermatitis.* December.

2-Butanamine

Immediate effects: Skin and lung irritation.

REFERENCE

Worthing, C. R. 1991. *The pesticide manual: A world compendium.* 9th ed. British Crop Protection Council.

Chloramben

Immediate effects: Skin and respiratory tract irritation.

REFERENCES

(1); (2); (3)

Chlorflurecol (Chlorflurenol)

Immediate effects: Eye irritation.

REFERENCE

(2)

Chlorine

Immediate effects: Eye and skin irritation; mucous membrane erosion.

REFERENCE

(1)

Chlorofluorocarbons

Immediate effects: Humming in ears; tingling; slurred speech; tremors; apprehension; EKG changes; amnesia; partial loss of consciousness; irregular heartbeat.

REFERENCE

EPA, Office of Water. 1989. Drinking water health advisory for dichloromethane (Freon 12).

Cholecalciferol

Mode of action: Moves bone calcium to plasma in rodents.
Immediate effects: Excessive calcium and phosphate in blood.

REFERENCE

(2)

Cinmethylin

Immediate effects: Eye and skin irritation.

REFERENCE

Humburg, N. E., et al. 1989. *Herbicide handbook of the Weed Science Society of America,* 6th ed. Champaign, IL: WSSA.

Clofentezine

Immediate effects: Eye and skin irritation.

REFERENCE

Agriculture Canada. 1989. *Decision document, clofentezine.*

Clomazine

Immediate effects: ?

Clopyralid

Immediate effects: Eye damage, skin irritation.

REFERENCE

Farm chemicals handbook. 1991. Willoughby, OH: Meister Publishing Co.

Credazine

Immediate effects: Related to diphenyl ethers, may act similarly.

REFERENCE

Corbett, J. R. 1974. *The biochemical mode of action of pesticides.*

Cyclohexane

Immediate effects: Skin irritant; central nervous system dysfunction; narcotic (brief anesthesia); diarrhea; vascular damage and circulatory collapse; convulsions; tissue destruction.

REFERENCE

(1)

Cyclohexanone

Immediate effects: Eye irritation; narcosis; central nervous system depression; respiratory arrest; dermatitis.

REFERENCES

(1); (3)

Dalapon

Immediate effects: Eye and skin irritation; conjunctivitis; corneal damage; gastrointestinal disturbances; vomiting; respiratory irritation; lethargy.
Environmental effects: Can enter groundwater.

REFERENCES

(1); (3)

Daminozide

Immediate effects: Eye and skin irriation.

REFERENCE

California Pesticide Safety Information Series-21, December 1990.

Dazomet

Immedate effects: Skin irritation: clonic convulsions.

REFERENCE

(1)

Dichloroethyl Ether

Immediate effects: Eye irritation; tearing; mucous membrane irritation; nausea; vomiting; respiratory irritation, with cough; pulmonary lesions; bronchitis; death from respiratory collapse. Suspect kidney congestion; brain congestion.

REFERENCE

Hallenbeck, W. H., and K. M. Burns. 1985. *Pesticides and human health.* New York: Springer-Verlag.

Dikegulac Sodium

Immediate effects: ?

Dimethylamine

Immediate effects: Nasal damage from inhalation. Skin and mucous membrane irritation.

REFERENCE

Budavari, S. M., et al. 1989. *The Merck index,* 11th ed. Rahway, NJ: Merck and Co.

Diphenylamine

Immediate effects: Protracted gastroenteritis, diarrhea; anorexia; emaciation; hyperthermia; blood damage (methoglobinemia). Death may be delayed two to three weeks after lethal dose.

REFERENCE

(1)

Disparlure

Immediate effects: No known mammalian toxicity; lure for male gypsy moths.

Long-term toxicity: Contact causes the body, clothes, and associated objects to retain the chemical for many years, making the individual and possessions very attractive to male gypsy moths.

REFERENCE

Ceameron, E. A., 1983. Apparent long-term bodily contamination by Disparlure, the gypsy moth (*Porthetria dispar*) attractant. *J. Chemical Ecology,* 9(1):33–37.

Dodine

Immediate effects: Irritates eyes, skin, gastrointestinal tract; nausea; vomiting; diarrhea.

REFERENCE

(3)

Drazoxolon

Immediate effects: No marked irritant effect on skin and eyes, but may be some sensitization. At high exposure, convulsions. Death comes from respiratory failure, heart damage.

REFERENCE

(1)

Ethofumesate

Mode of action: ?
Immediate effects: Eye and skin irritation.

REFERENCE

Farm chemicals handbook. 1991. Willoughby, OH: Meister Printing Co.

Ethoxyquin

Immediate effects: Skin irritant; depression; reversible liver changes.

REFERENCE

(1)

Ethylene Dibromide (EDB)

Immediate effects: Eye irritation, mucous membrane irritation; gastroenteritis; decreased appetite; headache; sleeplessness; dizziness.

REFERENCES

(1); (2); (3)

Fenoxaprop-ethyl

Mode of action: ?
Immediate effects: Skin and eye irritant.

REFERENCE

Worthing, C. R. 1991. *The pesticide manual: A world compendium,* 9th ed. British Crop Protection Council.

Flamprop-Isopropyl

Immediate effects: "No case of human intoxication recorded."

REFERENCE

(1)

Flurecol-Butyl (flurenol)

Immediate effects: ?

Fluridone

Immediate effects: Skin and eye irritation.

REFERENCE

Humburg, N. E., et al. 1989. *Herbicide handbook of the Weed Science Society of America,* 6th ed. Champaign, IL: WSSA.

Glyphosate

Mode of action: Affects enzyme system.
Immediate effects: Eye and skin irritation, gastrointestinal pain, vomiting, diarrhea, swelling lungs, pneumonia, dizziness, headaches, blurred vision, fever, weakness, allergic reactions.
Environmental effects: Depletion of animal species by reduction of plants needed for food and shelter, (EPA identified 76 endangered species that may be jeopardized, including 73 plants, a toad, and a beetle). Has been found in groundwater; in surface water affects fish directly and by temperature change.

REFERENCE

(2); (3)
Corbert, J. R., et al. 1984. *The biochemical mode of action of pesticides.* London, UK: Academic Press.
Cox, Caroline. 1991. Glyphosate. *J. Pest. Reform,* Summer 1991. 35–38, (includes additional references not cited here).
Hietanen, E., et al. 1983. *Acta Pharmacol. Toxicol.* 53:103-112.
Holtby, L. B. 1989. Changes in the temperature regime of a valley-bottom tributary of Carnation Creek, British Columbia, oversprayed with the herbicide Roundup (glyphosate), In Reynolds, P. E. (ed.), *Proceedings of the Carnation Creek Herbicide Workshop.* Sault. Ste. Marie, Ontario, Canada: Forest Pest Management Institute.
SAIF Corporation. 1987. Occupational Disease Referral (claim number deleted). Industrial Hygiene, Loss Control Section.
Santillo, D., et al. 1989. *J. Wild. Manage.* 53(1):64-71,164-172.
Sawada, Y., et al. 1988. *Lancet.* 1(8580):299.

U.S. EPA. 1980. *Summary of reported incidents involving glyphosate (isopropylamine salt)*. Report No. 375. Washington, DC: Health Effects Branch, EPA.

U.S. EPA. 1986. *Guidance for the reregistration of pesticide products containing glyphosate*. Washington, DC: Office of Pesticides Programs.

U.S. EPA. 1992. *EPA Pesticide in Ground Water Database: A Compilation of Monitoring Studies: 1971-1991 National Summary*. Office of Pesticides and Toxic Substances, EPA.

Hydramethylnon

Immediate effects: Eye irritant.

REFERENCE

Technical information bulletin, Amdro. 1980. American Cyanamid Co.

Imazapyr

Immediate effects: Eye and skin irritation.

REFERENCE

Humburg, N. E., et al. 1989. *Herbicide handbook of the Weed Science Society of America*, 6th ed. Champaign, IL: WSSA.

d-Limonene

Immediate effects: Eye and skin irritation, sensitizer. Blood, albumin in urine. Vomiting; nausea; salivation; muscle tremors; staggering; imbalance; hypothermia. Effects especially severe in cats.

REFERENCE

(1)

MBT

Mode of action: Reacts in the body to "the corresponding dithiocarbamate."

Immediate effects: Salivation; dilation of blood vessels; convulsions.

REFERENCE

(1)

Methoprene

Environmental effects: Risk to aquatic invertebrates if it reaches water.

REFERENCE

EPA Registration Eligibility Document. 1991. *Methoprene*.

Naphthaleneacetic Acid

Immediate effects: Difficulty breathing; imbalance; lethargy; prostration; gastrointestinal damage. The ethyl ester is an eye irritant.

REFERENCE

(1)

Naptalam

Immediate effects: Irreversible eye damage; if ingested, avoid alcohol.

REFERENCE

U.S. EPA. 1985. *Naptalam Pesticide Fact Sheet*. Office of Pesticide Programs.

Nitrapyrin

Immediate effects: "Get immediate medical aid . . . wash eyes and skin, do not induce vomiting."

REFERENCE

Farm chemicals handbook. 1991. Willoughby, OH: Meister Publishing Co.

Norflurazon

Immediate effects: ?

Oxadiazon

Immediate effects: Eye and skin irritation; depressed motor activity.

REFERENCE

(1)

Picloram

Mode of action: Uncouples oxidative phosphorylation.

Immediate effects: Irritation of skin, eye, respiratory tract; nausea; diarrhea; vaginal bleeding; dermatitis; hair loss (alopecia); depression; prostration; imbalance; tremors; convulsions. No known antidote.

REFERENCE

(3)

Potassium Bromide

Immediate effects: Vomiting; irritability; imbalance; confusion; skin rash; mania; hallucinations; other neurological signs; sensory disturbances; coma. Iodine-sensitive people should not take internally.

REFERENCES

(1)

Joyce, D. A., et al. 1985. Renal failure and upper urinary tract obstruction after retrograde pyelography with potassium bromide solution. *Human Toxicology* 4:481-490.

Potassium Permanganate

Immediate effects: Eye and skin irritation; esophagal stricture.

REFERENCE

Kochbar, R. 1986. Potassium permanganate induced oesophageal stricture. *Human Toxicology* 5:393-394.

Propionic Acid

Immediate effects: Allergic skin reaction; desquamation of gastric muscosa; bleeding.

REFERENCE

(1)

Soap

Immediate effects: Eye and skin irritation.

REFERENCE

(1)

Sodium Chlorate

Immediate effects: Eye, skin, and mucous membrane irritation; destruction of red blood cells; irregular heartbeat; low urine; phosphate loss; low blood oxygen.

REFERENCE

(3)

Sodium Omadine

Immediate effects: ?

Sodium Hyperchorite

Immediate effects: Corrodes skin and mucous membranes; pulmonary edema; esophagal closing; toxemia; shock; perforation; hemorrhage; infection; obstruction.

REFERENCE

(1)

ORGANOCHLORINE (CHLORINATED HYDROCARBONS)

aldrin	chlorfenethol
Bandane	chlorobenzilate
benzene hexachloride	chloroform
bithionol	chloroneb
carbon tetrachloride	chloropicrin
chlorbenside	chloropropylate
chlordane	4-CPA
chlordecone	D-D

DBCP	heptachlor
DDD	hexachlorobenzene
DDE	hexachlorophene
DDT	lindane
dicamba	methoxychlor
dichloropropane	methylene chloride
dichloropropene	mirex
dicofol	paradichlorobenzene
dieldrin	PCB
dienochlor	PCNB
endosulfan	pentachlorophenol
endrin	tetrachloroethylene
epichlorohydrin	tetradifon
ethylan	toxaphene
ethylene dichloride	triclopyr
fenac	

Mode of action: Interfere with transmissions of nerve impulses across axons disrupting primarily the central nervous system.

Immediate effects: Convulsions (may occur for several days after exposure); uncoordination; induces rapid metabolism of drugs and naturally occurring steroid hormones; hypersensitivity of skin or face and extremities; headache; dizziness; nausea; vomiting; tremors; confusion; muscle weakness; involuntary eye movements; slurred speech; pain in chest and joints; skin rash; labored breathing; central nervous stimulation followed by depression; diarrhea; brain wave disturbances; headache; hyperthermia; hypertension; salivation; sweating.

Long-term effects: Cumulative; transfers through placenta to fetus; found in mothers' milk. Carcinogens; liver and kidney damage; suspect teratogens; suspect mutagens; fetotoxins; aplastic anemia; "reproductive effects"; testicular damage; eye damage; affects hormone levels; central nervous system damage; bladder, kidney, lung, and thyroid damage; blood and spleen damage; anemia; recurrent asthma; irregular heartbeat; atrophy of adrenal cortex; behavior changes even in young of mother exposed at low levels during pregnancy; embryotoxin; decreased fertility immunotoxin; abnormal brain waves; increased mortality in young; lung damage; teratogens; porphyria cutanea tarda; sleep disturbance; hallucinations; anemia.

Environmental effects: Bioaccumulate; persistent; many are volatile, travelling long distances in the atmosphere and settling in distant locations; decreased fertility in birds; egg-shell thinning in birds; groundwater contaminants.

REFERENCES

(1); (2); (3)
Colburn, T.E., et al. 1990. *Great Lakes, great legacy?* Washington, DC: The Conservation Foundation and Institute for Research on Public Policy.

Hallenbeck, W. H., and K. M. Burns. 1985. *Pesticides and human health.* New York: Springer-Verlag.

Stickel, L. F. 1968. *Organochlorine pesticides in the environment.* Special Scientific Report-Wildlife no. 119. Washington, DC: U.S. Fish and Wildlife Service.

U.S. EPA. 1990. National pesticide survey phase 1 report. Office of water. PB91-125765.

ORGANOPHOSPHATE
(and related organophosphorus compounds)

acephate	ethion
acephate-met	ethoprop
Akton	etrimfos
azinphos-ethyl	famphur
azinphos-methyl	fenamiphos
bensulide	fenitrothion
Bomyl	fensulfothion
Bromophos	fenthion
Bromophos-ethyl	fonofos
carbophenothion	fosetyl-al
chlorfenvinphos	CC 6506
chlormephos	isazophos
chlorphoxim	isofenphos
chlorpyrifos	leptophos
coumaphos	malathion
crotoxyphos	methidathion
crufomate	methyl parathion
cyanophenphos	mevinphos
cyanophos	monocrotofos
cythioate	naled
DEF	omethoate
demeton	oxydemeton-methyl
demeton-methyl	parathion
dialifor	phorate
diamidfos	phosalone
diazinon	phosmet
dicapthon	phosphamidon
dichlofenthion	phoxim
dichlorvos	pirimiphos-ethyl
dicrotophos	pirimiphos-methyl
dimefox	ronnel
dimethoate	sulfo TEPP
dioxabenzofos	sulprofos
dioxathion	temephos
disulfoton	TEPP
ditalimfos	terbufos
DMPA	tetrachlorvinphos
edifenphos	triaziphos
EPN	trichlorfon
ethephon	vamidothion

Mode of action: Acetocholinesterase inhibitor, damaging nerve function, except for glyphosate.

Immediate effects: Behavioral disturbances; uncoordination; muscle twitching; headache; nausea; dizziness; anxiety; irritability; loss of memory; sleep pattern change; restlessness; weakness; tremor; abdominal cramps; diarrhea; sweating; salivation; tearing; excessive nasal discharge; blurred vision; constriction of pupil; slowed heartbeat; confusion; incontinence; hypertension.

Long-term effects: Delayed neurotoxicity [". . . tingling and burning sensations in the limb extremities followed by weakness in the lower limbs and atoxic. This progresses to paralyses, which, in several cases, affect the upper limbs also. . . . Recovering is seldom complete in adults; with the passage of time the clinical picture changes from flaccid to an epastic type paralysis" WHO (1986, p. 59)]; some are cumulative; persistent anorexia; weakness; malaise; nerve damage via destruction of myelin sheath around nerve fibers; carcinogens; mutagens; fetotoxins; hormonal inhibition; eye damage; suspect mutagens; suspect carcinogens; sterility and impotence; embryotoxins; suspect teratogens; immunotoxins; indication of bone marrow damage and aplastic anemia; kills white blood cells; sperm and other reproductive abnormalities; suspect viral enhancers; ulcers; abnormal brain waves; reduced protein synthesis in fetus; liver damage; kidney damage; suppressed antibody reproduction; decreased auditory attention, visual memory, problem solving, balance, and dexterity.

Environmental effects: Responsible for the deaths of large numbers of birds on turf and in agriculture; affect breeding success in birds; embryotoxins in birds; can change feeding habits in birds; para-Nitrophenol, a transformation product of parathion, is a groundwater contaminant.

REFERENCES

(3)

Bennett, R. S. 1989. Role of dietary choices in the ability of bobwhite to discriminate between insecticide treated and untreated food. *Environ. Toxicol. Chem.* 8:731–738.

Duffy, F. H., et al. 1980. Long-term effects of the organophosphate sarin on EEGs in monkeys and humans. *Neurotoxicology* 1:667–689.

Hallenbeck, W. H., and K. M. Burns. 1985. *Pesticides and human health.* New York: Springer-Verlag.

Hoffman, D. J. 1981. Effects of malathion, diazinon, and parathion on mallard embryo development and cholinesterase activity. *Environmental Research* x:xxx–xxx.

Ishikawa, S. 1971. Eye disease induced by organic phosphorous insecticides. *Acta. Soc. Opthamol. Jap.* 75:841–855.

Rosenstock, L., et al. 1991. Chronic central nervous system effects of acute organophosphate pesticide intoxication. *The Lancet* 338:223–227.

Tamura, O., et al. 1975. Organophosphorous pesticides as cause of myopia in school children: an epidemiological study. *Jap. J. Opthalmol.* 19:250–253.

U.S. EPA. 1990. National pesticide survey. Office of water. PB91-125765.

White, D. H., et al. 1979. Parathion causes secondary poisoning in laughing gull breeding colony. *Bull. Environ. Contam. Toxicol.* 23:281–284.

World Health Organization (WHO). 1986. Environmental health criteria no. 63 organophosphorus insecticides: A general introduction. Geneva: WHO.

OIL, PETROLEUM

fuel oil	stoddard solvents
kerosene	other names:
mineral oil	dormant oil
mineral spirits	miscible oil
naphtha	summer oil
naphthalene (see phenols)	

Immediate effects: Skin and mucous membrane irritation; burning sensation in chest; headache; ringing in the ear; nausea; weakness; restlessness; uncoordination; confusion; eye irritation; disorientation; blue coloration of extremities due to reduced oxygen in blood; aspiration toxicity is high and can occur during vomiting after an oral ingestion; death due to respiratory failure.

Long-term effects: Brain damage; carcinogens; mutagens; anemia; liver damage; lung damage; central nervous system depression; kidney damage.

REFERENCE

(1)

PHENOL

carbolic acid
creosote
cresylic acid
fluorodifen
naphthalene (major component of creosote)

Mode of action: General protoplasmic poison toxic to all cells, corrosive. Penetrates by dermal exposure rapidly, from solution or vapor.

Immediate effects: By ingestion, causes areas of dead tissue on face, mouth, and esophagus. Edema of larynx and lungs, sometimes with reduced pain because of damage to nerves (demyelination, other destruction of nerve fibers that transmit sensations). Depression of central nervous system; hypothermia; heart and circulation depression; coma; respiratory arrest; kidney failure. Repeated lower exposures cause diarrhea, dark urine, burning and sores in mouth. Additional symptoms for naphthalene include skin and eye irritant, anorexia, gastrointestinal disturbances, disorientation, red blood cell destruction. Babies exposed from clothing (mothballs) can develop kernicterus, with high levels of bilirubin in blood, giving a yellow color, severe neural symptoms, and widespread destructive changes.

REFERENCES

(1); (3)

Chlorophenol

dichlorophen
O-benzyl-p-chlorophenol

Immediate effects: Irritation of nose, throat, and eyes; sweating; weakness; dizziness; anorexia; nausea; increased body temperature; muscle spasms; tremor; labored breathing; abdominal pain; vomiting; restlessness; excitement; mental confusion; intense thirst; rapid heartbeat.

Long-term effects: Dermatitis; chloracne via dioxin contaminants; weight loss; carcinogens; teratogens; fetotoxins; embryotoxins; immunotoxins; aplastic anemia; nerve and liver damage; kidney and spleen damage.

REFERENCES

(1); (2); (3)

Nitrophenol

binapacryl	dinoseb
dinex	dinoterb
dinocap	dinoterb acetate
dinitrophenol	dinitrocresol

Mode of action: Destruction of cell membrane; cell respiration resulting in increased metabolism of carbohydrate and fat stores consumption.

Immediate effects: Yellow staining of skin and hair from dermal exposure; sweating; thirst; fever; headache; confusion; malaise; weakness; warm flushed skin; increased heartbeat; quick shallow breathing; most severe occupational dermal poisonings occur while laboring in hot environments.

Long-term effects: Liver, kidney, and nervous system damage; restlessness; apprehension; anxiety; manic behavior indicating brain damage; weight loss; dehydration; cumulative; teratogens; male sterility.

Environmental effects: Groundwater contaminants.

REFERENCES

(1); (2); (3)

PHENOXY

acilfluorfen	erbon
bifenox	fluazifop-butyl
CNP	MCPA
2,4-D	mecoprop
2,4-DB	silvex
dichlorprop	2,4,5-T
diclofop-methyl	

Mode of action: Act as synthetic growth hormones in plants; in animals it is poorly understood.

Immediate effects: Skin and mucous membrane irritation; dizziness with prolonged inhalation; vomiting; chest pain; diarrhea; headache; confusion; muscular stiffness; unconsciousness; increased acidity of blood; hyperventilation; nerve damage; brain wave changes; eye irritation; swelling of extremities; incontinence; sweating; stupor; respiratory depression.

Long-term effects: Carcinogens; heart; liver, and kidney damage; delayed fetal development; suspect mutagens; teratogens; fetotoxins; anorexia; ulceration of mouth and throat; immunotoxin; nerve damage. Several pesticides in this class are contaminated with dioxins. See dibenzodioxin/dibensofuran class.

Environmental effects: Groundwater contaminants.

REFERENCES

(1); (3)

Sjöden, P. O., and U. Söderberg. 1978. Phenoxyacetic acids: sublethal effects. In *Clorinated phenoxy acids and their dioxins.* C. Ramel, ed. Royal Swedish Academy of Sciences, Ecological Bulletin No. 27.

PHTHALATE

captafol	dimethyl phthalate
captan	endothall
DCPA	folpet
dibutyl phthalate	

Mode of action: Interfere with cell respiration.

Immediate effects: Skin, eye, and respiratory tract irritants; hypothermia; irritability; listlessness; blood in urine; death due to heart or lung failure; convulsions; may depress central nervous system.

Long-term effects: Skin sensitizers; anorexia; carcinogens; mutagens; teratogens; fetotoxins; immunotoxins; testicular atrophy.

Environmental effects: ?

REFERENCES

(1); (2); (3)

PYRETHROID

allethrin	flucythrinate
barthrin	fluvalinate
bifenthrin	τ-fluvalinate
bioallethrin	kadethrin
bioresmethrin	karate
cismethrin	permethrin
cyfluthrin	phenothrin
λ|cyhalothrin	resmethrin
cypermethrin	S-bioallethrin
d-cis,trans-allethrin	synthetic pyrethrum,
deltamethrin	pyrethrins
dimethrin	tefluthrin
esbiothrin	tetramethrin
fenpropathrin	tetramethrin (1R)-isomers
fenvalerate	tralomethrin

Mode of action: Pyrethroids inhibit sodium and potassium conduction in nerve cells and block nerve impulse transmission. Many times pyrethroids are mixed with piperonyl butoxide in formulations.

Immediate effects: Symptoms similar to DDT poisoning. T-syndrome: tremors; exaggerated startled response; hyperthermia. CS-syndrome: excessive writhing and salivation; decreased startle response; increase in adrenalin and blood sugar. Other possible effects: convulsions; diarrhea; headache; vomiting; labored breathing; excessive nasal mucous discharge; irritability; sweating; sudden swelling of face, eyelids, lips, mouth, and throat tissues. Hay fever-like symptoms; elevated pulse.

Long-term effects: Suspect mutagens; suspect teratogens; suspect carcinogens; immunotoxins; decreased hormone release from brain; some may be cumulative.

Environmental effects: Highly toxic to fish, bees, and aquatic anthropods.

REFERENCES

(2)

Hallenbeck, W. H., and K. M. Burns. 1985. *Pesticides and human health.* New York: Springer-Verlag.

PYRIMIDINE

dimethirimol	fenarimol
ethirimol	

Mode of action: May affect enzymes, including some related to sexual activity.

Immediate toxicity: ?

REFERENCES

Farm chemicals handbook. 1991. Willoughby, OH: Meister Printing Co.

Hirsch, K. S., et al. 1986. Studies to elucidate the mechanism of fenarimol-induced infertility in the male rat. *Toxicology and Applied Pharmacology* 86:391–399.

QUATERNARY AMMONIUM

benzalkonium chloride

Immediate toxicity: Severe skin and eye irritation; mucous membrane damage; gastrointestinal erosion; pulmonary edema; paralysis of skeletal muscles; cardiomuscular collapse; depression of central nervous system; convulsions.

REFERENCE

(1)

QUINONE

anthraquinone   dichlone
chloranil

Mode of action: ?
Immediate effects: Skin irritation and sensitization; diarrhea; depression of central nervous system; coma; death.
Long-term effects: Suspect mutagens; liver damage.
Environmental effects: ?

REFERENCE

(1)

THIOCARBAMATE

amobam	mancozeb
butylate	maneb
cartap	metam sodium
CDEC	metiram
cycloate	nabam
diallate	thiram
disulfiram	triallate
EPTC	vernolate
ethiolate	zineb
ferbam	ziram

Mode of action: Inhibits acetaldehyde dehydrogenase, which is essential in conversion of acetaldehyde to acetic acid.
Immediate effects: Skin, eye, and respiratory tract irritation; skin sensitization; hyperactivity; central nervous system depression; blood diarrhea; general weakness. Thiram is the methyl analog of disulfiram, used in drug therapy for alcoholics. In combination with alcohol, disulfiram quickly induces flushing, restlessness, anxiety, headache, nausea, vomiting, hyperventilation, constriction sensation in the neck, chest pain, sweating, thirst, weakness, vertigo, and possible circulatory collapse, coma, and death. These reactions may occur when thiram and alcohol exposure coincide.
Long-term effects: Protein-deficient animals are more susceptible to toxicity of some thiocarbamates; carcinogens; mutagens; delayed neurotoxicity; "testicular and ovarian effects"; kidney damage; sperm damage; teratogen; fetotoxin; anemia. The ethylene thiourea (ETU), a transformation product of some thiocarbamates, is characterized as a carcinogen, mutagen, teratogen, and goiterogen (thyroid damage).
Environmental effects: ETU, is a groundwater contaminant.

REFERENCES

(1); (2); (3)

TRIAZINE

ametryn	dipropetryn
anilazine	ethiozin
atrazine	hexazinone
aziprotryne	promaton
chlorinated isocyanurates	promatryn
cyanazine	propazine
cyprazine	simazine
cyromazine	terbutryn
desmetryn	

Mode of action: May disturb the metabolism of vitamins.
Immediate effects: Skin and eye irritation; nausea; vomiting; diarrhea; muscular weakness; salivation.
Long-term effects: Carcinogens; suspect mutagens; immunotoxin; adrenal damage; kidney and urinary tract stone formation; teratogens; lung damage; suspect fetotoxins; liver and kidney damage; disturbances in sperm production.
Environmental effects: Groundwater contaminants.

REFERENCES

(2); (3)

Hallenbeck, W. H., and K. M. Burns. 1985. *Pesticides and human health.* New York: Springer-Verlag.

TRIAZOLE

amitrole   triadimefon
flusilazole

Mode of action: Inhibition of liver enzymes.
Immediate effects: ?
Long-term effects: Carcinogens; suspect mutagens; may affect growth rate; goiter producing; fetotoxins; liver damage.

REFERENCE

(1)

URACIL

bromacil   terbacil

Immediate toxicity: ?

UREA

benzthiazuron   diuron
chlorbromuron   fenuron
chlorfluazuron   flufenoxuron
chlorimuron   fluometuron
chloroxuron   linuron
chlorsulfuron   monolinuron
cycluron   monuron
cymoxanil   siduron
diflubenzuron   tebuthiuron

Immediate effects: Anemia; skin and eye irritation; diarrhea; nausea; vomiting; pulmonary edema and congestion.
Long-term effects: Reduces oxygen carrying capacity of blood; suspect mutagens; suspect teratogens; anemia; growth inhibition; poisoning potential increased with protein deficient diet; lung, liver, kidney, and spleen damage; carcinogens.

REFERENCE

(1)

REFERENCES

Chapter Six

References

COMMONLY USED SOURCES

The sources listed here are those that we used with frequency when gathering information for this guide. The following section comprises the citations that support the statements on the charts in Chapter 6. To avoid repeating full citations, we use numbers to refer back to these 58 sources. For example, the citations for acephate begin as

acephate
(1) *33*
(2) *7*
(3) Rattner, B.A., et al. 1984. *Arch. Environ. Contam. Toxicol.* 13:483–491.

The first citation refers to *33*, which on the list of commonly used sources is:

Spencer, E. Y. 1982. *Guide to the chemicals used in crop protection,* 7th ed. Publication 1092. London, Ontario, Canada: Research Branch, Agriculture Canada.

When dealing with EPA Reregistration Standards and Special Review Technical Support Documents we put the name of the chemical and year that appear on the document next to the numbers *40* or *41*.

Sources marked with an asterisk should appear in your local library. If you cannot find them there, you might suggest that the librarian obtain them.

1. Atkins, E. L., E. A. Greywood, and R. L. MacDonald. 1975. *Toxicity of pesticides and other agricultural chemicals to honey bees.* Leaflet 2287. Laboratory Studies. Division of Agricultural Sciences, University of California.
2. Ben-Dyke, R., D. M. Sander, and D. N. Noakes. 1970. Acute toxicity data for pesticides. *World Rev. Pest Control* 9:119–127.
3.* Budavari, S., M. J. O'Neil, A. Smith, and P. E. Heckelman. 1989. *The Merck index: An encyclopedia of chemicals, drugs, and biologicals,* 11th ed. Rahway, NJ: Merck and Co.
4. *EXTOXNET: Extension Toxicology Network.* 1989. Cooperative Extension Offices of Cornell University, the University of California, Michigan State University, and Oregon State University.
5.* *Farm chemicals handbook.* 1991. Willoughby, OH: Meister.

6. Gaines, T. B. 1960. The acute toxicity of pesticides to rats. *Toxicol. Appl. Pharmacol.* 2:88–89.
7. Gaines, T. B., and R. E. Linder. 1986. Acute toxicity of pesticides in adults and weanling rats. *Fundam. Appl. Toxicol.* 7:299–308.
8. Gaines, T. B. 1969. Acute toxicity of pesticides. *Toxicol. Pharmacol.* 14(3):515–534.
9.* Gosselin, R. E., R. P. Smith, and H. C. Hodge. 1984. *Clinical toxicology of commercial products.* Baltimore, MD: Williams and Wilkins.
10.* Hallenbeck, W. H., and K. M. Cunningham Burns. 1985. *Pesticides and human health.* New York: Spring-Verlag.
11.* Hayes, W. J. 1982. *Pesticides studied in man.* Baltimore, MD: Williams and Wilkins.
12.* Hayes, W. J., and E. R. Laws. 1991. *Handbook of pesticide toxicology, vol. 1: General principles.* San Diego, CA: Academic.
13.* Hayes, W. J., and E. R. Laws. 1991. *Handbook of pesticide toxicology, vol. 2: Classes of pesticides.* San Diego, CA: Academic.
14.* Hayes, W. J., and E. R. Laws. 1991. *Handbook of pesticide toxicology, vol. 3: Classes of pesticides.* San Diego, CA: Academic.
15. Hill, E. F., and M. B. Camardese. 1986. *Lethal dietary toxicities of environmental contaminants and pesticides to coturnix.* Technical Report 2. Washington, DC: U.S. Department of the Interior, Fish and Wildlife Service.
16. Hill, E. F., R. G. Health, J. W. Spann, and J. D. Williams. 1975. *Lethal dietary toxicities of environmental pollutants to birds.* Technical Report 2. Washington, DC: U.S. Department of the Interior, Fish and Wildlife Service.
17. Howard, P. H., E. M. Michalenko, W. F. Jarvis, D. K. Basu, G. W. Sage, W. M. Meylan, J. A. Beauman, and D. A. Gray. 1991. *Handbook of environmental fate and exposure data for organic chemicals, vol. 3.* Chelsea, MI: Lewis.
18. Howard, P. H., ed. 1991. *Handbook of environmental fate and exposure data for organic chemicals, vol. 2. Solvents.* Chelsea, MI: Lewis.
19.* Hudson, E. F., R. K. Tucker, and M. A. Haegele. 1984. *Handbook of toxicity of pesticide to wildlife.* Resource Publication 153. Washington, DC: U.S. Department of the Interior, U.S. Fish and Wildlife Service.
20.* Humburg, N. E., S. R. Colby, E. R. Hill, L. M. Kitchen, R. G. Lym, W. J. McAvoy, and R. Prasad.

1989. *Herbicide handbook of the Weed Science Society of America,* 6th ed. Champaign, IL: Weed Science Society of America.

21.* Johnson W. W., and M. T. Finley. 1980. *Handbook of acute toxicity of chemicals to fish and aquatic invertebrates.* Resource Publication 137. Washington, DC: U.S. Department of the Interior, U.S. Fish and Wildlife Service.

22. Martin, H. 1972. *Pesticide manual: Basic information on the chemicals used as active components of pesticides.* British Crop Protection Council.

23.* Menzie, C. M. 1969–80. *Metabolism of Pesticides—4 Vols.* Special Scientific Reports—Wildlife Nos. 127, 184, 212, 232. Washington, DC: U.S. Department of the Interior, U.S. Fish and Wildlife Service.

24.* Morgan, D. P. 1989. *Recognition and management of pesticide poisonings,* 4th ed. EPA-540/9-88-001. Washington, DC: U.S. Environmental Protection Agency.

25. Moriya, M., T. Ohta, K. Watanabe, T. Miyazawa, K. Kato, and Y. Shirasu. 1983. Further mutagenicity studies on pesticides in bacterial reversion assay systems. *Mutat. Res.* 115:185–216.

26.* National Fire Protection Association. 1986. *Fire protection guide on hazardous materials,* 9th ed. Boston, MA.

27.* National Institute for Occupational Safety and Health. 1990. *Pocket guide to chemical hazards.* Washington, DC: U.S. Department of Health and Human Services, Centers for Disease Control.

28.* National Institute for Occupational Safety and Health. 1991. *Registry of toxic effects of chemical substances.* Cincinnati, OH: U.S. Department of Health and Human Services.

29.* National Toxicology Program. 1989. *Fifth annual report on carcinogens: 1989 summary.* U.S. Department of Health and Human Services, Public Health Service.

30. Pimentel, D. 1971. *Ecological effects of pesticides on non-target species.* Washington, DC: Executive Office of the President, Office of Science and Technology.

31. Schafer, E. W., W. A. Bowles, and J. Hurlburt. 1983. The acute oral toxicity, repellency, and hazard potential of 998 chemicals to one or more species of wild and domestic birds. *Arch. Environ. Contam. Toxicol.* 12:355–382.

32.* Smith, G. J. 1987. *Pesticide use and toxicology in relation to wildlife: Organophosphorus and carbamate compounds.* Resource Publication 170. Washington, DC: U.S. Department of the Interior, U.S. Fish and Wildlife Service.

33. Spencer, E. Y. 1982. *Guide to the chemicals used in crop protection,* 7th ed. Publication 1092.

London, Ontario, Canada: Research Branch, Agriculture Canada.

34.* Thomson, W. T. 1985–1986. *Agricultural chemicals—Book I, Insecticides.* Fresno, CA: Thomson.

35.* Thomson, W. T. 1986. *Agricultural chemicals—Book II, Herbicides.* Fresno, CA: Thomson.

36.* Thomson, W. T. 1986. *Agricultural chemicals—Book III, Fumigants, growth regulators, repellents, and rodenticides.* Fresno, CA: Thomson.

37.* Thomson, W. T. 1988. *Agricultural chemicals—Book IV, Fungicides.* Fresno, CA: Thomson.

38. U.S. Environmental Protection Agency. 1990. *Suspended, cancelled, and restricted pesticides.* Washington, DC: Office of Pesticide Products.

39. U.S. Environmental Protection Agency. *Guidance for the reregistration of pesticide products containing * as the active ingredient.* Washington, DC: Office of Pesticide Programs.

40. U.S. Environmental Protection Agency. *Special review position document.* Washington, DC: Office of Pesticide Programs.

41. Weast, R. C., M. J. Astle, and W. H. Beyer. 1988. *CRC Handbook of chemistry and physics: A ready-reference book of chemical and physical data,* 68th ed. Boca Raton, FL: CRC Press.

42. Wiswesser, W. J. 1976. *Pesticide index,* 5th ed. College Park, MD: The Entomological Society of America.

43. World Health Organization. 1974. *IARC Monographs on the evaluation of the carcinogenic risk of chemicals to humans. Some organochlorine pesticides,* vol. 5. Lyon, France: International Agency for Research on Cancer.

44. World Health Organization. 1975. *IARC Monographs on the evaluation of the carcinogenic risk of chemicals to humans. Some aziridines, N-, S-, O-mustards and selenium,* vol. 9. Lyon, France: International Agency for Research on Cancer.

45. World Health Organization. 1976. *IARC Monographs on the evaluation of the carcinogenic risk of chemicals to humans. Some naturally occurring substances,* vol. 10. Lyon, France: International Agency for Research on Cancer.

46. World Health Organization. 1976. *IARC Monographs on the evaluation of the carcinogenic risk of chemicals to humans. Some carbamates, thiocarbamates, and carbazides,* vol. 12. Lyon, France: International Agency for Research on Cancer.

47. World Health Organization. 1977. *IARC Monographs on the evaluation of the carcinogenic risk of chemicals to humans. Some fumigants, the herbicides, 2,4-D and 2,4,5-T, chlorinated dibenzodioxin and miscellaneous industrial*

chemicals, vol. 15. Lyon, France: International Agency for Research on Cancer.

48. World Health Organization. 1978. *IARC Monographs on the evaluation of the carcinogenic risk of chemicals to humans. Some aromatic amines and related nitor compounds—hair dyes, coloring agents, and miscellaneous industrial chemicals, vol. 16.* Lyon, France: International Agency for Research on Cancer.

49. World Health Organization. 1978. *IARC Monographs on the evaluation of the carcinogenic risk of chemicals to humans. Some N-nitroso compounds, vol. 17.* Lyon, France: International Agency for Research on Cancer.

50. World Health Organization. 1978. *IARC Monographs on the evaluation of the carcinogenic risk of chemicals to humans. Polybrominated biphenyls, vol. 18.* Lyon, France: International Agency for Research on Cancer.

51. World Health Organization. 1979. *IARC Monographs on the evaluation of the carcinogenic risk of chemicals to humans. Some halogenated hydrocarbons, vol. 20.* Lyon, France: International Agency for Research on Cancer.

52. World Health Organization. 1980. *IARC Monographs on the evaluation of the carcinogenic risk of chemicals to humans. Some Nonnutritive sweetening agents, vol. 22.* Lyon, France: International Agency for Research on Cancer.

53. World Health Organization. 1982. *IARC Monographs on the evaluation of the carcinogenic risk of chemicals to humans. Some aromatic amines, anthraquinones, and nitroso compounds, and inorganic fluorides used in drinking-water and dental preparations, vol. 27.* Lyon, France: International Agency for Research on Cancer.

54. World Health Organization. 1983. *IARC Monographs on the evaluation of the carcinogenic risk of chemicals to humans. Miscellaneous pesticides, vol. 30.* Lyon, France: International Agency for Research on Cancer.

55. World Health Organization. 1983. *IARC Monographs on the evaluation of the carcinogenic risk of chemicals to humans. Some food additives, feed additives, and naturally occurring substances, vol. 31.* Lyon, France: International Agency for Research on Cancer.

56. World Health Organization. 1985. *IARC Monographs on the evaluation of the carcinogenic risk of chemicals to humans. Allyl compounds, aldehydes, epoxides and peroxides, vol. 36.* Lyon, France: International Agency for Research on Cancer.

57. World Health Organization. 1986. *IARC Monographs on the evaluation of the carcinogenic risk of chemicals to humans. Some halogenated hydrocarbons and pesticides exposures, vol. 41.* Lyon, France: International Agency for Research on Cancer.

58.* Worthing, C. R. 1991. *The pesticide manual: A world compendium,* 9th ed. British Crop Protection Council.

REFERENCES FOR CHARTS: Numbered references refer to list of Commonly Used Sources.

abamectin
(1) Moye, H.A., et al. 1987. *J. Agric. Food Chem.* 35:859-864.
(2) Wislocki, P.G., et al. 1988. *Toxicol. Appl. Pharmacol.* 94:238-245.
(3) *3*

acephate
(1) *33*
(2) *7*
(3) Rattner, B.A., et al. 1984. *Arch. Environ. Contam. Toxicol.* 13:483-491.
(4) *39. Acephate. 1987.*
(5) U.S. Environmental Protection Agency Science Advisory Panel Hearing on February 12, 1986. Comment made by Dr. Richter of Chevron.
(6) *33*
(7) Geen, G.H., et al. 1984. *J. Environ. Sci. Health.* [B]19(2):131-155.
(8) Frank, R., et al. 1984. *J. Econ. Entom.* 77(5):1110-1115.
(9) Bouchard, D.C., et al. 1982. *J. Econ. Entomol.* 75(5):921-923.
(10) *23*
(11) Zinkl, J.G., et al. 1981. *Arch. Environ. Contam. Toxicol.* 10:185-192.
(12) Merle, R.L. 1979. *Effects of Sevin-4-oil, Dimilin, and Orthene on forest birds in northeastern Oregon.* Research Paper PSW- 148. U.S. Department of the Interior. Fish and Wildlife Service, Pacific Southwest Forest and Range Experiment Station.

acephate-met
(1) Davis, et al. 1974. *J. Econ. Entomol.* 67:766-768.
(2) *7*
(3) U.S. Environmental Protection Agency. 1990. *Tox Oneliner: Methamidophos.* Office of Pesticide Programs. p. 9.
(4) *32*
(5) *5*
(6) *1*
(7) Juarez, L.M., et al. 1989. *Bull. Environ. Contam. Toxicol.* 43:302-309.
(8) Hixson, E.J. 1984. *Unpublished Study No. 82-671-01.* Mobay Chem. Co.

acetochlor
(1) *58*
(2) U.S. Environmental Protection Agency. 1991. *List of Chemicals Evaluated for Carcinogenic Potential.* Health Effects Division.

acetonitrile
(1) *3*
(2) Willhite, C.C. 1983. *Teratology.* 27:313-325.

acifluorfen
(1) *20*
(2) U.S. Environmental Protection Agency. 1989. *Acifluorfen Health Advisory Summary.* Office of Water.

acrolein
(1) *20*
(2) *9*
(3) Curren, R.D., et al. 1988. *Mutat. Res.* 209:17-22.
(4) Hales, B.F. 1982. *Cancer Res.* 42:3016-3021.
(5) Au, W., et al. 1980. *Cytogenet. Cell Genet.* 29:108-116.
(6) *19*

(7) *30*

acrylonitrile
(1) *3*
(2) Hogy, L.L., et al. 1986. *Cancer Res.* 46:3932-3938.
(3) Solomon, J.J., et al. 1984. *Chem. Biol. Interact.* 51:167-190.
(4) Myhr, B., et al. 1985. *Prog. Mutat. Res.* 5:555-569.
(5) Doherty, P.A. 1982. *Toxicol. Appl. Pharmacol.* 64:456-464.

Akton
(1) Bull, D.L., et al. 1982. *J. Agric. Food Chem.* 30:150-155.
(2) *5*
(3) *19*
(4) *21*
(5) *1*

alachlor
(1) *20*
(2) *5*
(3) *40*
(4) *39. Alachlor. 1984.*
(5) Lin, M.F. et al. 1987. *Mutat. Res.* 188:241-250.
(6) Georgian, L., et al. 1983. *Mutat. Res.* 116:341-348.
(7) *19*
(8) *21*
(9) *1*

aldicarb
(1) U.S. Environmental Protection Agency. 1988. *Aldicarb: Special Review Technical Support Document.* Office of Pesticide Programs. p. II-19.
(2) Maitlen, J.C., et al. 1982. *J. Agric. Food Chem.* 30:589-592.
(3) Wagner, S.L. 1983. *Clinical Toxicology Of Agricultural Chemicals.* Noyes Data Corporation. p. 250.
(4) Boyd, D., et al. 1990. *J. Toxicol. Environ. Health.* 30(3):209-221.
(5) *19*
(6) U.S. Environmental Protection Agency. 1984. *Aldicarb Chemical Information Fact Sheet.* Office of Pesticide Programs. p. 6.
(7) *5*
(8) *1*

aldicarb sulfone
(1) U.S. Environmental Protection Agency. 1988. *Aldicarb: Special Review Technical Support Document.* Office of Pesticide Programs. p. II-2.

aldicarb sulfoxide
(1) U.S. Environmental Protection Agency. 1988. *Aldicarb: Special Review Technical Support Document.* Office of Pesticide Programs. p. II-2.

aldrin
(1) *Review Of The Toxicity Of Aldrin.* 1987. Report of the Ministerial Committee. Public Health Service. South Australian Health Commission.
(2) *39. Aldrin. 1986.*
(3) *5*
(4) *43*
(5) Ottolenghi, A.D., et al. 1974. *Teratology.* 9:11-16.
(6) Reuber, M.D. 1977. Arch. Toxicol. 38:163-168.
(7) *30*
(8) *1*
(9) California Department of Food and Agriculture. 1978. *Report on Environmental Assessment of Pesiticide Regulatory*

Programs. State Component. Vol. 2. p. 3.2-15

allethrin
(1)   *3*
(2)   World Health Organization. 1989. *Environmental Health Criteria 87*. International Programme on Chemical Safety. p. 14.
(3)   Miyamoto, U. 1976. *Environ. Health Perspect.* 14:15-28.
(4)   *2*
(5)   Herrera, A., et al. 1988. *Mutagenesis.* 3(6):509-514.
(6)   Descotes, J. 1988. *Immunotoxicology of Drugs and Chemicals.* Elsevier.
(7)   *39. Allethrin. 1988.*
(8)   *16*
(9)   *58*
(10) *21*

d-cis/trans-allethrin
(1)   World Health Organization. 1989. Environmental Health Criteria 87: *Allethrins - Allethrin, d-Allethrin, Bioallethrin, and S-Bioallethrin*. International Programme on Chemical Safety. p. 21.
(2)   *39. Allethrin Stereoisomers. 1988.*

allyl alcohol
(1)   *5*
(2)   Dunlop, M.K., et al. 1958. *A.M.A. Archives of Industrial Health.* 18:303-311.
(3)   Lutz, D., et al. 1982. *Mutat. Res.* 93:305-315.
(4)   Lijinsky, W., et al. 1980. *Teratogenesis. Carcinog. Mutagen.* 1:259-267.
(5)   Brenner, D.E., et al. 1987. *Cancer Res.* 47:3259-3265.
(6)   Smith, P.F., et al. 1987. *Toxicol. Appl. Pharmacol.* 87:509-522.

aluminum phosphide
(1)   *3*
(2)   *23*

ametryn
(1)   *20*
(2)   *5*
(3)   *30*
(4)   *1*

aminocarb
(1)   Sundaram, K.M.S., et al. 1984. *J. Agric. Food Chem.* 32:1138-1141.
(2)   *19*
(3)   *33*
(4)   Kingsbury, P.D., et al. 1981. *The Environmental Impact of Nonyl Phenol and the Matacil Formulation Part 2: Terrestrial Ecosystem*. Forest Pest Management Institute. Canadian Forestry Service. Dept. of the Environment.
(5)   Gualindi, G. 1987. *Mutat. Res.* 178:33-41.
(6)   Sundaram, K.M.S., et al. 1979. *J. Environ. Sci Health.* [B]14(6):589-602.
(7)   *1*

aminonitrofen
(1)   Hurt, S.S.B., et al. 1983. *Toxicology.* 29:1-37.

amitraz
(1)   *5*
(2)   *39. Amitraz. 1987.*
(3)   U.S. Environmental Protection Agency. 1979. *Amitraz (BAAM): Position Document 3*. Office of Pesticide Programs.

(4)   *22*
(5)   *1*

amitrole
(1)   *20*
(2)   *5*
(3)   Yanagisawa, K., et al. 1987. *Mutat. Res.* 183:89-94.
(4)   *40. Amitrole. 1984.*
(5)   U.S. Environmental Protection Agency. 1986. *Neoplasia Induced by Inhibition of Thyroid Gland Function (Guidance for Analysis and Evaluation)*. Office of Pesticide Programs.
(6)   *19*
(7)   *22*
(8)   *1*

ammonium methanearsenate
(1)   *5*

ammonium salt of imazapyr
(1)   *58*

ammonium thiocyanate
(1)   *22*
(2)   *42*
(3)   *9*

amobam
(1)   *42*

AMS
(1)   *22*
(2)   *20*
(3)   *1*

anilazine
(1)   *39. Anilazine. 1983.*
(2)   *30*
(3)   *1*
(4)   *Pesticide Information Manual.* 1966. Northeastern Regional Pesticide Coordinators. Cooperative Extension Service.

anthraquinone
(1)   *36*
(2)   *5*
(3)   Nishio, A. and E.M. Uyeki. 1983. *Cancer Res.* 43:1951-1956.
(4)   Cesaroxe, C.F., et al. 1982. *Archives of Toxicology Supplement.* 5:355-359.
(5)   *15*

antimycin A
(1)   *36*
(2)   *21*
(3)   U.S. Department of the Interior. 1969. *Investigations in Fish Control*. Fish and Wildlife Service, Bureau of Sport Fisheries and Wildlife.

antimycin A3
(1)   *36*
(2)   *3*

ANTU
(1)   *36*
(2)   *54*

Aramite
(1)   *Pesticide Information Manual.* 1969. Northeastern Regional Pesticide Coordinators Cooperative Extension Service.
(2)   *3*
(3)   *43*

(4) *31*
(5) *21*
(6) *1*

Arosurf
(1) U.S. Environmental Protection Agency. 1984. *Arosurf* MSF Chemical Information Fact Sheet. Office of Pesticide Programs.

arsenic
(1) Woolson, E.A. 1975. *Arsenical Pesticides*. American Chemical Society. pp.40-41.
(2) *9*
(3) World Health Organization. 1980. *IARC Monographs on the Evaluation of the Carcinogenic Risk of Chemicals to Humans: Some Metals and Metallic Compounds*. 23 International Agency for Research on Cancer. pp. 39-115.
(4) Enterline, P.E., et al. 1987. *Am. J. Epidemiol.* 125:929-938.
(5) Poma. K., et al. 1983. *Mutat. Res.* 113:293-294.
(6) Tseng, W.P. 1977. *Environ. Health Perspect.* 19:109-119.
(7) Leonard, A. 1980. *Mutat. Res.* 75:49-62.

arsenic acid
(1) *5*
(2) Nakamuro,K., et al. 1981. *Mutat. Res.* 88:73-80.

arsenic pentoxide
(1) Ohno, H., et al. 1982. *Mutat. Res.* 104:141-145.
(2) Nakamuro, K., et al. 1981. *Mutat. Res.* 88:73-80.
(3) Kamboj, V.G., et al. 1964. *J. Reprod. Fertil.* 7:21-28.

arsenic trioxide
(1) *9*
(2) Pershagen, G., et al. 1984. *Environ. Res.* 34:227-241.
(3) Nagymajtenyi, L., et al. 1985. *J. Appl. Toxicol.* 5(2):61-63.
(4) Nakamuro, K., et al. 1981. *Mutat. Res.* 88:73-80.

asulam
(1) *20*
(2) 39. Asulam. 1987.
(3) *5*
(4) *1*

atrazine
(1) Behki, R.M. et al. 1986. *J. Agric. Food Chem.* 34(4): 746-749.
(2) Capriel, P., et al. 1985. *J. Agric. Food Chem.* 33(4):546-569.
(3) *7*
(4) Harrison-Biotech. 1983. *Control of Vegetation on Utility and Railroad Rights-of-Way*. 1983. Commonwealth of Massachusetts Generic Environmental Impact Report. p. II-19.
(5) U.S. Environmental Protection Agency. 1990. *Atrazine Environmental Fact Sheet*. Office of Pesticide Programs.
(6) Hoar, S.K., et al. 1986. *J. Am. Med. Assoc.* 256(9):1141-1147.
(7) Butler, M.A., et al. 1989. *Bull. Environ. Contam. Toxicol.* 43:797-804.
(8) Pino, A., et al. 1988. *Mutat. Res.* 209:145-147.
(9) Descotes, J. 1988. *Immunotoxicology of Drugs and Chemicals*. Elsevier.
(10) *9*
(11) *30*
(12) *1*
(13) O'Brein, M.H. 1986. *Atrazine*. Northwest Coalition for Alternatives to Pesticides. Eugene, OR.

auramine
(1) U.S. Department of Health & Human Services. 1983. *Third Annual Report on Carcinogens: Summary September, 1983*. National Toxicology Program. p. 26.
(2) Seiler, P. 1977. *Mutat. Res.* 46:305-310.
(3) *28*

Avitrol 100
(1) Takahashi, K., et al. 1979. *Gann.* 70:799-806.
(2) Andoh, T., et al. 1975. *Cancer Research.* 35:521-527.
(3) *19*

Avitrol 200
(1) 39. Avitrol. 1980.

azacosterol
(1) *5*

azinphos-ethyl
(1) *5*

azinphos-methyl
(1) *23*
(2) *5*
(3) 39. Azinphos-methyl. 1986.
(4) Milman, H.A., et al. 1978. *J. Environ. Pathol. Toxicol.* 1:829-840.
(5) National Cancer Institute. 1973. *Bioassay of Azinphos-methyl for Possible Carcinogenicity*. Technical Report Series No. 69. National Institutes of Health.
(6) Gilot-Delhalle, J., et al. 1983. *Mutat. Res.* 117:139-148.
(7) Alam, M.T., et al. 1976. *Caw. J. Genet. Cytol.* 18:665-671.

aziprotryne
(1) *35*
(2) *5*

azobenzene
(1) *42*
(2) Goodman, D.G., et al. 1984. *J. Natl. Canc. Inst.* 73(1):265-270.
(3) National Cancer Institute. 1979. *Bioassay of Azobenzene for Possible Carcinogenicity*. Technical Report Series No. 154. National Institutes of Health.
(4) Bus, J.S., et al. 1987. *Food Chem. Toxicol.* 25(8):619-626.
(5) Sina, J.F., et al. 1983. *Mutat. Res.* 113:357-391.
(6) *9*

azoxybenzene
(1) Tsunenari, S. 1973. *Japanese Journal of Legal Medicine.* 27:123-133.

B naphthylamine
(1) *16*

Bacillus thuringiensis (Berliner)
(1) Bio Integral Resource Center. *IPM Practitioner.* April 1980. 2(4):2.
(2) *5*
(3) *3*
(4) *19*

barban
(1) *30*
(2) *20*
(3) *1*

barthrin
(1) *2*

basic copper carbonate
(1) *37*

basic copper chloride
(1) 58
(2) 39. Group II Copper Compounds. 1987.
(3) 30
basic cupric acetate
(1) 28
benalaxyl
(1) 58
benazolin
(1) 5
(2) 33
(30) 58
bendiocarb
(1) 39. Bendiocarb. 1987.
(2) 32
(3) 33
(4) BFC Chemicals, Wilmington, DE. Label Instructions.
benefin
(1) Mrak, E.M. 1969. *Report of the Secretary's Commission of Pesticides and Their Relationship to the Environmental Health*. U.S. Department of Health, Education, and Welfare. p. 70.
(2) 20
(3) 1
benomyl
(1) 23
(2) 3
(3) 5
(4) 39. Benomyl. 1987.
(5) Pekkanen, T., et al. 1981. *The Toxicity of Benomyl. A Report by the Finnish National Board of Health Toxicology Expert Group.*
(6) Carter, S.D., et al. 1984. *J. Toxicol. Environ. Health.* 13(1): 53-68.
(7) Schafer, E.W. 1972. *Toxicol. Appl. Pharmacol.* 21:315-330.
(8) 21
(9) 4
(10) 1
(11) U.S. Environmental Protection Agency. 1977. *Rebuttable Presumption Against Reregistration and Continued Registration of Pesticide Products Containing Benomyl.* Federal Register, Tuesday, December 6, 1977. p. 61791.
bensulide
(1) 20
(2) 5
(3) 1
bentazon
(1) 20
(2) *Pesticide Science.* 1972. 3:(2). p. 242a.
(3) El-Mahdi, M.M., et al. 1988. *Arch. Exper. Vet. Med.* 42:261-266.
benthiocarb
(1) 20
(2) Schimmel, S.C., et al. 1983. *J. Agric. Food Chem.* 31(1):104-113.
(3) 21
bentranil
(1) 22

benzadox
(1) 5
(2) Mullison, W.R., et al. 1979. *Herbicide Handbook.* 4th ed. Weed Science Society of America, p. 54.
benzalkonium chloride
(1) 5
(2) 39. Alkyl Benzylammonium Chloride. 1985.
(3) 28
benzene
(1) Suzuki, S., et al. 1989. *Mutat. Res.* 223:407-410.
(2) 9
(3) Crespi, C.L., et al. 1985. *Prog. Mutat. Res.* 5:497-501.
(4) Henderson, R.F., et al. 1989. *Environ. Health Perspect.* 82:9-18.
(5) 29
(6) Crespi, C.L., et al. 1985. *Prog. Mutat. Res.* 5:497.
(7) Kuna, R.A. 1981. *Toxicol. Appl. Pharmacol.* 57:1-7.
(8) Green, J.D., et al. 1978. *Toxicol. Appl. Pharmacol.* 46:9-18.
(9) Aksoy, M. 1989. *Environ. Health Perspect.* 82:193-198.
benzene hexachloride
(1) 30
(2) 3
(3) Crosby, A.D., et al. 1986. *Vet. Hum. Toxicol.* 28(6):569-571.
(4) 43
(5) 29
(6) Babu, K.A., et al. 1981. *Bull. Environ. Contam. Toxicol.* 26:508-512.
(7) Nigam, S.K., et al. 1982. *Arch. Environ. Health.* 37(3):156-158.
(8) Sherman, J.D. 1978. *Aplastic Anemia Resulting from Adjacent Agricultural Use of Pesticide.* Paper presented before the Society for Occupational and Environmental Health. December 12, 1978.
(9) 19
(10) 21
(11) 30
(12) 1
(13) California Department of Food and Agriculture. 1978. *Report on Environmental Assessment of Pesticide Regulatory Programs.* Vol. 2. p. 3.2-25a.
(14) Anderegg, B.N., et al. 1977. *J. Agric. Food Chem.* 25(4):923-9.
1,2,4-benzentriol
(1) Kawanishi, S., et al. 1989. *Cancer Res.* 49:164-168.
benzthiazuron
(1) 5
6-benzyladenine
(1) 5
(2) Farrow, M.G., et al. 1976. *Experientia.* 32:29-30.
O-benzyl-p-chlorophenol
(1) 5
bifenox
(1) 20
(2) Kruger, P.J., et al. 1974. *Bifenox: A Selective Weed Killer.* Proceedings 12th British Weed Control Conference. p. 839-845.
(3) 5
(4) 39. Bifenox. 1981.

bifenthrin
(1) U.S. Environmental Protection Agency 1988. *Bifenthrin Pesticide Fact Sheet.* Office of Pesticide Programs.
(2) *58*
binapacryl
(1) *5*
(2) *22*
(3) *21*
(4) *1*
bioallethrin
(1) *58*
(2) World Health Organization. 1989. *Environmental Health Criteria 87: Allethrins - Allethrin, d-Allethrin, Bioallethrin, S-Bioallethrin.*International Programme on Chemical Safety.
(3) *40. Allethrin Stereoisomers. 1988.*
S-bioallethrin
(1) *58*
(2) World Health Organization. 1989. *Environmental Health Criteria 87: Allethrins - Allethrin, d-Allethrin, Bioallethrin, S-Bioallethrin.* International Programme on Chemical Safety. p. 38.
bithinol
(1) *42*
blasticidin S
(1) *5*
(2) Yamashita, M., et al. 1986. *Vet. Hum. Toxicol.* 29(1):8-11.
(3) *37*
Bomyl
(1) *33*
(2) *3*
(3) *5*
(4) *30*
(5) *1*
borax
(1) *20*
(2) *30*
boric acid
(1) *3*
(2) Linder, R.E., et al. 1990. *J. Toxicol. Environ. Health* 31:133-146.
brodifacoum
(1) ICI Americas, Inc. 1982. *Environmental Impact of Talon (Brodifacoum) Rodenticide.* Health and Environmental Affairs.
(2) *5*
bromacil
(1) *20*
(2) *15*
(3) Anderson, K.J., et al. 1972. *J. Agric. Food Chem.* 20(3):649-658.
(4) *1*
bromadiolone
(1) *3*
(2) *5*
bromethalin
(1) *5*
(2) Van Lier, R.B. 1988. *Fundam. Appl. Toxicol.* 11:664-672.
bromophos
(1) *34*

(2) *22*
(3) *3*
(4) *33*
(5) Nehez, M., et al. 1986. *Regul. Toxicol. Pharmacol.* 6:416-421.
(6) *30*
bromophos-ethyl
(1) *5*
(2) *22*
(3) Muacevic, G. 1973. *Toxicol. Appl. Pharmacol.* 25:180-189.
bromopropylate
(1) Wheeler, W.B., et al. 1973. *J. Environ. Qual.* 2(1):115-117.
(2) *5*
(3) *1*
(4) *58*
bromoxynil
(1) *23*
(2) *20*
(3) *38*
(4) *1*
bronopol
(1) *3*
bufencarb
(1) *23*
(2) *22*
(3) *15*
(4) *1*
butachlor
(1) *20*
(2) *5*
(3) Lin, M.F., et al. 1987. *Mutat. Res.* 188:241-250.
butam
(1) Mullison, W.R., et al. 1979. *Herbicide Handbook.* 4th Edition. Weed Science Society of America. Champaign, IL. pp. 78-79.
(2) *22*
2-butanamine
(1) *33*
(2) Lijinsky, W., et al. 1988. *Cancer Res.* 48:6648-6652.
(3) Lijinsky, W. 1983. *J. Natl. Cancer Inst.* 70(5):959-963.
(4) *5*
(5) *1*
buthidiazole
(1) *5*
(2) *33*
butoxycarboxim
(1) *34*
(2) *5*
butoxy polypropylene glycol
(1) *5*
butralin
(1) Kearney, P.C., et al. 1974. *J. Agric. Food Chem.* 22(5):856-859.
(2) *5*
(3) *3*
(4) *33*
(5) *20*
butylate
(1) *20*

(2) *39. Butylate. 1983.*
(3) *1*
cadmium
(1) Yoshikawa, H., et al. 1974. *Japanese J. Ind. Health.* 16:212.
(2) Fox, M.R.S. 1987. *J. Anim. Sci.* 65:1744-1752.
(3) Eisler, R. 1985. *Cadmium Hazards to Fish, Wildlife, and Invertebrates: A Synoptic Review.* Biological Report 85 (1.2). U.S. Fish and Wildlife Service.
(4) Friberg, L. 1984. *Environ. Health Perspect.* 54:1-12.
(5) Nordberg, M. 1984. *Environ. Health Perspect.* 54:13-20.
(6) Whelton, B.D., et al. 1988. *J. Toxicol. Environ. Health.* 24 (3):321-343.
(7) Chung, J., et al. 1986. *Arch. Environ. Health.* 41(5):319-323.
(8) Engel, D.W., et al. 1979. *Environ. Health Perspect.* 28:81-88.
cadmium chloride
(1) *37*
(2) Kostial, K., et al. 1979. *Environ. Health Perspect.* 28:89-95.
(3) *29*
(4) Kanematsu, N., et al. 1980. *Mutat. Res.* 77:109-116.
(5) Samarawickrama, G.P., et al. 1979. *Environ. Health Perspect.* 28:245-249.
(6) Levin, A.A., et al. 1981. *Toxicol. Appl. Pharmacol.* 58:297-306.
(7) Gabbiani, D. B., et al. 1967. *Exp. Neurol.* 18:154-160.
(8) Goyer, R.A., et al. 1989. *Toxicol. Appl. Pharmacol.* 101:232-244.
(9) Rao, P.V.V.P., et al. 1989. *J. Toxicol. Environ. Health.* 26 (3):327-348.
(10) Zenic, H., et al. 1982. *J. Toxicol. Environ. Health.* 9:367.
(11) Stacey, N.H. 1986. *J. Toxicol. Environ. Health.* 18(2):293-300.
(12) Blakley, B.R. 1985. *Can. J. Comp. Med.* 49(1):104-108.
(13) Kotsonis, F.N., et al. 1977. *Toxicol. Appl. Pharmacol.* 41:667-680.
(14) Suzuki, K.T., et al. 1983. *J. Toxicol. Environ. Health.* 11(4-6):727-737.
(15) Suzuki, K.T., et al. 1983. *J. Toxicol. Environ. Health.* 11(4-6):713-736.
(16) Burton, D.T., et al. 1990. *Bull. Environ. Contam. Toxicol.* 44:776-783.
cadmium oxide
(1) *29*
cadmium sebacate
(1) *5*
cadmium succinate
(1) *5*
(2) *37*
cadmium sulfate
(1) Sina, J.F., et al. 1983. pp. 357-391.
(2) Oberly, T.J., et al. 1982. *J. Toxicol. Environ. Health.* 9:367-376.
(3) Kuczuk, M.H., et al. 1984. *Teratology.* 29:427-435.
calcitriol
(1) *14*
calcium acid methanearsenate
(1) *5*
calcium arsenate
(1) Woolson, E.A. 1977. *Environ. Health Perspect.* 19:73-81.

(2) *3*
(3) *22*
(4) *42*
(5) *1*
calcium chlorate
(1) *3*
(2) Zaloga, G.I., et al. 1987. *Ann. Emerg. Med.* 16(6):637-639.
calcium cyanamide
(1) *22*
calcium cyanide
(1) *3*
calcium propanearsenate
(1) *42*
camphor
(1) *9*
(2) Jimenez, J.F., et al. 1983. *Gastroenterology.* 84:394-398.
captafol
(1) Wolfe, N.L., et al. 1976. *J. Agric. Food and Chem.* 24(5):1041-1045.
(2) *5*
(3) *3*
(4) Ito, N., et al. 1988. 9(3):387-394.
(5) Ito, N., et al. 1984. *Gann.* 75: 853-865.
(6) Bridges, B.A. 1975. *Mutat. Res.* 32: 3-34.
(7) Collins, T.F.X. 1972. *Fd. Cosmet. Toxicol.* 10:363-371.
(8) U.S. Environmental Protection Agency. 1985. *Captafol-Special Review Position Document 1.* Office of Pesticide Programs.
(9) *33*
(10) *21*
captan
(1) *40. Captan. 1985.*
(2) *30*
(3) *5*
(4) Reuber, M.D. 1989. *J. Environ. Pathol. Toxicol. Oncol.* 9(2):127-143.
(5) Ito, N., et al. 1988. *Carcinogenesis.* 9(3):387-394.
(6) *NCI Bioassay of Captan for Possible Carcinogenicity. 1971.* National Cancer Institute Carcinogenesis Technical Report Series. No. 15. National Institutes of Health.
(7) Tezuka, H., et al. 1980. *Mutat. Res.* 78:177-191.
(8) Robens, J.F. 1970. *Toxicol. Appl. Pharmacol.* 16:24-34.
(9) Descotes, J. 1988. *Immunotoxicology of Drugs and Chemicals.* Elsevier.
(10) *1*
(11) Deep, I.W., et al. 1965. *Phytopathology.* 55:212-216.
carbanolate
(1) *7*
(2) *1*
carbaryl
(1) *30*
(2) *Pesticide Information Manual.* 1966. Northeastern Regional Pesticide Coordinators. Cooperative Extension Service.
(3) *19*
(4) National Institute of Occupational Safety and Health. 1989. *Criteria for a Recommended Standard. Occupational Exposure to Carbaryl.* No. 77-107.

(5)   U.S. Environmental Protection Agency. 1980. *Carbaryl Decision Document.* Office of Pesticides and Toxic Substances.

(6)   39. Carbaryl. 1984.

(7)   National Institute of Occupational Safety and Health. 1976. *Criteria for a Recommended Standard. Occupational Exposure to Carbaryl.* pp. 68-72.

(8)   Richmond, M.L.et al. 1979. *Effects of Sevin-4-Oil, Dimilin, and Orthene on Forest Birds in Northeastern Oregon.* Research Paper PSW-I48. Pacific Southwest Forest and Range Experiment Station.

(9)   Moulding, J.D. I976. *The Auk.* 93(4): 692-707.

(10) 30

(11) California Department of Food and Agriculture. 1978. *Report on Environmental Assessment of Pesticide Regulatory Programs. Vol. 2. State Component.*

(12) Epstein, S.S., et al. 1972. *The Mutagenicity of Pesticides.* MIT Press.

(13) Ghosh, P.S., et al. 1990. *Biomed. Environ. Sci.* 3(1):106-112.

(14) Little, E.E., et al. 1990. *Arch. Environ. Contam. Toxicol.* 19:380-385.

carbendazim

(1)   22

(2)   58

(3)   39. Thiophanate-methyl. 1986.

(4)   Speakman, J.B., et al. 1981. *Mutat. Res.* 88:45-51.

(5)   Carter, S.D., et al. 1987. *Biol. Reprod.* 37:709-717.

(6)   Evenson, D.P., et al. 1987. *J. Toxicol. Environ. Health.* 20:387-399.

(7)   5

(8)   34

(9)   Stringer, A., et al. 1973. *Pesticide Science.* 4:165-170.

carbofuran

(1)   34

(2)   Eislerl, R. 1985. *Carbofuran Hazards to Fish, Wildlife, and Invertebrates: A Synoptic Review. 1985.* U.S. Fish and Wildlife Service. Biological Report 85(1.3). p. 13.

(3)   U.S. Environmental Protection Agency. 1989. *Carbofuran Special Review: Technical Support Document.* Office of Pesticide Programs. p.IV-1.

(4)   39. Carbofuran. 1989.

(5)   Descotes, J. 1988. *Immunotoxicology of Drugs and Chemicals.* Elsevier.

(6)   19

carbon disulfide

(1)   33

(2)   9

(3)   13

carbon monoxide

(1)   Stewart, R.D., et al. 1976. *JAMA.* 235(4):398-401.

(2)   National Institute for Occupational Safety and Health. *Occupational Health Guidelines for Methylene Chloride.* Department of Labor. 1978.

(3)   Barrowcliff, D.F., et al. 1979. *J. Soc. Occup. Med.* 29:12-14.

carbon tetrachloride

(1)   5

(2)   51

(3)   29

(4)   9

(5)   Reuber, M.D., et al. 1970. *J. Natl. Cancer Inst.* 44 (2): 419-423.

(6)   National Institute for Occupational Safety and Health. 1975. *Criteria for a Recommended Standard: Occupational Exposure to Carbon Tetrachloride.* Center for Disease Control. pp.21-22.

(7)   Barlow, S.M., et al. 1982. *Reproductive Hazards of Industrial Chemicals: An Evaluation of Animal and Human Data.* Academic Press. pp. 200-211.

(8)   9

carbophenothion

(1)   5

(2)   3

(3)   33

(4)   21

(5)   30

(6)   1

carboxin

(1)   4

(2)   Lane, J.R. 1970. *J. Agric. Food Chem.* 18(3):409-412.

(3)   5

(4)   3

(5)   Uniroyal Chemical. *Technical data sheet Vitavax Fungicide.* Bethany, CT.

(6)   1

cartap

(1)   5

(2)   42

CDAA

(1)   Beste, C.E. (Ed.). 1983. *Herbicide Handbook.* 5th Edition. Weed Science Society of America. Champaign, IL. pp.88-92.

(2)   5

CDEC

(1)   22

(2)   30

(3)   5

(4)   29

(5)   Mullison, W.R., et al. 1979. Herbicide Handbook. 4th ed. Weed Science Society of America. pp.95-98.

(6)   1

chloramben

(1)   20

(2)   5

(3)   39. Chloramben. 1981.

(4)   9

(5)   U.S. Enviromental Protection Agency. 1989. *Chloramben. Health Advisory Summary.* Office of Water.

chloranil

(1)   5

(2)   38

(3)   Innes, J.R.M., et al. *J. Natl. Canc. Inst.* 42:1101-1114.

chlorbenside

(1)   *Pesticide Information Manual.* 1966. Northeastern Regional Pesticide Coordinators. Cooperative Extension Service.

(2)   42

chlorbromuron

(1)   33

chlordane

(1)   39. Chlordane. 1986.

(2) U.S. Environmental Protection Agency. 1985. *Chlordane: Health Advisory.* Office of Drinking Water. pp. 3-5.
(3) *30*
(4) *58*
(5) *9*
(6) U.S. Environmental Protection Agency. 1987. *Project Summary: Carcinogenicity assessment of chlordane and heptachlor/heptachlor epoxide.* Office of Health and Environmental Assessment.
(7) Khasawinah, A.M. and J.F. Grutsch. 1989. *Regulatory Toxicology and Pharmacology.* 10:95-109.
(8) Balsh, K.J., et al. 1987. *Bull. Environ. Contam. Toxicol.* 39: 434-442.
(9) Lundholm, C.E. 1988. *Comparative Biochemistry and Physiology [C]:Comparative Pharmacology and Toxicology.* 89C(2):361-368.
(10) Takamiya, K. 1990. *Bull. Environ. Contam. Toxicol.* 44:905-909.
(11) Descotes, J. 1988. *Immunotoxicology of Drugs and Chemicals.* Elsevier.
(12) *19*
(13) *21*
(14) Kawano, M., et al. 1988. *Environ. Science. Tech.* 22:792-797.

chlordecone
(1) Sterrett, F.S., et al. 1977. *Environment.* 19(2):30-37.
(2) *5*
(3) *3*
(4) *58*
(5) Linder, R.E., et al. 1983. *J. Toxicol. Environ. Health.* 12:183-192.
(6) *10*
(7) *29*
(8) Reuber, M.D. 1978. *J. Toxicol. Environ. Health.* 4:895-911.
(9) Kavlock, R.J., et al. 1985. *Teratogenesis Carcinog. Mutagen.* 5:3-13.
(10) Chernoff, N., et al. 1976. *Toxicol. Appl. Pharmacol.* 38:189-194.
(11) *9*
(12) Vzodinma, J.E., et al. 1984. *J. Environ. Bio.* 5(2):81-88.
(13) Huang, E.S., et al. 1986. *Toxicol. Appl. Pharmacol.* 82:62-69.
(14) *19*
(15) *30*
(16) *1*
(16) *Fisheries and Wildlife Research.* 1978. Scott, T.G. et al (eds.) U.S. Fish and Wildlife Service. p. 28.
(17) Larson, P.S., et al. 1979. *Toxicol. Appl. Pharmacol.* 48:29-41.

chlordimeform
(1) 39. Chlordimeform.
(2) *5*
(3) Ahmed, A.K. 1986. *Toxicology of Chlordimeform: With Particular Reference to Its Carcinogenicity. A Summary Report.* Natural Resources Defense Council. New York, NY. p. 11.
(4) *22*

chlorfenethol
(1) *5*

(2) *3*

chlorfenson
(1) *14*
(2) *3*
(3) Association of American Pesticide Control Officials, Inc. 1966. *Pesticide Chemicals Official Compendium.*

chlorfensulphide
(1) *22*

chlorfenvinphos
(1) *34*
(2) *33*
(3) *3*
(4) *5*
(5) *19*

chlorfluazuron
(1) *The Agrochemicals Handbook.* 1990. 2nd Edition. Update 5. The Royal Society of Chemistry.

chlorflurecol
(1) *20*
(2) *5*

chlorimuron
(1) *58*

chlorinated isocyanurates
(1) 39. Chlorinated Isocyanurates. 1988.

chlorinated naphthalenes
(1) *9*

chlorine
(1) *Chlorine and Hydrogen Chloride. 1976.* The National Research Council. National Academy of Sciences. p. 117.
(2) *3*
(3) Mattice, J.S. and H.E. Zittel. 1976. *Journal Water Pollution Control Federation.* 48:2284-2297.
(4) *27*
(5) Brungs, W.A. 1973. *Journal Water Pollution Control Federation.* 45(10):2180-2193.

chlormephos
(1) *5*
(2) *34*
(3) *22*

chlormequat
(1) *36*
(2) *58*

2-chloroacrolein
(1) Marsolen, P.J. and J.E. Casida. 1982. *J. Agric. Food Chem.* 30(4):627-631.

4-chloroaniline
(1) U.S. Environmental Protection Agency. 1988. *Dimilin (p-chloroaniline)-Interim quantitative risk assessment from a NTP rat and mouse oncogenicity study.* Health Effects Division, Office of Pesticide Programs.
(2) Dost, F.M., et al. 1985. *Toxicological Examination of Dimilin.* Extension Service Toxicology Information Program. Oregon State University.

chlorobenzilate
(1) 40. Chlorobenzilate.
(2) *22*
(3) *54*

(4)   National Cancer Institute. 1978. *Bioassay of chlorobenzilate for Possible Carcinogenicity*. *Carcinogenesis Technical Report Series*. No. 75. National Institutes of Health.
(5)   *1*

chlorofluorocarbons
(1)   Mrak, E.M. 1969. *Reports of the Secretary's Commission on Pesticides and their Relationship to Environmental Health*. Parts I & II. U.S. Department of Health, Education, & Welfare.

chloroform
(1)   Pericin, C., et al. 1979. *Archives of Toxicology*. Supplement 2: 371-373.
(2)   *51*
(3)   Lundberg, I., et al. 1986. *Environ. Res.* 40:411-420.
(4)   Marty, M. 1989. *Health Effects of Chloroform*. California Department of Health Services. pp.3-9 to 3-11.
(5)   National Cancer Institute. 1976. *Report on Carcinogenesis Bioassay of Chloroform*. National Institutes of Health.
(6)   Morimoto, K., et al. 1983. *Environ. Res.* 32:72-79.
(7)   Chu, I., et al. 1982. *J. Environ. Sci. Health.* B17(3):205-224.
(8)   U.S. Department of Health, Education, and Welfare. 1974. *Criteria for a Recommended Standard: Occupational Exposure to Chloroform*. Public Health Service. Center for Disease Control. National Institute for Occupational Safety and Health. pp. 19-21.

chloroneb
(1)   *39. Chloroneb. 1980.*
(2)   *38*

chlorophacinone
(1)   *5*
(2)   *33*

chloropicrin
(1)   *5*
(2)   Garry, V.F., et al. 1990. *Teratogenesis Carcinog. Mutagen.* 10:21-29.
(3)   Association of American Pesticide Control Officials, Inc. 1966. *Pesticide Chemicals Official Compendium*. Topeka, KS. p.238.

chloropropylate
(1)   *5*
(2)   *22*
(3)   *91*
(4)   *30*
(5)   *1*

chlorothalonil
(1)   *39. Chlorothalonil. 1984.*
(2)   Northover, J., et al. 1980. *J. Agric. Food Chem.* 28(5):971-974.
(3)   *5*
(4)   Liden, C. 1990. *Contact Dermatitis.* 22: 206-211.
(5)   U.S. Environmental Protection Agency. 1991. *List of Chemicals Evaluated for Carcinogenic Potential*. Health Effects Division. Office of Pesticide Programs.
(6)   National Cancer Institute. 1978. *Bioassay of chlorothalonil for possible carcinogenicity*. Carcinogenesis Technical Report Series. No. 41. National Institutes of Health.
(7)   Prior, L.R. 1985. *Amicus Journal.* 7(8):8-9.
(8)   *22*
(9)   *1*

(10) *37*
4-chloro-o-toluidine
(1)   *39. Chlordimeform.*

chloroxuron
(1)   *20*
(2)   *5*
(3)   *30*
(4)   *1*

chlorphoxim
(1)   *33*

chlorpropham
(1)   *20*
(2)   *5*
(3)   *39. Chlorpropham. 1987.*

chlorpyrifos
(1)   Odenkirchen, E.W., et al. 1988. *Chlorpyrifos Hazards to Fish, Wildlife, and Invertebrates: A Synoptic Review.* Biological Report 85 (1.13). U.S. Fish and Wildlife Service.
(2)   *19*
(3)   *5*
(4)   *32*
(5)   Berteau, P.E., et al. 1978. *Bull. Environ. Contam. Toxicd.* pp.113-120.
(7)   *9*
(8)   Court upholds award for bull's impotence. *The Washington Post.* April 26, 1985.
(9)   *30*
(10) *1*
(11) World Health Organization. 1986. *Organophosphorus Insecticides: A General Introduction. Environmental Health Criteria 63.* International Programme on Chemical Safety.
(12) DOW Chemical. *Dursban Insecticides Technical Information Technical Data Sheet.* Agricultural Products Department. Midland, MI.
(13) *23*
(14) *23*

chlorsulfuron
(1)   *20*
(2)   *5*

cholecalciferol
(1)   *36*
(2)   *14*
(3)   U.S. Environmental Protection Agency. 1984. *Chemical Information Fact Sheet for Vitamin D_3.* Office of Pesticide Programs.

chromated copper arsenate
(1)   U.S. Consumer Product Safety Commission. *Summary of Health Sciences memoranda regarding skin cancer risk from dislodgeable arsenic on pressure treated playground equipment wood.* Memorandum of January 26, 1990. From: Brian C. Lee To: Elaine A. Tyrrell.

cinmethylin
(1)   *20*
(2)   *5*

cismethrin
(1)   *9*

citronella
(1)   *33*

clofentezine
(1) *34*
(2) *5*
(3) Agriculture Canada. 1989. *Decision Document: Clofentezine (Miticide)*. Food Production and Inspection Branch. Pesticides Directorate.
(4) *3*

clomazone
(1) *20*
(2) *35*
(3) U.S. Environmental Protection Agency. January 29, 1986. Memorandum: *Request for "8 Point Summary" for Command*. To: J. Yowell, Registration Division. From: C. Gregorio, Hazard Evaluation Division. Office of Pesticide Programs.
(4) U.S. Environmental Protection Agency. 1984. *Enviromental Effects Branch Review: 2-(chlorophenyl)-methyl-4-4-dimethyl-3-isoxazolidinone.*
(5) U.S. Environmental Protection Agency. Nov. 9, 1988. Memorandum: *Command: Review of Possible Toxicology Concern of an Impurity of Command*. To: R. Taylor, Registration Division. From: W. Phang, Hazard Evaluation Division. Office of Pesticide Programs.

clopyralid
(1) *20*
(2) *20*

CNP
(1) *5*

copper acetoarsenite
(1) *30*

copper ammonium carbonate
(1) *37*

copper ammonium sulfate
(1) *21*

copper bis(3-phenylsalicylate)
(1) *9*

copper chelate
(1) *20*

copper-ethylenediamine complex
(1) *20*

copper hydroxide
(1) *58*
(2) 39. Group II copper compounds. 1987.

copper linoleate
(1) *37*

copper napthenate
(1) *58*
(2) *5*
(3) Grove, S.L. 1987. *Copper Napthenate: An Alternative Wood Preservative*. Moorey Chemicals, Inc.

copper oxide
(1) *58*
(2) *5*
(3) *37*

copper oxychloride
(1) *58*
(2) 40. *Group II Copper Compounds. 1987.*
(3) *30*

copper oxychloride sulfate
(1) *39.* Group II copper compounds. 1987.
(2) *1*

copper sulfate
(1) *5*
(2) *20*
(3) *4*
(4) Robison, S.H., et al. 1984. *Mutat. Res.* 131:173-181.
(5) Casto, B.C., et al. 1979. *Cancer Res.* 39:193-198.

copper sulfate monohydrate
(1) *37*

copper tea complex
(1) *20*

copper zinc chromate
(1) *9*

coumafuryl
(1) *14*

coumaphos
(1) *58*
(2) U.S. Environmental Protection Agency. 1985. E.P.A. *Chemical Profile: Coumaphos*. Chemical Emergency Preparedness Program. (3)   *19*
(4) *21*
(5) World Health Organization. 1986. *Environmental Health Criteria 63. Organophosphorus Insecticides: A General Introduction*. International Programme on Chemical Safety.

4-CPA
(1) *30*
(2) *9*

credazine
(1) *33*

creosote (coal tar)
(1) U.S. Department of Health and Human Services. 1985. *Fourth Annual Report on Carcinogens*. Public Health Service. NTP 85-002. p.205.
(2) 40. Creosote. 1985.
(3) National Institute for Occupational Safety and Health. 1977. *Criteria for a Recommended Standard......Occupational Exposure to Coal Tar Products*. U.S. Department of Health, Education, and Welfare. Publication No. 78-107.

creosote (wood tar)
(1) Letizia, C., et al. 1982. *Food Chem. Toxicol.* 20:697-701.
(2) Bos, R.P., et al. 1983. *Exposure to mutagenic aromatic hydrocarbons of workers creosoting wood*. IARC Scientific Publications No. 59. International Agency for Research on Cancer.
(3) Vogelbein, W.K., et al. 1990. *Cancer Res.* 50:5978-5986.

cresylic acid
(1) *3*
(2) *30*

crotoxyphos
(1) *5*
(2) *22*
(3) *32*
(4) *21*
(5) *1*

crufomate
(1) *5*
(2) *33*

(3)   *19*
cryolite
(1)   *30*
(2)   U.S. Environmental Protection Agency. 1987. *Toxicology Chapter of the Cryolite Registration Standard.* Prepared by W. Woodrow. Toxicology Branch. Hazard Evaluation Division. p.5.
(3)   *39. Cryolite. 1988.*
(4)   *21*
(5)   *1*
(6)   *5*
cyanazine
(1)   *20*
(2)   U.S. Environmental Protection Agency. 1984. *Pesticide Fact Sheet.* Cyanazine. Fact Sheet Number 41. Office of Pesticide Programs.
(3)   U.S. Environmental Protection Agency. 1988. *Cyanazine Special Review Technical Support Docunment.* Office of Pesticide Programs.
(4)   *19*
(5)   *21*
(6)   *1*
cyanophenphos
(1)   *3*
(2)   Enan, E.et al. 1987. *J. Environ. Sci. Health.* B22(2):149-170.
(3)   *34*
        Soliman, S.A., et al. 1986. *J. Environ. Sci. Health.* B21(5):401-411.
cyanaphos
(1)   *58*
(2)   *34*
(3)   World Health Organization. 1986 *Environmental Health Criteria 63. Organophosphorus Insecticides: A General Introduction.* International Programme on Chemical Safety.
cycloate
(1)   *20*
(2)   *33*
(3)   *58*
(4)   *1*
cyclohexane
(1)   *3*
cyclohexanone
(1)   World Health Organization. 1989. *IARC Monographs on the Evaluation of the Carcinogenic Risk of Chemicals to Humans.* Vol. 47. International Agency for Research on Cancer.pp. 157-169.
cycloheximide
(1)   *39. Cycloheximide. 1982.*
(2)   *37*
(3)   *9*
(4)   Garberg, P., et al. 1988. *Mutat. Res.* 203:155-176.
(5)   Wangenheim, J., et al. 1988. *Mutagenesis.* 3(3):193-205.
(6)   Ritter, E.J. 1984. *Fundam. Appl. Toxicol.* 4:352-359.
(7)   Lary, J.M., et al. 1982. *Teratology.* 25:345-349.
(8)   *19*
(9)   *1*
(10) Descotes, J. 1988. *Immunotoxicology of Drugs and Chemicals.* Elsevier.

cycloprate
(1)   *5*
cycluron
(1)   *33*
cyfluthrin
(1)   U.S. EPA. 1987. *Pesticide Fact Sheet: Cyfluthrin.* Office of Pesticide Programs.
(2)   Mobay Corporation. 1990. *Material Safety Data Sheet: Tempo 20 Wettable Powder-0.1% End-use Dilution.* Agricultural Chemicals Division. Kansas City, MO.
(3)   U.S. EPA. 1989. *Tox One-Liner: Cyfluthrin.* Office of Pesticide Programs.
(4)   *48*
(5)   Mobay Corporation. 6/30/89. *Material Safety Data Sheet. Tempo.* Agricultural Chemicals Division. Kansas City, MO.
(6)   *1*
λ-cyhalothrin
(1)   *58*
(2)   Jacques, A., et al. 1984. Cyhalothrin. *Analytical Methods For Pesticides and Plant Growth Regulators.* 8:9-13.
cyhexatin
(1)   *7*
(2)   *33*
(3)   *1*
cymoxanil
(1)   *58*
cypermethrin
(1)   United Nations Environment Programme. March, 1990. *News About Chemicals: Cypermethrin.* IRPTC Bulletin - Journal of the International Register of Potentially Toxic Chemicals.
(2)   *5*
(3)   *33*
(4)   World Health Organization. 1989. *Environmental Health Criteria 82. Cypermethrin.*
(5)   Puig, M. et al 1989. *Mutagenesis.* 4:72-74.
(6)   Descotes, J. 1988. *Immunotoxicology of Drugs and Chemicals.* Second Updated, Edition. Elsevier.
(7)   Qadri, S.H. et al 1987. *J. Appl. Toxicol.* 7(6):367-371.
(8)   Edwards, R., et al. 1986. *Toxicol. Appl. Pharmacol.* 84:512-522.
(9)   Coats, J.R., et al. 1979. *Bull. Environ. Contam. Toxicol.* 23:250-255.
(10) Clarke, J.R., et al. 1989. *Environ. Toxicol. Chem.* 8:393-401.
cyprazine
(1)   *5*
(2)   *33*
(3)   Hilton, J.L., et al. 1979. *Herbicide Handbook of the Weed Science Society of America.* Weed Science Society of America. pp.126-129.
(4)   *42*
cyromazine
(1)   *58*
(2)   1984. *Science.* 225(4659):295.
(3)   U.S. Environmental Protection Agency. 1989. Memorandum: *Review submission of "weight-of-evidence" evaluation of cyromazine potential to cause developmental effects, from Ciba-Geigy.* To: P. Hutton/M. Mendelsohn, Insecticide-Rodenticide Branch. Registration Division. From: S.C.

Dapson, Review Section I. Toxicology Branch–Herbicide, Fungicide, Antimicrobial Support/HED.

(4)   5

cythioate

(1)   58

(2)   3

2,4-D

(1)   Newton, M. et al. 1990. *J. Agric. Food Chem.* 38(2):574-583.

(2)   20

(3)   3

(4)   4

(5)   U.S. Environmental Protection Agency. 1989. *Pesticide Fact Sheet: 2,4-Dichlorophenoxy acetic acid.* Office of Pesticide Programs.

(6)   Hoar, S.K. et al. 1986. *JAMA.* 256(9):1141-1147.

(7)   Shearer, R.W. 1985. *Health effects of 2,4-D herbicide.* Unpublished report. Issaquah Health Research Institute. Washington.

(8)   Kappas, A. 1988. *Mutat. Res.* 204:615-621.

(9)   Zhao, Y. et al. 1987. *J. Toxicol. Environ. Health* 20:11-26.

(10) Blakley, P.M. et al. 1989. *Teratology* 39:237-241.

(11) 39. 2,4-Dichlorophenoxyacetic acid(2,4-D). 1988.

(12) 47

(13) Descotes, J. 1988. *Immunotoxicology of Drugs and Chemicals.* Elsevier.

(14) 19

(15) 58

(16) 1

(17) 30

(18) 1976. *Science.* 193:239-240.

(19) 24

(20) Hayes, H.M. et al. 1991. *J. Natl. Cancer Inst.* 83:1226-1231.

2,4-DB

(1)   58

(2)   U.S. Environmental Protection Agency. 1989. *Fact Sheet. 2,4-DB.* Office of Pesticide Programs.

(3)   1

2,4-DB sodium

(1)   58

4,4-dichloroazobenzene

(1)   Hurt, S.S.B., et al. 1983. *Toxicology.* 29:1-37.

dalapon

(1)   Worthing, C.R. (Ed.) 1983. *Pesticide Manual: Basic Information on the Chemicals used as Active Components of Pesticides.* British Crop Protection Council.

(2)   58

(3)   7

(4)   21

(5)   1

(6)   Kenaja, E.E. 1974. Residue Review. 53:109-151.

daminozide

(1)   5

(2)   33

(3)   39. Daminozide. 1984.

(4)   Toth, B. et al. 1977. *Cancer Res.* 37:3497-3500.

(5)   58

(6)   1

dazomet

(1)   22

(2)   30

(3)   58

(4)   20

(5)   1

(6)   9

DBCP

(1)   5

(2)   29

(3)   39. DBCP. 1978.

(4)   Rosenkranz, H.A. 1975. *Bull. Environ. Contam. Toxicol.* 14:8-12.

(5)   Kluwe, W.M. et al. 1983. *Toxicol. Appl. Pharmacol.* 71:294-298.

(6)   11

(7)   30

(8)   1

DCPA

(1)   20

(2)   5

(3)   39. DCPA. 1988.

(4)   1

D-D

(1)   Cohen, D.B., et al. 1983. *Water Quality and Pesticides 1,2-dichloropropane/1,3-dichloropropene.* California State Water Resources Control Board Toxic Substances Control Program. Vol. 3. Special Projects Report No. 83-85P.

(2)   5

(3)   43

(4)   De Lorenzo, F., et al. 1977. *Cancer Res.* 36:1915-1917.

(5)   Torkelson, T.R., et al. 1977. *Am. Ind. Hyg. Assoc. J.* 38:217-223.

DDD

(1)   World Health Organization. 1989. *Environmental Health Critera 83. DDT and its Derivatives-Environmental Aspects.* International Programme on Chemical Safety.

(2)   5

(3)   9

(4)   National Cancer Institute. 1978. *Bioassay of DDT, TDE, and p,p'-DDE for possible Carcinogenicity.* National Institutes of Health. Technical Report Series No. 131.

(5)   43

(6)   National Institute for Occupational Safety and Health. 1978. *Special Occupational Hazard Review for DDT.* U.S. Department of Health, Education, and Welfare.

(7)   19

(8)   21

(9)   1

DDE

(1)   30

(2)   6

(3)   Clark, D.R. 1976. *Bull. Environ. Contam. Toxicol.* 15:1-8.

(4)   National Cancer Institutes. 1978. *Bioassay of DDT, TDE, AND p,p'-DDE for possible Carcinogenicity.* National Institutes of Health.

(5)   43

(6)   21

(7)   1

(8) Anderson, D.W. et al. 1976. *Environ. Pollut.* 10:183-200.
(9) Longcore, J.R. et al. 1971. *Bull. Environ. Contam. Toxicol.* 6(6): 485-490.
(10) Ohlendorf, H.M. et al. 1977. *Organochlorine residues and eggshell thinning in anhingas and waders.* Patuxent Wildlife Research Center. U.S. Department of the Interior. Fish and Wildlife Service.

DDT
(1) U.S. Environmental Protection Agency. 1975. *DDT - A Review of Scientific and Economic Aspects of the Decision to Ban its Use as a Pesticide.*
(2) *58*
(3) *11*
(4) *29*
(5) U.S. Department of Health, Education and Welfare, National Institute for Occupational Safety and Health. 1978. *Special Occupational Hazard Review for for DDT.*
(6) *9*
(7) National Cancer Institute. 1978. *Bioassay of DDT, TDE, and p,p'-DDE for possible Carcinogenicity. Technical Report Series No. 131.* National Institutes of Health.
(8) Descotes, J. 1988. *Immunotoxicology of Drugs and Chemicals.* Elsevier.
(9) *19*
(10) *21*
(11) *1*
(12) California Department of Food and Agriculture. 1978. *Report on Environmental Assessment of Pesticide Regulatory Programs.* State Component. Vol. 2. p.3.2-9.
(13) World Health Organization. 1989. *Environmental Health Criteria 83 DDT and its Derivatives-Environmental Aspects.*
(14) Fleet, R.R., et al. 1978. *Bull. Environ. Contam. Toxicol.* 7(4):383-388.
(15) Wurstor, C.F. 1968. *Science.* 159(3822):1474-1475.
(16) Niethammer, K.R., et al. 1984. *Arch. Environ. Contam. Toxicol.* 13:63-74.
(17) *30*

deet
(1) *33*
(2) *2*
(3) *3*

DEF
(1) Eichelberger, J.W. et al. 1971. *Environ. Science Tech.* 5(6):541-544.
(2) *5*
(3) *32*
(4) U.S. Environmental Protection Agency. 1978. *DEF Pesticide Fact Sheet.* Office of Pesticide Programs.
(5) *19*
(6) *21*
(7) *30*
(8) *1*
(9) Abou-Donia, M.B. et al. 1986. *Toxicol. Appl. Pharmacol.* 82:461-473.
(10) Little, E.E. et al. 1990. *Arch. Environ. Contam. Toxicol.* 19:380-385.
(11) California Department of Food and Agriculture. 1982. *DEF and Folex. Pesticides Safety Information Series F-7.* Worker Health and Safety Unit.

delta⁴ tetra-tetrahydrophthalimide
(1) *28*

deltamethrin
(1) *5*
(2) *33*

demeton
(1) *30*
(2) *33*
(3) *3*
(4) *34*
(5) Dzwondowski, A. et al. 1986. *Arch. Toxicol.* 58:152-156.
(6) Chen, H.H. et al. 1981. *Mutat. Res.* 88:307-316.
(7) Budrean, C.H. et al. 1973. *Toxicol. Appl. Pharmacol.* 24:324-332.
(8) *19*
(9) *21*
(10) *1*

demeton-methyl
(1) *33*
(2) *58*
(3) *30*
(4) *34*

desbromoleptophos
(1) Abou-Donia, M.B., et al. 1978. *Toxicol. Appl. Pharmacol.* 45:280.

desmedipham
(1) *20*
(2) *58*
(3) *5*
(4) *32*
(5) *1*

desmetryn
(1) *5*
(2) *58*

dialifor
(1) Freed, V.H. et al. 1979. *J. Agric. Food Chem.* 27(4):706-708.
(2) *3*
(3) *33*
(4) Robens, J.F. 1970. *Toxicol. Appl. Pharmacol.* 16:24-34.
(5) *1*

diallate
(1) *20*
(2) *5*
(3) *40*
(4) *46*
(5) *32*
(6) *1*

diamidfos
(1) *5*
(2) *32*

diatomaceous earth
(1) *9*
(2) *1*

diazinon
(1) *32*
(2) Eisler, R. 1986. *Diazinon Hazards to Fish, Wildlife, and Invertebrates: A Synoptic Review.* U.S. Fish and Wildlife Service. Biol. Rep. 85.

(3) 5
(4) 33
(5) McGregor, D.B. 1988. *Environ. Mol. Mutagen.* 12:85-154.
(6) Kiraly, J., et al. 1979. Environ. Contam. Toxicol. 8:309-319.
(7) Crammer, J.S., et al. 1978. *J. Environ. Pathol. Toxicol.*
     2:357-369.
(8) *11*
(9) California Department of Food and Agriculture. 1978. *Report
     on Environmental Assessment of Pesticide Regulatory
     Programs State Component. Vol.* 2. p. 3.3-31.
(10) World Health Organization. 1986. *Environmental Health
     Criteria 63 Organophosphorus Insecticides: A General
     Introduction.* International Program on Chemical Safety. p.
     72.
(11) *19*
(12) *21*
(13) 39. Diazinon. 1988.
(15) Northeastern Regional Pesticide Coordinators. 1966.
     Pesticide Information Manual. Cooperative Extension
     Service.

dibutyl phthalate
(1) *33*
(2) Agarwal, D.K., et al. 1985. *J. Toxicol. Environ. Health.*
     16:61-69.
(3) Shiota, K., et al. 1982. *Environ. Health Perspect.* 45:65-70.
(4) *9*

dicamba
(1) *20*
(2) *30*
(3) *6*
(4) *7*
(5) *21*

dicapthon
(1) Freed, V.H., et al. 1979. *J. Agric. Food Chem.* 27(4):706-
     708.
(2) *33*
(3) *32*

dichlobenil
(1) *20*
(2) *58*
(3) U.S. Environmental Protection Agency. 1991. *List of
     Chemicals Evaluated for Carcinogenic Potential.* Health
     Effects Division, Office of Pesticide Programs.
(4) *1*

dichlofenthion
(1) *5*
(2) *7*
(3) *22*
(4) Schafer, E.W. 1972. *Toxicol. Appl. Pharmacol.* 21:315-330.
(5) *21*

dichlofluanid
(1) *33*
(2) *5*
(3) *58*
(4) *1*

dichlone
(1) *5*
(2) *9*
(3) *30*

(4) *42*
dichloran
(1) *58*
(2) Morpurgo, G. et al. 1979. *Environ. Health Perspect.* 31:81-
     95.
(3) *11*
(4) *19*
(5) *1*

dichlormid
(1) *20*
(2) *58*

2,4-dichlorobenzamide
(1) 39. Dichlobenil. 1987.

dichloroethyl ether
(1) *3*
(2) *42*
(3) *44*
(4) *9*

dichlorophen
(1) *5*
(2) *30*
(3) Kader, H.A. et al. 1976. *Comparative Physiology and
     Ecology.* 1(3):78-82.

1,2-dichloropropane
(1) Cohen, D.B. et al. 1983. *Water Quality and Pesticides: 1,2-
     dichloropropane (1,2-D) and 1,3-dichloropropene.* California
     State Water Resources Control Board. Vol. 3. Special
     Projects Report No. 83-85P.
(2) Howard, P.H. ed. 1990. *Handbook of Environmental Fate
     and Exposure Data for Organic Chemicals.* Lewis Publishers.
(3) U.S. Environmental Protection Agency. 1989. *Health
     Advisory Summary. 1,2-Dichloropropane.* Office of Water.

dichloropropene
(1) Cohen, D.B. et al. 1983. *Water Quality and Pesticides: 1,2-
     dichloropropane (1,2-D), 1,3-dichloropropene (1,3-D).*
     California State Water Resources Control Board. Vol. 3.
     Special Projects Report No. 83-85P.
(2) *5*
(3) *29*
(4) Yang, R.S.H. et al. 1986. *J. Toxicol. Environ. Health.* 18:
     377-392.
(5) Albrecht, W.N. 1987. *Arch. Environ. Health.* 42(5):292-296.
(6) *21*
(7) Bently, R.E. 1975. *Acute Toxicity of M-3993 to Bluegill
     (Lepomis macrochirus) and Rainbow Trout (Salmo gairdneri).*
     Unpublished Report. Bionomics, E.G.& G. Wareham,
     Massachusetts.
(8) *1*

dichlorprop
(1) Thompson, D.G., et al. 1984. *J. Agric. Food Chem.* 32:578-
     581.
(2) *30*
(3) U.S. Environmental Protection Agency. 1989. *2,4-DP
     Pesticide Fact Sheet.* Office of Pesticide Programs.
(4) *33*
(5) *20*
(6) U.S. Environmental Protection Agency. 1986. *Dichlorprop
     Tox One-Liner.* Office of Pesticide Programs.
(7) *57*

(8)   *58*
(9)   *5*

dichlorvos

(1)   World Health Organization. 1989. *Environmental Health Criteria 79: Dichlorvos.* International Programme on Chemical Safety.
(2)   *39. DDVP. 1987.*
(3)   *51*
(4)   *5*
(5)   Sasinovich, L.M. 1968. *Gig. Sanit.* (translated from Russian). 33(12):361-366.
(6)   Chan, P.C. 1989. *Toxicology and Carcinogenesis Studies of Dichlorvos in F344/N Rats and B6C3F1 Mice.* Technical Report No. 342. National Toxicology Program.
(7)   U.S. EPA. 1987. Memorandum: Subject: *DDVP Cover Memo.* To: J. Edwards, Reg. Div. Thru: A.S. Rispin, HED. Office of Pesticide Programs.
(8)   Reeves, J.D. 1982. *Am. J. Pediatr. Hematol. Oncol.* 4(4):438-439.
(9)   Civen, M. et al. 1977. *Biochem. Pharmacol.* 26:1901-1907.
(10)   Reeves, J.D. et al. 1980. *Clin. Res.* 28(1):106A.
(11)   World Health Organization. 1986. *Environmental Health Criteria 63. Organophosphorus Insecticides: A General Introduction.* International Programme on Chemical Safety.
(12)   *32*
(13)   *21*

diclofop-methyl

(1)   *58*
(2)   *20*
(3)   *35*

dicofol

(1)   *58*
(2)   *5*
(3)   *8*
(4)   National Cancer Institute. 1978. *Bioassay of Dicofol for Possible Carcinogenicity. Technical Report Series No. 90.* National Institutes of Health.
(5)   *19*
(6)   *21*
(7)   *30*
(8)   *1*

dicrotophos

(1)   *58*
(2)   Lee, P.W. et al. 1989. *J. Agric. Food Chem.* 37:1169-1174.
(3)   *40. Dicrotophos. 1982.*
(4)   Breau, A.P. 1985. *J. Toxicol. Environ. Health.* 16:403-413.
(5)   *19*
(6)   *21*
(7)   *30*

dicryl

(1)   Association of American Pesticide Control Officials, Inc. 1966. *Pesticide Chemicals Official Compendium.*
(2)   *5*
(3)   *2*

dicumarol

(1)   *3*
(2)   *9*

dieldrin

(1)   World Health Organization. 1989. *Environmental Health Criteria 91. Aldrin and Dieldrin.* International Programme on Chemical Safety.
(2)   *5*
(3)   Heinz, G.H. et al. 1981. *Diagnostic brain residues of dieldrin: some new insights.* Avian and Mammalian Wildlife Toxicology: Second Conference. D.W. Lamb and E.E. Kenaga. Eds., American Society for Testing and Materials. pp. 72-92.
(4)   Zabik, M.E. et al. 1973. *Arch. Environ. Health.* 27:25-31.
(5)   National Cancer Institute. *Bioassay of Dieldrin for possible Carcinogenicity 1978.* Technical Report Series. No. 22. National Institutes of Health.
(6)   Epstein, S.S. 1975. *Sci. Total Environ.* 4:1-52.
(7)   Ottolenghi, A.D. et al. 1974. *Teratology.* 9:11-16.
(8)   Murphy, D.A. et al. 1970. *Wildlife Management.* 34:887-903.
(9)   Hamilton, H.E. et al. 1978. *Environ. Res.* 17:155-164.
(10)   Descotes, J. 1988. *Immunotoxicology of Drugs and Chemicals.* Elsevier. New York.
(11)   Joy, R.M. 1976. *Toxicol. Appl. Pharmacol.* 35:95-106.
(12)   *19*
(13)   *21*
(14)   *30*
(15)   *1*
(16)   California Department of Food and Agriculture. 1978. *Report on Environmental Assessment of Pesticide Regulatory Programs Vol. 2.* State Component. p. 3.2-15.
(17)   Fergin, T.J. et al. 1977. *Arch. Environ. Contam. Toxicol.* 6:213-219.
(18)   Lehner, P.N. 1969. *Nature.* 224:1218-1219.
(19)   *28*

dienochlor

(1)   Quistad, G.B. 1983. *J. Agric. Food Chem.* 31:621-624.
(2)   *5*
(3)   U.S. Environmental Protection Agency. *Dienochlor Tox One-Liner.* Office of Pesticide Programs.
(4)   *58*
(5)   *33*
(6)   *Pentac WP Data Sheet No. 710.* 1978. Hooker Specialty Chemicals Division. Niagara Falls, NY.

diethatyl

(1)   *58*
(2)   *5*
(3)   *20*

diethyl fumarate

(1)   *28*

difenacoum

(1)   *33*

difenzoquat

(1)   *58*
(2)   *20*
(3)   *39. Difenzoquat Methyl Sulfate. 1988.*
(4)   *5*

diflubenzuron

(1)   Smuker, R.A. 1988. *Environmental Residues of Dimilin as a Consequence of Aerial Spraying for Gypsy Moth Control.* Maryland Department of Agriculture.

(2) U.S. Environmental Protection Agency. 1986. *Dimilin Product Label.*

(3) 39. Diflubenzuron. 1985.

(4) U.S. Envrionmental Protection Agency. 1989. *Diflubenzuron Pesticide Fact Sheet.* Office of Pesticide Programs.

(5) Dost, F.M. et al. 1985. *Toxicological Examination of Dimilin.* Extension Service Toxicology Information Program. Oregon State University.

(6) Tester, P.A. et al. 1981. *Marine Ecology - Progress Series.* 5:297-302.

dihydrosafrole

(1) *45*

(2) Reuber, M. 1981. *Fd. Cosmet. Toxicol.* 19:130-131.

dikegulac sodium

(1) *5*

(2) *58*

(3) *33*

dimefox

(1) *5*

(2) *58*

(3) *33*

dimethirimol

(1) *58*

(2) *5*

(3) *3*

dimethoate

(1) *32*

(2) World Health Organization. 1989. *Environmental Health Criteria 90-Dimethoate.* International Programme on Chemical Safety. p.18.

(3) *58*

(4) Reuber, M.D. 1984. *Environ. Res.* 34:193-211.

(5) *25*

(6) Courtney, K.D. et al. 1985. *J. Environ. Sci. Health.* B20(4): 373-406.

(7) Stieglitz, R. 1974. *Acta Haemat.* 52:70-76.

(8) Descotes, J. 1988. *Immunotoxicology of Drugs and Chemicals.* Elsevier.

(9) Beusen, J.M. et al. 1989. *Bull. Environ. Contam. Toxicol.* 42: 126-133.

(10) *1*

dimethrin

(1) *33*

(2) *21*

(3) *30*

dimethyl phthalate

(1) *5*

(2) U.S. Environmental Protection Agency. 1985. *EPA Chemical Profile: Dimethyl Phthalate.* Chemical Emergency Preparedness Program.

(3) Agarwal, D.K. et al. 1985. *J. Toxicol. Environ. Health.* V.16:61-69.

(4) Seed, J.L. 1982. *Environ. Health Perspect.* 45:111-114.

dimethyl sulfoxide

(1) *5*

(2) Walles, A.S. et al. 1984. *Carcinogenesis.* 5(3):319-323.

(3) *31*

dinex

(1) *9*

(2) *1*

(3) California Department of Food and Agriculture. 1978. *Report on Environmental Assessment of Pesticide Regulatory Programs. Vol. 2.* State Component. p. 3.2-20.

dinitramine

(1) *58*

(2) *5*

(3) *21*

dinitrocresol

(1) *30*

(2) *58*

(3) *3*

(4) *19*

(5) *21*

dinitrophenol

(1) *5*

(2) *3*

(3) California Department of Food and Agriculture. 1982. *Pesticide Safety Information Series F-6 Dinitrophenol.* Worker Health and Safety Unit.

(4) *31*

dinobuton

(1) *58*

(2) *5*

(3) *42*

dinocap

(1) *58*

(2) Mittelstaedt, W. and F. Fuhr. 1984. *J. Agric. Food Chem.* 32(5):1151-1155.

(3) *3*

(4) U.S. Environmental Protection Agency. November 7, 1984. *Environmental News: EPA notified of voluntary suspension of pesticide Dinocap.* Office of Public Affairs.

(5) *30*

(6) *1*

dinoseb

(1) *20*

(2) *30*

(3) *58*

(4) U.S. Environmental Protection Agency. 1986. *Pesticide Fact Sheet: Dinoseb-Special Review/Emergency Suspension.* Office of Pesticide Programs.

(5) U.S. Environmental Protection Agency. 1986. Federal Register Notice, October 14. 51(198). p. 36637.

(6) Giavini, E. et al. 1989. *Bull. Environ. Contam. Toxicol.* 43 (2):215-219.

(7) U.S. Environmental Protection Agency. 1987. *Decision and Final Order Modifying Final Suspension of Pesticide Products Which Contain Dinoseb.* Office of the Administrator.

(8) Call, D.J. et al. 1984. *J. Environ. Qual.* 13(3):493-498.

(9) *19*

dinoterb

(1) *22*

dinoterb acetate

(1) *5*

dioxabenzofos

(1) *58*

(2) *5*

(3)   World Health Organization. *Organophosphorus Insecticides: A General Introduction. Environmental Health Criteria 63.* International Programme on Chemical Safety.

dioxacarb
(1)   *58*
(2)   World Health Organization. 1986. *Environmental Health Criteria 64. Carbamate Pesticides: A General Introduction.*
(3)   *1*

1,4-dioxane
(1)   International Agency for Research on Cancer, World Health Organization. 1976. *IARC Monographs on the Evaluation of Carcinogen Chemicals to Humans.* Vol. 11. p.247-255.

dioxathion
(1)   *32*
(2)   *5*
(3)   U.S. Environmental Protection Agency. 1985. *EPA Chemical Profile: Dioxathion.* Chemical Emergency Preparedness Program.
(4)   Mortelmans, K. et al. 1986. *Environ. Mutagen.* 8(Suppl.7):1-119.
(5)   *21*
(6)   *1*

diphacinone
(1)   *37*
(2)   *58*

diphenamid
(1)   *20*
(2)   *39. Diphenamid. 1987.*
(3)   *21*
(4)   *1*

diphenyl
(1)   *5*
(2)   *3*
(3)   *9*

diphenylamine
(1)   *5*
(2)   Katayama, S., et al. 1982. *J. Natl. Cancer Inst.* 68(5):867-871.
(3)   Helleman, A.H., et al. 1984. *Cancer Lett.* 22:211-218.
(4)   Lenz, S.D., et al. 1990. *Vet. Pathol.* 27:171-178.

dipropetryn
(1)   *58*
(2)   *5*
(3)   U.S. Environmental Protection Agency. 1985. *Chemical Information Fact Sheet for Dipropetryn.* Office of Pesticide Programs.

dipropylnitrosamine
(1)   *49*

diquat
(1)   *58*
(2)   *5*
(3)   *39. Diquat. 1986.*
(4)   Bus, J.S. et al. 1975. *Toxicol. Appl. Pharmacol.* 33:450-460.
(5)   Spalding, D.J.M. et al. 1989. *Toxicol. Appl. Pharmacol.* 101:319-327.
(6)   Selypes, A. et al. 1980. *Bull. Environ. Contam. Toxicol.* 25:513-517.
(7)   *20*
(8)   *19*

(9)   *21*

disparlure
(1)   *3*

disulfiram
(1)   *28*
(2)   Thompson, P.A. et al. 1985. *J. Appl. Toxicol.* 5(1):1-10.

disulfoton
(1)   Gohre, L., et al. *J. Agric. Food Chem.* 34(4):709-713.
(2)   *30*
(3)   *5*
(4)   *58*
(5)   *25*
(6)   *19*
(7)   *21*

ditalimfos
(1)   *5*
(2)   *33*

diuron
(1)   Dynamac Corporation. June 8, 1982. *Diuron: Task 2: Topical Discussions.* Environ Control Divisions. Rockville, MD.
(2)   *30*
(3)   *20*
(4)   *5*
(5)   *11*
(6)   Seiler, J.P. 1978. *Mutat. Res.* 58:353-359.
(7)   Khera, K.S. et al. *Bull. Environ. Contam. Toxicol.* 22:522-529.
(8)   *19*
(9)   *21*
(10) *1*

DMPA
(1)   *3*
(2)   *30*
(3)   Francis, B.M., et al. 1980. *J. Environ. Sci. Health* B15 (4):313-331.

dodine
(1)   *5*
(2)   *39. Dodine. 1987.*
(3)   *58*
(4)   *1*
(5)   *30*

drazoxolon
(1)   *58*
(2)   Clark, D.G., et al. 1969. *Fd. Cosmet. Toxicol.* 7:481-491.
(3)   *9*
(4)   *5*

DSMA
(1)   E.A. Woolson. 1975. *Arsenical Pesticides.* American Chemical Society. p. 65.
(2)   *20*
(3)   Fowler, B.A. (Ed.). 1983. *Biological and Environmental Effects of Arsenic.* Elsevier Publishers. B.V.
(4)   *5*

edifenphos
(1)   *23*
(2)   *58*

emetic
(1)   *39. Paraquat Dichloride. 1987.*

endod

(1) Lemma, A., et al. 1984. *Phytolacco dodecandra* (Endod). Final Report of the International Scientific Workshop, Lusaka, Zambia. March 1983. Zambian National Council for Scientific Research.

(2) Stabaeus, J.K. et al. 1990. *Vet. Hum. Toxicol.* 2(3):212-216.

endosulfan

(1) Ranga Rao, D.M., et al. 1980. *J. Agric. Food Chem.* 28(6):1099-1101.

(2) *8*

(3) *39.* Endosulfan. 1982.

(4) Reuber, M.D. 1981. *Sci. Total Environ.* 20:23-47.

(5) Sobti, R.C., et al. 1983. *Arch. Toxicol.* 52:221-231.

(6) Dzwonkowska, A., et al. 1986. *Arch. Toxicol.* 58:152-156.

(7) Anand, M., et al. 1987. *J. Toxicol.-Cut. & Ocular Toxicol.* 6(3): 161-171.

(8) Banaerjee, B.D., et al. 1986. *Arch. Toxicol.* 59(4):279-284.

(9) Daniel, C.S., et al. 1986. *Toxicol. Lett.* 32(1-2):113-118.

(10) *19*

(11) *21*

(12) *1*

(13) Vardia, H.K., et al. 1984. *Arch. Hydrobiol.* 100(3):395-400.

(14) Mane, U.H., et al. 1984. *Toxicol. Lett.* 23(2):147-155.

(15) Pandy, A.C. 1988. *Ecotoxicol. Environ. Safety.* 15(2):221-225.

(16) Kulshrestha, S.K., et al. 1984. *Toxicol. Lett.* 20:93-98.

(17) Dmoch, J. 1988. *Acta. Hortic.* 219:15-20.

(18) Naqvi, S.M., et al. 1990. *J. Environ. Sci. Health. B.* 25(4):511-526.

(19) *34*

endothall

(1) Reinert, L.H., et al. 1988. *Arch. Environ. Contam. Toxicol.* 17:195-199.

(2) *5*

(3) *7*

(4) U.S. Environmental Protection Agency. 1983. *Endothall Tox One-Liner.* Office of Pesticide Programs.

(5) *21*

(6) *30*

(7) Eller, L. 1969. *Trans. Am. Fisheries Soc.* 98(1): 52-59.

endrin

(1) *30*

(2) *5*

(3) Reuber, M.D. 1979. *Sci. Total Environ.* 12:101-135.

(4) Reuber, M.D. 1978. *Exp. Cell Biol.* 46:129-145.

(5) Kavlock, R.J., et al. 1985. *Teratogenesis Carcinog. Mutagen.* 5:3-13.

(6) Chernoff, N., et al. 1979. *Toxicology.* 13:155-165.

(7) Mehrotra, B.D. et al. 1989. *Toxicology.* 54:17-29.

(8) *19*

(9) *21*

(10) Scott, T.G. et al (eds.). 1978. *Fisheries And Wildlife Research.* U.S. Fish and Wildlife Service. p. 37.

(11) *1*

(12) California Department of Food and Agriculture. 1978. *Report on Environmental Assessment of Pesticide Regulatory Programs. Vol. 2. State Component.*

(13) Spann, J.W., et al. 1979. *Environ. Toxicol. Chem.* 5:755-759.

(14) Fleming, Q.J. et al. 1982. *Journal of Wildlife Management.* 46(2):462-468.

epichlorohydrin

(1) Wester, P.W., et al. 1985. *Toxicology.* 36:325-339.

(2) Laskin, S., et al. 1980. *J. Natl. Cancer Inst.* 65(4):751-755.

(3) Burlinson, B. 1989. *Carcinogenesis.* 10(8):1425-1428.

(4) DeFlora, S., et al. 1987. *Cancer Res.* 47:4740-4745.

(5) Toth, G.P., et al. 1989. *Fund. Appl. Toxicol.* 13:16-25.

(6) John, J.A., et al. 1983. *Toxicol. Appl. Pharmacol.* 68:415-423.

EPN

(1) *39.* EPN. 1987.

(2) Mrak, E.M. et al. 1969. *Report of the Secretary's Commission on Pesticides and their Relationship to Environmental Health.* U.S. Department of Health, Education and Welfare.

(3) Abou-Donia, M.B., et al. 1983. *Toxicol. Appl. Pharmacol.* 68:54-65.

(4) *19*

(5) *21*

(6) *1*

(7) *34*

EPTC

(1) Smith, A.E., et al. 1982. *Bull. Environ. Contam. Toxicol.* 29(2):243-247.

(2) *20*

(3) *2*

(4) Schafer, E.W. 1972. *Toxicol. Appl. Pharmacol.* 21:315-330.

(5) *21*

erbon

(1) *5*

(2) Hilton, J.L., et al. 1974. *Herbicide Handbook.* 3rd ed. Weed Science Society of America.

(3) *1*

esbiothrin

(1) World Health Organization. 1989. *Environmental Health Criteria 87: Allethrins Allethrin, d-Allethrin, Bioallethrin, S-Bioallethrin.* International Programme on Chemical Safety. p. 38.

(2) *40.* Allethrins: Allethrin, D-Allethrin, Bioallethrin, S-Bioallethrin. 1989.

etacelasil

(1) *58*

(2) *5*

ethalfluralin

(1) *58*

(2) *20*

(3) U.S. Environmental Protection Agency. 1985. *Ethalfluralin Fact Sheet.* Office of Pesticide Programs.

(4) *39*

ethephon

(1) *58*

(2) *20*

(3) *1*

(4) *39*

ethiofencarb

(1) *58*

ethiolate
(1) Hilton, J.L., et al. 1974. *Herbicide Handbook*. 3rd ed. Weed Science Society of America. p. 188.

ethion
(1) *32*
(2) Fierberg, F.E., et al. 1985. *J. Agric. Food Chem.* 31(4):704-709.
(3) *58*
(4) *5*
(5) *39.* Ethion. 1989.
(6) *19*
(7) *21*
(8) *1*
(9) *30*

ethiozin
(1) *5*

ethirimol
(1) *58*

ethofumesate
(1) *58*
(2) Fisions Corp. 1978. *Nortron Technical Information.* Agricultural Chemicals Division. Bedford, MA.

ethoprop
(1) *58*
(2) *39.* Ethoprop. 1988.
(3) *19*

ethoxyquin
(1) *5*
(2) Fukushima, S., et al. 1984. *Cancer Lett.* 23:29-37.
(3) Epstein, S.S., et al. 1970. *Toxicol. Appl. Pharmacol.* 16:321-334.
(4) Rannug, A., et al. 1984. *Prog. Clin. Biol. Res.* 141:407-419.

ethylan
(1) *3*
(2) *5*
(3) *30*
(4) *1*

ethyl benzene
(1) Ungvàry, G. and E. Tátrai. 1985. *Arch. Toxicol.* Suppl. 8:425-430.

ethyl formate
(1) *5*
(2) *42*
(3) *1*

ethyl hexanediol
(1) *5*
(2) *42*

ethylene dibromide
(1) Ali, S.M., et al. 1984. *Ethylene Dibromide (EDB): A Water Quality Assessment.* California State Water Resources. Special Projects Report No. 84-8SP.
(2) *40*
(3) *58*
(4) *4*
(5) Brown, A.F. 1984. *J. Environ. Health.* 46(5):220-225.
(6) Kitchin, K.T., et al. 1986. *Biochem. Biophys. Res. Comm.* 141(2):723-727.

(7) Smith, R.F., et al. 1983. *Neurobehav. Toxicol. Teratol.* 5:579-585.
(8) Short, R.D., et al. 1979. *Toxicol. Appl. Pharmacol.* 49:97-105.
(9) *3*
(10) Westlake, G.E., et al. 1981. *Br. Poult. Sci.* 22:355-364.
(11) Landau, M., et al. 1984. *Bull. Environ. Contam. Toxicol.* 33:127-132.
(12) Darnerud, P.O., et al. 1989. *J. Toxicol. Environ. Health.* 26:209-22
(13) U.S. Department of Health, Education, and Welfare, National Institute Of Safety and Health. 1977. *Criteria For A Recommended Standard...Occupational Exposure To Ethylene Dibromide.*

ethylene dichloride
(1) *3*
(2) *51*
(3) Storer, R.D., et al. 1984. *Cancer Res.* 44:4267-4271.
(4) *11*
(5) Igwe, O.J., et al. 1986. *Toxicol. Appl. Pharmacol.* 86:286-297.

ethylene oxide
(1) *11*
(2) *56*
(3) Dunkelberg, H. 1982. *Br. J. Cancer.* 46:924-933
(4) Kelsey, K., et al. 1988. *Cancer Res.* 48:5045-5050.
(5) Tucker, J.D., et al. 1986. *Teratogenesis. Carcinog. Mutagen.* 6:15-21
(6) Kuzuhara, S., et al. 1983. *Neurology.* 33:377-380.
(7) Mori, K., et al. 1989. *Toxicol. Appl. Pharmacol.* 101:299-309.

ethylene thiourea
(1) *9*
(2) *40.* Terrazole. 1980.
(3) National Toxicology Program. 1989. *Report on Perinatal Toxicity and Carcinogenicity Studies of ETU in F/344 Rats and B6C3F1 Mice (feed studies).* U.S. Department of Health and Human Services. Public Health Service. National Institutes of Health.
(4) McGregor, D.B., et al. 1988. *Environ. Mol. Mutagen.* 12:85-154.
(5) Chernoff, N., et al. 1979. *J. Toxicol. Environ. Health.* 5(5):821-834.
(6) Stula, E.F. and W.C. Krauss. 1977. *Toxicol. Appl. Pharmacol.* 41:35-55.
(7) Daston, G.P., et al. 1987. *Fundam. Appl. Toxicol.* 9:415-422.

ethyl mercaptan
(1) Fairchild, E.J. and H.E. Stokinger. 1958. *Amer. Ind. Hyg. Assoc. J.* 19:171-189.

etridiazol
(1) *58*
(2) *33*
(3) *39.* Terrazole. 1980.

etrimfos
(1) Bowman, M.C., et al. 1978. *J. Agric. Food Chem.* 26:(1):35-42.
(2) *3*
(3) *5*

(4) *58*
EXD
(1) *33*
(2) *5*
famphur
(1) *7*
(2) *19*
(3) Franson, J.C. et al. 1985. *J. Wildlife Diseases.* 21(3):318-320.
(4) Hill, E.F. et al. 1980. *J. Wildlife Mgmt.* 44(3):676-681.
fenac
(1) *20*
(2) *21*
fenaminosulf
(1) *30*
(2) *9*
(3) *33*
(4) *39.* Fenaminosulf. 1983.
(5) Tezuka, H. et al. 1980. *Mutat. Res.* 78:177-191.
(6) National Cancer Institute. 1978. *Bioassay of Formulated Fenaminosulf for Possible Carcinogenicity. Report No. 101.* National Institutes of Health.
(7) *31*
(8) *21*
(9) *1*
fenamiphos
(1) Krause, M. et al. 1986. *J. Agric. Food Chem.* 34(4):717-720.
(2) *32*
(3) *58*
(4) *1*
fenarimol
(1) *58*
(2) *3*
(3) U.S. EPA. 1984b. Memorandum: *A two-generation reproduction study with fenarimol in guinea pigs: reponse to EPA comments date August 14, 1984.* From: W. Dykstra, Tox. Branch, HED. To: H. Jacoby, Prod. Mngr., Office of Pesticide Programs.
(4) U.S. EPA. 1986a. Memorandum: *Fenarimol in/on apples.* From: W. Dykstra, Tox. Branch, HED. To: H. Jacoby, Prod. Mngr., Office of Pesticide Programs.
(5) Hirsch, E.S. et al. 1986. *Toxicol. Appl. Pharmacol.* 86:391-399.
fenbutatin-oxide
(1) *58*
(2) *19*
(3) *1*
fenitrothion
(1) *32*
(2) *13*
(3) *25*
(4) Center for Disease Control lends credence to pesticide, chemical causes of Reye's syndrome. 1977. *Pesticide and Toxic Chemical News.* 5(9):13-14.
(5) Lehotzky, K., et al. 1989. *Neurotoxicol. Teratol.* 11:321-324.

(6) World Health Organization. 1986. *Environmental Insecticides: A General Introduction.* International Programme on Chemical Safety. p.71.
(7) *19*
(8) *21*
(9) *1*
(10) California Department of Food and Agriculture. 1978 *Report on Environmental Assessment of Pesticide Regulatory Programs. Vol. 2.* State Component.
fenoxaprop-ethyl
(1) Dynamac Corporation. 1986. *Final Report: Task I: Review & Evaluation of Individual Studies, Task II: Environmental Fate & Exposure Assessment.* Rockville, MD.
(2) *58*
(3) *5*
(4) *20*
(5) U.S. Environmental Protection Agency. 1987. *Fenoxaprop-ethyl Pesticide Fact Sheet.* Office of Pesticide Programs.
fenpropathrin
(1) Mikami, N. et al. 1985. *J. Agric. Food Chem.* 33:980-987.
(2) Tikahashi, N. et al. 1985. *Pesticide Science* 16:119-131.
(3) *58*
fensulfothion
(1) *32*
(2) *3*
(3) *19*
(4) *58*
(5) U.S. Environmental Protection Agency. 1985. *Chemical Information Fact Sheet for Fensulfothion.* Office of Pesticide Programs.
(6) *1*
fenthion
(1) Gohre, K. et al. 1986. *J. Agric. Food Chem.* 34:709-713.
(2) *32*
(3) *58*
(4) *14*
(5) National Cancer Institute. 1979. *Bioassays of Fenthion for possible Carcinogenicity.* Technical Report Series. National Institutes of Health.
(6) Cherniack, M.G. 1988. *Neurotoxicology.* 9(2):249-272.
(7) Budreau, C.H., et al. 1973. *Toxicol. Appl. Pharmacol.* 24:324-332.
(8) Misra, U.K., et al. 1988. *Arch. Toxicol.* 61:496-500.
(9) Imai, H. 1974. *Nippon Ganka Gakki Zasshi.* 81(8)925-932.
(10) Imai, H. 1974. *Nippon Ganka Gakki Zasshi.* 78(4)163-172.
(11) *19*
(12) U.S. Environmental Protection Agency. 1987. *Data Evaluation Record: Fenthion.* Health Effects Division.
(13) *21*
(14) *1*
fentin hydroxide
(1) U.S. Environmental Protection Agency. 1984. *Triphenyl Tin Hydroxide Pesticide Fact Sheet.* Office of Pesticide Programs.
(2) *5*
(3) *31*
(4) *58*

(5) *1*

fenuron

(1) *20*

(2) Seiler, J.P. 1978. *Mutat. Res.* 58:353-359.

(3) *5*

fenvalerate

(1) *5*

(2) Chatterjee, K.K., et al. 1982. Mutat. Res. 105:101-106.

(3) Puig, M., et al. 1989. *Mutagenesis.* 4: 72-74.

(4) Clark, J.R., et al. 1989. *Environ. Toxicol. Chem.* 8:393-401.

(5) Bradbury, S.P., et al. 1989. *Environ. Toxicol. Chem.* 8:373-380.

ferbam

(1) *30*

(2) *58*

(3) *25*

(4) *46*

(5) *9*

(6) Minor, J.L. et al. 1974. *Toxicol. Appl. Pharmacol.* 29:120.

(7) Quinto, E.D.M., et al. 1989. *Mutat. Res.* 224:406-408.

(8) *1*

(9) Serio, R., et al. 1984. *Toxicol. Appl. Pharmacol.* 72:333-

ferrous sulfate

(1) *3*

flamprop-isopropyl

(1) *5*

fluazifop-butyl

(1) Negre, M., et al. 1988. *J. Agric. Food Chem.* 36:1319-1322.

(2) *35*

(3) *58*

(4) *20*

fluchloralin

(1) *5*

(2) *58*

flucythrinate

(1) *58*

(2) U.S. Environmental Protection Agency. 1984. Memorandum: *Environmental fate review of flucythrinate.* From: S.M. Creeger, Exposure Assessment Branch, HED. To: T. Gardner/Heyward, Prod. Mgr., Reg. Div. Office of Pesticide Programs.

(3) *5*

(4) U.S. Environmental Protection Agency. 1984. *Tox One-Liner: 2AA Pay-Off®.* Office of Pesticide Programs.

(5) Clark, J.R., et al. 1989. *Environ. Toxicol. Chem.* 8:393-401.

fluenethyl

(1) *5*

(2) *42*

(3) *1*

flufenoxuron

(1) *58*

(2) *5*

fluometuron

(1) *58*

(2) *20*

(3) National Cancer Institute. 1980. *Bioassays of Fluometuron for Possible Carcinogenicity. Technical Report Series.* National Institutes of Health.

(4) *54*

(5) *1*

fluoroacetamide

(1) *14*

(2) *31*

fluorocitrate

(1) Peters, R.A. 1973. *Am. J. Clin. Nutr.* 27:750-759.

(2) Cater, D.B. and R.A. Peters. 1961. *Br. J. Exp. Path.* 42:278-289.

(3) *1*

(4) *9*

fluorodifen

(1) *33*

(2) *7*

(3) *1*

flurecol-butyl

(1) *33*

(2) *5*

(3) *58*

fluridone

(1) *58*

(2) *20*

(3) *5*

(4) *Sonar₂ Specimen Label.* 1987. Lilly Corporate Center. Indianapolis.

flusilazole

(1) *58*

(2) *5*

fluvalinate

(1) U.S. Environmental Protection Agency. 1986. *Fluvalinate Pesticide Fact Sheet.* Office of Pesticide Programs.

(2) *5*

r-fluvalinate

(1) *58*

folpet

(1) *7*

(2) 39. Folpet. 1987.

(3) U.S. Environmental Protection Agency. 1989. *Folpet Tox One-Liner.* Office of Pesticide Programs.

(4) Robens, J.F. 1970. *Toxicol. Appl. Pharmacol.* 16:24-34.

(5) *19*

(6) *21*

(7) *1*

(8) *30*

(9) *33*

fomesafen

(1) *20*

(2) *5*

(3) *58*

fonofos

(1) *32*

(2) U.S. Environmental Protection Agency. 1985. *Chemical Information Fact Sheet for Fonofos.* Office of Pesticides Programs.

(3) *3*

(4) *58*

formaldehyde
(1) *58*
(2) *39. Formaldehyde and Paraformaldehyde. 1988.*
(3) *29*
(4) Cosma, G.N. and A.C. Marchok. 1988. *Toxicology.* 51:309-320.
(5) Strubelt, O., et al. 1989. *J. Toxicol. Environ. Health.* 27:351-366.

formetanate hydrochloride
(1) *58*
(2) *32*
(3) *1*

formic acid
(1) *3*

N-formyl-chloro-o-toluidine
(1) *39. Chlodimeform.*

fosamine ammonium
(1) *20*
(2) Envirologic Data. 1984. *Risk Assessment of Fosamine Ammonium for Maine Department of Transportation and Maine Board of Pesticide Control.* Portland, ME.
(3) *58*
(4) Hoffman, D.J. 1988. *Environ. Toxicol. Chem.* 7:69-75.

fosetyl-al
(1) U.S. Environmental Protection Agency. 1990. *Reregistration Eligibility Document Aluminum Tris (o-Ethylphosphonate) (Refered to as Fosetyl-Al).* Office of Pesticide Programs.
(2) *58*

furamethrin
(1) Miyamoto, V. 1976. *Environ. Health Perspect.* 14:15-28.

GC 6506
(1) *42*

glyphosate
(1) Rueppel, M.L., et al. 1977. *J. Agric. Food Chem.* 25(2):517-528.
(2) Edwards, W.M. 1980. *J. Environ. Quality.* 9:661-665.
(3) *33*
(4) *39. Glyphosate. 1986.*
(5) Vigfusson, N.V. et al. 1980. *Mutat. Res.* 79(l):53-57.
(6) *20*
(7) Envirologic Data. 1984. *Risk Assessment of Glyphosate for Maine Board of Pesticide Control.*
(8) Boyle, W., et al. 1974. *J. Heredity.* 65:250.

heptachlor
(1) *39*
(2) *51*
(3) *58*
(4) National Research Council. 1982. *An Assessment of the Health Risks of Seven Pesticides used for Pest Control.* National Academy Press. pp. 19-21.
(5) *9*
(6) Reuber, M.D. 1987. *J. Environ. Pathol. Toxicol. Oncol.* 7(3):85-114.
(7) *4*
(8) *13*
(9) *21*
(10) *30*
(11) *19*

(12) *1*
(13) Bresch, H. & U. Arendt. 1977. *Environ. Res.* 13:121-128.

heptachlor epoxide
(1) National Research Council. 1982. *An Assessment of the Health Risks of Seven Pesticides used for Pest Control.* National Academy Press. pp. 19-21.
(2) *9*
(3) Reuber, M.D. 1987. *J. Environ. Pathol. Toxicol. Oncol.* 7(3):85-114.

heptachlorodibenzo-p-dioxin
(1) Kerkvliet, N.J., et al. 1987. *Toxicol. Appl. Pharmacol.* 87:18-31.

hexachlorobenzene
(1) *51*
(2) U.S. EPA. 1978. *Hexachlorobenzene Sources/Effects Review (Phase I Report).* Office of Pesticide Programs.
(3) Arnold, D.L., et al. 1985. Two-generation chronic toxicity study with hexachlorobenzene in the rat. pp. 405-410. in *Hexachlorobenzene: Proceedings Of An International Symposium.* IARC Scientific Publication No. 77. World Health Organization.
(4) Erturk, E., et al. 1985. Oncogenicity of HCB. In *Hexachlorobenzene: Proceedings of an International Symposium.* IARC Scientific Publication No. 77. World Health Organization. pp. 417-423.
(5) Cabral, J.R.P., et al. 1977. *Nature.* 269:510-511.
(6) Courtney, K.D., et al. 1976. *Toxicol. Appl. Pharm.* 35:238-256.
(7) *14*
(8) Peters, H.A., et al. 1985. Neurotoxicity of HCB-induced porphyria turcica. In *Hexachlorobenzene: Proceedings of an International Symposium.* IARC Scientific Publication No. 77. World Health Organization. pp. 575-579.
(9) Bleavins, M.R., et al. 1984. *J. Toxicol. Environ. Health.* 14:363-377.
(10) Kleiman de Pisarev, D.L., et al. 1990. *Biochem. Pharmacol.* 39(5):817-825.
(11) Andrews, J.E., et al. 1989. *Fundam. Appl. Toxicol.* 12(2):242-251.
(12) Sherwood, R.L., et al. 1989. *Toxicol. Indus. Health.* 5(3):451-461.
(13) Vos, J. 1985. Immunotoxicity of hexachlorobenzene. pp. 347-356 in *Hexachlorobenzene: Proceedings of an International Symposium.* IARC Scientific publication No. 77. World Health Organization.
(14) *21*
(15) Bleavins, M.R., et al. 1984. *Arch. Environ. Contam. Toxicol.* 13:357-365.

hexachlorodibenzo-p-dioxin
(1) World Health Organization. 1987. *Environmental Health Criteria 71: Pentachlorophenol.* International Programme on Chemical Safety.
(2) Holsapple, M.P., et al. 1984. *J. Pharmacol. Exp. Ther.* 231 (3):518-526.

hexachlorodibenzofuran
(1) Palmer, F.H., et al. 1988. *Chlorinated dibenzo-p-dioxin and dibenzofuran contamination in California from chlorophenol wood preservative uses.* Report No. 88-

5WQ. Division of Water Quality. State Water Resource Control Board.
(2)   Hébert, C.D., et al. 1990. *Toxicol. Appl. Pharmacol.* 102(2):362-377.
(3)   Birnbaum, L.S., et al. 1987. *Toxicol. Appl. Pharmacol.* 90(2):206-216.

hexachlorophene
(1)   *7*
(2)   *51*
(3)   Kimmel, C.A., et al. *Arch. Environ. Health.* 28(1):43-48.
(4)   *3*
(5)   Thompson, J.P., et al. 1987. *J. Am. Vet. Med. Asssoc.* 190(10):1311-1312.
(6)   *3*

hexazinone
(1)   Roy, D.N., et al. 1989. *J. Agric. Food Chem.* 37:443-447.
(2)   *20*
(3)   *39.* Hexazinone. 1988.

hydramethylnon
(1)   *58*
(2)   *5*
(3)   U.S. Environmental Protection Agency. 1991. *List of Chemicals Evaluated for Carcinogenic Potential.* Health Effects Division, Office of Pesticide Programs.

hydrazoic acid
(1)   Humburg, N.E. et al. 1979. *Herbicide Handbook.* 4th Edition. Weed Science Society of America.

hydrogen chloride
(1)   *3*

hydrogen cyanide
(1)   *13*
(2)   *5*
(3)   *3*

hydrogen sulfide
(1)   *9*

N-hydroxymethyl methiocarb sulfoxide
(1)   *39.* Methiocarb. 1987.

4-hydroxy-2,5,6-trichloroisophtalonitrile
(1)   *39.* Methiocarb. 1987.

hypochlorous acid
(1)   *39.* Chlorinated Isocyanurates. 1988.

imazamethabenz
(1)   *58*

imazapyr
(1)   *58*
(2)   *20*

ioxynil
(1)   *58*
(2)   *33*

iprodione
(1)   *58*
(2)   *3*
(3)   *33*
(4)   *Technical Bulletin.* Rhône Poulenc Ag Company. Research Triangle Park, N.C.

isazophos
(1)   *5*

isofenphos
(1)   Racke, K.D., et al. 1987. *J. Agric. Food Chem.* 35(1):94-99.
(2)   *58*
(3)   *5*
(4)   Catz, A., et al. 1988. *J. Neurol. Neurosurg. Psychiatry.* 51:1338-1340.
(5)   Rubin, H. 1984. *The Nation.* May 19, 1984. p. 599-602.
(6)   *32*
(7)   Wilson, B.W., et al. 1984. *Bull. Environ. Contam. Toxicol.* 33:386-394.

isopropalin
(1)   *58*
(2)   *20*
(3)   *Technical Report on Paarlan.* Elano Products Company. Indianapolis, IN.

isoxaben
(1)   U.S. Environmental Protection Agency. 1988. *Isoxaben Pesticide Fact Sheet.* Office of Pesticide Programs.

kadethrin
(1)   *58*

kerosene
(1)   *28*
(2)   Blackburn, G.R. et al. 1986. *Cell Biol. Toxicol.* 2(1):63-83.
(3)   *9*

12-ketoendrin
(1)   Stickel, W.H., et al. 1979. Endrin versus 12-ketoendrin in birds and rodents. *Avian and Mammalian Wildlife.* pp. 61-68.

lead arsenate
(1)   *30*

leptophos
(1)   *13*
(2)   *7*
(3)   *22*
(4)   *1*

d-limonene
(1)   National Toxicology Program. 1990. *Toxicology and Carcinogenesis Studies of d-Limonene in F344/N Rats and B6C3F Mice.* U.S. Dept. of Health and Human Services. Public Health Service.
(2)   Kodama, R. 1977. *Oyo Yakuri Kenyukai.* 13(6):863-873.
(3)   Tsuji, M. 1975. *Oyo Yakuri Pharmacometrics.* 9(5):775-808.
(4)   Evans, D.L. et al. 1987. *J. Toxicol. Environ. Health.* 20:51-66.
(5)   Hooser, S.B. 1990. *Veterinary Clinics of North America: Small Animal Practice.* 20(2):383-385.

lindane
(1)   *30*
(2)   *13*
(3)   Aparicio-Lopez, P., et al. 1988. *Pesticide Biochemistry and Physiology.* 31:109-119.
(4)   *40.* Lindane. 1983.
(5)   *43*
(6)   Cooper, R.L., et al. 1989. *Toxicol. Appl. Pharmacol.* 99:384-394
(7)   Chowdhury, A.R., et al. 1987. *Bull. Environ. Contam. Toxicol.* 38:154-156.

(8) Descotes, J. 1988. *Immunotoxicology of Drugs and Chemicals.* Elsevier.

(9) Sanfeliu, C., et al. 1989. *Neurotoxicology.* 10:727-742.

(10) Suñol, C., et al. 1988. *Toxicology.* 49:247-252.

(11) *9*

(12) U.S. Environmental Protection Agency. 1977. Rebuttable Presumption Against Registration and Continued Registration of Pesticide Products containing Lindane. *Federal Register.* 42(33): 9816-9947.

(13) *19*

(14) *21*

(15) *30*

(16) *Report on Environmental Assessment of Pesticide Regulation Programs.* 1978. State Component. Volume 2. Dept. of Food and Agriculture.

(17) *1*

(18) Anderegg, B.N., et al. 1977. *J. Agric. Food Chem.* 25(4):923-928.

(19) Sauter, E.A., et al. 1972. *Poult. Sci.* 51:71-76.

(20) Braunbeck, T., et al. 1990. *Ecotoxicol. Environ. Safety.* 19(3):355-372.

(21) Segault, D.M., et al. 1981. *Environ. Res.* 24:250-258.

linuron

(1) *20*

(2) U.S. Environmental Protection Agency. August 17, 1988. Linuron; preliminary determination to conclude the special review; Notice. *Federal Register.* 53(159):31262-31268.

(3) *1*

(4) Wuu, K.D., et al. 1967. *Cytologia.* 32:31-41.

malaoxon

(1) *4*

(2) *40. Malathion.* 1988.

malathion

(1) *30*

(2) *28*

(3) Walter, Z., et al. 1980. *Hum. Genet.* 53:375-381.

(4) Lechner, D.M.W., et al. 1984. *J. Toxicol. Environ. Health.* 14:267-278.

(5) Duffy, F,. et al. 1979. *Toxicol. Appl. Pharmacol.* 47:161-176.

(6) Milby, T.H,. et al. 1964. *Arch. Environ. Health.* 9:434-437.

(7) National Cancer Institute. 1979. *Bioassay of Malathion for Possible Carcinogenicity.* Technical Report Series No. 192. National Institutes of Health.

(8) Tamura, O., et al. 1975. *Jap. J. Opthalmol.* 19:250-253.

(9) Kurtz, P.L. 1976. *Behavioral and biochemical effects of malathion.* Study No. 51-051-73/6. U.S. Army Environmental Hygiene Agency.

(10) World Health Organization. 1986. Environmental Health Criteria 63. *Organophosphorus Insecticides: A General Introduction.* International Programme on Chemical Safety.

(11) *19*

(12) *30*

(13) California Department of Food and Agriculture. 1978. *Report on Environmental Assessment of Pesticide Regulatory Programs. Vol. 2.* State Component.

mancozeb

(1) *58*

(2) *39. Mancozeb.* 1987.

(3) *5*

(4) *1*

(5) Bristow, P.R., et al. 1981. *J. Amer. Hort. Sci.* 106(3):290-292.

maneb

(1) Nash, R.G., et al. 1980. *J. Agric. Food Chem.* 28:322-330.

(2) Rhodes, R.C. 1977. *J. Agric. Food Chem.* 25(3):528-533.

(3) *5*

(4) *37*

(5) *58*

(6) Larsson, K.S., et al. 1976. *Teratology.* 14:171-184.

(7) Petrova-Vergieva, et al. 1973. *Food Cosmetic Toxicol.* 11:239-244.

(8) *1*

MBT

(1) Guess, W.L. and R.K. O'Leary. 1969. *Toxicol. Appl. Pharmacol.* 14:221-231.

(2) *9*

MCPA

(1) Crosby, D.G., et al. 1985. *J. Agric. Food Chem.* 33:569-573.

(2) *39. MCPA.* 1989.

(3) *54*

(4) *1*

mecoprop

(1) *20*

(2) *40. Mecoprop (MCPP).* 1988.

(3) *59*

(4) *1*

melamine

(1) National Toxicology Program. 1983. *Carcinogenesis Bioassay of Melamine (CAS. No. 108-78-1) IN F344/N Rats and B6C3F1 Mice (FEED STUDY).* Technical Report Series # 245. National Institutes of Health.

mercuric chloride

(1) *33*

(2) Swensson, A., et al. 1963. *Occup. Health Rev.* 15(3):5-11.

(3) Casto, B.C., et al. 1979. *Cancer Res.* 39:193-198.

(4) McGregor, D.B., et al. 1988. *Environ. Mol. Mutagen.* 12:85-154.

(5) Montaldi, A., et al. 1985. *Environ. Mutagenensis.* 7:381-391.

(6) Steffek, A.J., et al. 1987. *Teratology.* 35:59A.

(7) Koos, B.J., et al. 1976. *Am. J. Obstet. Gynecol.* 126:390-409.

(8) Lille, F., et al. 1988. *Clin. Toxicol.* 26(1 & 2):103-116.

(9) Singer, R., et al. 1987. *Arch. Environ. Health.* 42(4):181-184.

(10) Daston, G.P., et al. 1986. *Toxicol. Appl. Pharmacol.* 85:39-48.

(11) Houser, M.T., et al. 1986. *Toxicol. Appl. Pharmacol.* 83(3):506-515.

(12) Aten, J,. et al. 1988. *Am. J. Pathol.* 133(1):127-137.

(13) Pelletier, L., et al. 1986. *J. Immunol.* 137(8):2548-2554.

(14) Hill, E.F., et al. 1987. *J. Toxicol. Environ. Health.* 20:105-116.

(15) Grissom, R.E., Jr., et al. 1985. *Arch. Environ. Contam. Toxicol.* 14:193-196.

mercurous chloride
(1)   9
(2)   *33*
mercury
(1)   Eisler, R. 1987. *Mercury hazards to fish, wildlife, and invertebrates: a synoptic review.* U.S. Fish and Wildlife Service. Biol. Rep. 85(1.10).
(2)   *13*
(3)   Miyamato, M.D. 1983. *Brain. Res.* 267:375-379.
(4)   Erfurth, E.M. 1990. *Br. J. Ind. Med.* 47:639-644.
(5)   *9*
(6)   Bridger, M.A. and J.P. Thaxton. 1983. *Arch. Environ. Contam. Toxicol.* 12(1):45-49.
metalaxyl
(1)   *39. Metalaxyl. 1988.*
(2)   *58*
(3)   *5*
metaldehyde
(1)   Iwata, Y., et al. 1982. *J. Agric. Food Chem.* 30:606-608.
(2)   *58*
(3)   Verschuuren, H.G., et al. 1975. *Toxicology.* 4:97-115.
metam-sodium
(1)   *58*
(2)   *20*
(3)   *5*
(4)   U.S. Environmental Protection Agency. 1990. *Phase 3 summary of MRID No. 41577101 metam-sodium teratogenicity, 83-3 teratology rat.*
(5)   U.S. Environmental Protection Agency. 1990. *Phase 3 summary of MRID No. 40330901 metam-sodium teratogenicity, 83-3(b) teratology rabbit.*
methidathion
(1)   *19*
(2)   *58*
(3)   *32*
(4)   Chen, H.H., et al. 1982. *Environ. Mutagen.* 4:621-624.
(5)   Sandberg, A.A. (ed.) 1987-*Sister Chromatid Exhange.* Wiley. Liss, Inc. New York, NY.
(6)   *1*
methiocarb
(1)   *32*
(2)   *14*
(3)   *58*
(4)   *39. Methiocarb. 1987.*
(5)   *1*
methiocarb sulfoxide
(1)   *39. Methiocarb. 1987.*
methomyl
(1)   *32*
(2)   *58*
(3)   *5*
(4)   *39. Methomyl. 1989.*
(5)   Hemavathy, K.C., et al. 1987. *Environ. Res.* 42:362-365.
(6)   Kaplan, A., et al. 1977. *Toxicol. Appl. Pharmacol.* 40:1-17.
(7)   *14*
(8)   *19*
(9)   *21*
(10) *1*

methoprene
(1)   *Altosid Mosquito Growth Regulator.* Technical Data Sheet. Zoecon Corporation. Palo Alto, CA.
(2)   *59*
(3)   *5*
(4)   *19*
(5)   *21*
(6)   U.S. Environmental Protection Agency. 1991. *Reregistration Eligibilty Document.* Office of Pesticide Programs.
(7)   *39. Methoprene. 1982.*
methoxychlor
(1)   Barger, J.H. 1984. *J. Econ. Entomol.* 77:794-797.
(2)   *5*
(3)   *51*
(4)   Reuber, M.D. 1980. *Environ. Health Perspec.* 36:205-219.
(5)   Khera, K.S., et al. 1978. *Toxicol. Appl. Pharmacol.* 45:435-444.
(6)   Tullner, W.W., et al. 1962. *J. Pharmacol. Exp. Therap.* 138:126.
(7)   Cooke, P.S., et al. 1990. *Biol. Reprod.* 42:585-596.
(8)   Gray, L.E., Jr., et al. 1989. *Prog. Clin. Biol. Res.* 302:193-205.
(9)   Harris, S.J. 1974. *J. Agric. Food Chem.* 22:969-973.
(10) *19*
(11) *21*
(12) *30*
(13) *1*
methyl bromide
(1)   U.S. Environmental Protection Agency. 1989. *Bromoethane. Drinking Water Health Advisory.* Office of Water.
(2)   Stein, E.R., et al. 1989. *J. Agric. Food Chem.* 37(6):1507-1509.
(3)   *5*
(4)   *57*
(5)   *13*
(6)   Honma, T., et al. 1985. *Toxicol. Appl. Pharmacol.* 81:183-191.
(7)   *9*
(8)   National Toxicology Program. 1990. *Toxicology and Carcinogenesis Studies of Methyl Bromide in B6C3F$_1$ Mice.* Department of Health & Human Services.
methyl isocyanate
(1)   *3*
(2)   *58*
(3)   Bucher, J.A. 1987. *Fundam. Appl. Toxicol.* 9:367-379.
methyl isothiocyanate
(1)   *3*
(2)   Vernot, E.H., et al. 1977. *Toxicol. Appl. Pharmacol.* 42:417-423.
methyl parathion
(1)   *23*
(2)   National Research Council. 1975. *Pest Control: An Assessment of Present and Alternative Technologies. Vol. III Cotton Pest Control.* National Academy Press.
(3)   *39. Methyl Parathion. 1986.*
(4)   Malhi, P.K., et al. 1987. *Mutat. Res.* 188:45-51.
(5)   Yu, Y.D., et al. 1984. *Environ. Sci. Res.* 31:842-843.

(6) Komeil, A.A., et al. 1988. *Teratology.* 38(2):21a.
(7) Gupta, R.C., et al. 1984. *Toxicol. Appl. Pharmacol.* 72:457-468.
(8) Street, J.C., et al. 1975. *Toxicol. Appl. Pharmacol.* 32:587-602.
(9) *32*
(10) *1*
(11) Fairbrother, A., et al. 1988. *Environ. Toxicol. Chem.* 7:499-503.
(12) Ghosh, P., et al. 1990. *Biomed. Environ. Sci.* 3(1):106-112.
(13) Little, E.E., et al. 1990. *Arch. Environ. Contam. Toxicol.* 19:380-385.
(14) Gomez-Arroyo, S., et al. 1988. *Cytologia.* 53:627-634.

methylene chloride
(1) *18*
(2) *13*
(3) *3*
(4) Alexeeff, G.V., et al. 1983. *J. Toxicol. Environ. Health.* 11:569-581.
(5) Kitchin, K.T., et al. 1989. *Teratogenesis Carcinog. Mutagen.* 9:61-69.
(6) National Toxicology Program. 1986. *Toxicology and Carcinogenesis of Dichloromethane (Methylene Chloride) (CAS No. 75-09-2) IN F344/N Rats and B6C3F$_1$ Mice (Inhalation Studies).* NTP TR Ser. No. 306. NIH Publication No. 86-2562.
(7) Briving, C., et al. 1986. *Scand. J. Work Environ. Health.* 12:216-220.
(8) Miller, L., et al. 1985. *Arch. Intern. Med.* 145:145-146.
(9) National Institute for Occupational Safety and Health. Sept. 1978. *Occupational Health Guidelines for Methylene Chloride.* U.S. Department of Health and Human Services. U.S. Department of Labor.
(10) Weiss, G. 1968. *Abstracts on Hygiene.* 43(9):1123.
(11) Burton, D.T., et al. 1990. *Bull. Environ. Contam. Toxicol.* 44:776-783.

methylthioacetate
(1) *39. Acephate. 1987.*

metiram
(1) *58*
(2) *5*
(3) *39. Metiram. 1988.*
(4) *1*

metolachlor
(1) Chesters, G,. et al. 1989. *Rev. Environ. Contam. Toxicol.* 110:1-74.
(2) *20*
(3) *5*

metribuzin
(1) *20*
(2) *5*
(3) *39. Metribuzin. 1985.*
(4) *58*
(5) *1*

mevinphos
(1) *32*
(2) *5*
(3) *39. Mevinphos. 1988.*

(4) Carricaburu, P., et al. 1981. *Toxicological European Research.* 3(2):87-91.
(5) *19*
(6) *21*
(7) *1*

mexacarbate
(1) *32*
(2) *5*
(3) *19*
(4) *21*
(5) *1*

MGK R11
(1) *58*
(2) U.S. Environmental Protection Agency. 1990. Notice of Intent to Remove the Active Ingredient Repellent R-11 from Reregistration List C, and to Cancel all Pesticide Product Registrations Containing Repellent R-11. *Federal Register.* 55(114):24052-24058.

milky spore disease
(1) *34*
(2) *5*

mineral oil
(1) Blackburn, G.R., et al. 1986. *Cell Biol. Toxicol.* 2(1):63-83.
(2) *Pesticide Toxic Chemical News.* 1983. 11(32):4-5.

mirex
(1) *9*
(2) *3*
(3) *5*
(4) Chambers, J.E., et al. 1982. *J. Agric. Food Chem.* 30(5):878-882.
(5) Smrek, A.L. 1977. *J. Agric. Food Chem.* 25(6):1321-1325.
(6) Grabowski, C.T. 1981. Project Summary. *Pesticide effects on prenatal cardiovascular physiology.* U.S. Environmental Protection Agency Health Effects Research Laboratory. EPA-600/51-80-032.
(7) *19*
(8) *21*
(9) *1*
(10) Eisler, R. 1985. *Mirex Hazards to Fish, Wildlife and Invertebrates: A Synoptic Review.* U.S. Fish and Wildlife Service. Biol. Rep. 85. (1.1). 42 pp.

monocrotophos
(1) *32*
(2) *58*
(3) Kumar, D.V,. et al. 1988. *Bull. Environ. Contam. Toxicol.* 41: 189-194.
(4) Bhunya, S.P,. et al. 1988. *Cytologia.* 53: 801-807.
(5) *19*
(6) *1*
(7) Nagabhushanam, R., et al. 1984. *Acta Physiol. Pol.* 35 (5-6): 551-557.

monofluoroacetic acid
(1) *23*

monolinuron
(1) *58*

monomethylformamide
(1) Saunders, D.G., et al. 1983. *J. Agric. Food Chem.* 31:237-265.

monuron
(1)   Freitag, D., et al. 1984. *J. Agric. Food Chem.* 32:203-207.
(2)   *3*
(3)   National Toxicology Program. 1988. *Toxicology and Carcinogenesis Studies of Monuron in F344/N Rats and B6C3F¹ Mice (Feed Studies).* Technical Report Series No. 266.
(4)   U.S. Environmental Protection Agency. 1975. *Substitute Chemical Program. Initial Scientific and Minieconomic Review of Monuron.* Office of Pesticide Programs.
(5)   *30*
(6)   *1*

MSMA
(1)   Fowler, B.A. 1983. *Biological and Environmental Effects of Arsenic.* Elsevier Publishers. B.V. p. 73.
(2)   E.A. Woolson. 1975. *Arsenical Pesticides.* American Chemical Society. p. 65.
(3)   *5*
(4)   *20*
(5)   Jaghabir, M.T.W., et al. 1988. *Bull. Environ. Contam. Toxicol.* 40:119-122.
(6)   World Health Organization. 1980. *IARC Monographs on the Evaluation of the Carcinogenic Risk of Chemicals to Humans.* Vol. 23. International Agency for Research on Cancer. p. 90.
(7)   *23*

nabam
(1)   *30*
(2)   Merck, et al. 1983. *The Merck Index.* Tenth edition. Merck and Co., Inc.
(3)   39. Nabam. 1987.
(4)   Birch, W.X., et al. 1986. *Arch. Environ. Contam. Toxicol.* 15:637-645.
(5)   *19*
(6)   Ghate, H.V. 1985. *Riv. Biol.* 78:288-291.

naled
(1)   *58*
(2)   *5*
(3)   Shiau, S.U., et al. 1981. *J. Agric. Food Chem.* 29(2):268-271.
(4)   *19*
(5)   *21*
(6)   *30*
(7)   *1*

naphthalene
(1)   U.S. Environmental Protection Agency. 1990. *Naphthalene. Drinking Water Health Advisory.* Office of Drinking Water.
(2)   *58*
(3)   *13*
(4)   *9*
(5)   Crider, J.Y., et al. 1982. *Bull. Environ. Contam. Toxicol.* 28:52-57.
(6)   Collier, T.K., et al. 1980. *Environ. Res.* 23:35-41.

naphthaleneacetic acid
(1)   *58*
(2)   *33*
(3)   39. Napthaleneacetic acid its, salts, Ester and Acetamide. 1981.

(4)   *5*

1-naphthol
(1)   *9*

naptha
(1)   McKee, R.H. et al. 1987. *Can. J. Physiol. Pharmacol.* 65:1793-1797.
(2)   Blackburn, G.R. 1986. *Cell Biol. Toxicol.* 291):63-83.
(3)   Cruzan, G. 1986. *Toxicol. Ind. Health.* 2(4):427-444.
(4)   *30*

naptalam
(1)   *20*
(2)   *58*
(3)   *1*
(4)   U.S. Environmental Protection Agency. 1985. *Naptalam Pesticide Fact Sheet.* Office of Pesticide Programs.
(5)   Epstein, S.S., et al. 1971. *The Mutagenicity of Pesticides.* MIT Press.

niclosamide
(1)   *58*
(2)   Espinosa-Aguirre, J.J., et al. 1989. *Mutat. Res.* 222:161-166.
(3)   Vega, S.G., et al. 1988. *Mutat. Res.* 204:269-276.

nicotine
(1)   *58*
(2)   *8*
(3)   *19*
(4)   *30*

nitrapyrin
(1)   39. Nitrapyrin. 1985.
(2)   *58*
(3)   Berdasco, N.M. 1988. *Fund. Appl. Toxicol.* 11:464-471.

nitrofen
(1)   Kale, S.P., et al. 1989. *Bull. Environ. Contam. Toxicol.* 42:544-547.
(2)   *5*
(3)   Hurt, S.S.B., et al. 1983. *Toxicology.* 29:1-37.
(4)   *9*
(5)   *29*
(6)   Francis, B.M. 1986. *J. Environ. Sci. Health.* Part B 21(4):303-317.
(7)   Kavlock, R.J., et al. 1983. *J. Toxicol. Environ. Health.* 11:1-13.

nitrosocarbaryl
(1)   Rickard, R.W., et al. 1984. *J. Toxicol. Environ. Health.* 14:279-290.
(2)   U.S. Environmental Protection Agency. 1980. *Carbaryl Decision Document.* Office of Pesticide Programs.

N-formyl-chloro-o-toluidine
(1)   39. Chlordimeform.

N-hydroxymethyl methiocarb sulfoxide
(1)   39. Methiocarb. 1987.

N-nitrososamide
(1)   U.S. Environmental Protection Agency. 1977. *Pronamide: Position Document 1.* Office of Pesticide Programs.

N-nitroso-di-n-propylamine
(1)   Afkham, J., et al. 1967. *Zeitschrift fur Krebsforschung.* 69:103-201.
(2)   Lijinsky, W., et al. 1983. *Cancer Letters.* 19:207-213.
(3)   *49*

(4) Martelli, A., et al. 1988. *Cancer Research.* 48:4144-4152.
(5) Brambilla, G., et al. 1987. *Cancer Res.* 47:3485-3491.

N-nitrosodimethylamine
(1) Brambilla, G., et al. 1987. *Cancer Research.* 47:3485-3491.
(2) 29
(3) Martelli, A., et al. 1988. *Cancer Research.* 48:4144-4152.

N-nitrosodiphenylamine
(1) Gorsdorf, S.,et al. 1984. *Mutagenesis.* 4:323-324.

N-nitrosodipropylamine
(1) 49

N-nitrosomethyl-n-butylamine
(1) Lijinsky, W. et al. 1988. *Cancer Res.* 48:6648-6652.
(2) Lijinsky, W. 1983. *J. Natl. Cancer Inst.* 70(5)959-963.

N-nitrosonornicotine
(1) 49

N-nitrosopendimethalin
(1) 49

N-nitroso propoxur
(1) Cid, et al. 1990.
(2) Gichner, T., et al. 1990. *Mut. Res.* 229: 37-41.

nonachlor
(1) 39. Chlordane. 1986.

norflurazon
(1) 20
(2) 58
(3) 39. Norflurazon. 1984.

nuclear polyhedrosis virus
(1) U.S. Environmental Protection Agency. 1990. *Reregistration Eligibility Document. Polyhedrosis of Heliothis Zea Nuclear Polyhedrosis Virus.* Office of Pesticide Programs.

octachlorodibenzo-p-dioxin
(1) Birnbaum, L.S., et al. 1988. *Toxicol. Appl. Pharmacol.* 93:22-30.
(2) World Health Organization. 1987. *Environmental Health Criteria 71: Pentachloropehnol.* International Programme on Chemical Safety.

omethoate
(1) 58
(2) 5
(3) Food & Agriculture Organization. 1981. *Plant Production and Protection Paper 42.* United Nations. Rome, Italy. pp. 375-379.

oryzalin
(1) Gaynor, J.D. 1985. *Can. J. Soil Science.* 65:587-592.
(2) 58
(3) 20
(4) 39. Oryzalin. 1987.

oxadiazon
(1) Ambrosi, D., et al. 1977. *J. Agric. Food Chem.* 25(4):868-872.
(2) 58
(3) 20
(4) Wang, G.M. 1984. *Regulatory Toxicol. Pharmacol.* 4:355-360.
(5) Wang, G.M. 1984. *Regulatory Toxicol. Pharmacol.* 4:361-371.
(6) 1

(7) Blacker, A.M. et al. 1988. *Summary of the Oxadiazon Dossier.* Rhone Poulenc. Research Triangle Park, N.C.

oxamyl
(1) 58
(2) 5
(3) 4
(4) 39. Oxamyl. 1987.
(5) 1

oxine-copper
(1) 58
(2) 25
(3) 30

oxychlordane
(1) 10
(2) Biros, F.J., et al. 1973. *Bull. Environ. Contam. Toxicol.* 10(5):257-260.

oxydemeton-methyl
(1) 32
(2) 58
(3) 5
(4) Gomez-Arroyo, S., et al. 1988. *Cytologia.* 53:627-634.
(5) Romero, P., et al. 1989. *Environ. Res.* 50:256-261.
(6) 39. Oxydemeton-methyl. 1987.
(7) 19
(8) 21
(9) 1
(10) 30

oxyfluorfen
(1) 20
(2) 5
(3) 58
(4) U.S. Environmental Protection Agency. 1981. *Oxyfluorfen Position Document. No. 1-2-3.* Office of Pesticide Programs.

oxygen analogue of azinphosmethyl
(1) California Department of Food and Agriculture. Worker Safety Unit-Dr. Keith Maddy.

oxytetracycline hydrochloride
(1) Morissey, R.E., et al. 1986. *Fundam. Appl. Toxicol.* 434-443.
(2) 40. Oxytetracycline, Oxytetracycline Hydrochloride, Oxytetra Cycline Calcium Complex. 1988.

paradichlorobenzene
(1) 5
(2) 12
(3) U.S. Environmental Protection Agency. 1985. *Ortho-, meta-, and paradichlorobenzene.* Health Advisory. Office of Drinking Water.
(4) 30

paraformaldehyde
(1) 40. Formaldehyde. 1988.

paraoxon
(1) 28

paraquat
(1) Eisler, R. 1990. *Paraquat Hazards to Fish, Wildlife, and Invertebrates: A Synoptic Review.* U.S. Fish and Wildlife Service. Biol. Rep. No. 85(1.22).
(2) 30

(3) U.S. Environmental Protection Agency. 1982. *Paraquat Decision Document*. Office of Pesticide Programs.

(4) Wang, T.C., et al. 1987. *Mutat. Res.* 188:311-321.

(5) Bus, J.S., et al. 1975. *Toxicol. Appl. Pharmacol.* 33:450-460.

(6) DeGori, N., et al. 1988. *Neuropharmacology.* 27(2):201-207.

(7) *24*

(8) Calò, M., et al. 1990. *J. Comp. Path.* 103:73-78.

(9) Lindquist, N.G., et al. 1988. *Neurosci. Lett.* 93:1-6.

(10) Barbeau, A., et al. 1986. Studies on MPTP, MPP⁺ and paraquat in frogs and in vitro. pp.85-103. *in* S.P. Markey, et al eds. *MPTP: A Neurotoxin Producing a Parkinsonian Syndrome.* Academic Press.

(11) *21*

(12) Carr, R.J.G., et al. 1986. *Appl. Environ. Pharmacol.* 52(5):1112-1116.

(13) Hoffman, D.J., et al. 1982. *Arch. Environ. Contam. Toxicol.* 11:79-86.

(14) Hoffman, D.J., et al. 1985. *Arch. Environ. Contam. Toxicol.* 14:495-500.

(15) Clark, M.W., et al. 1988. *Comp. Biochem. Physiol [C].* 39C(1):15-30.

(16) Vaishampayan, A. 1984. *Mutat. Res.* 138:39-46.

paraquat dichloride

(1) *58*

(2) *3*

(3) Haley, T.J. 1979. *Clinical Toxicol.* 14(1):1-46.

(4) *39. Paraquat Dichloride. 1987.*

(5) Grant, H.C., et al. 1980. *Histopathology.* 4:185-195.

(6) *19*

(7) Dial, N.A., et al. 1984. *Bull. Environ. Contam. Toxicol.* 33:592-597.

parathion

(1) *12*

(2) *39. Parathion. 1986.*

(3) Kimbrough, R.D., et al. 1968. *Arch. Environ. Health.* 16:239-247.

(4) Casale, G.P., et al. 1984. *Toxicol. Lett.* 23:239-247.

(5) *32*

(6) *21*

(7) *30*

(8) *1*

(9) California Department of Food and Agriculture. 1978. *Report of Environmental Assessment of Pesticide Programs.* State Component, Vol. 2.

(10) White, D.H., et al. 1983. *Bull. Environ. Contam. Toxicol.* 31:93-97.

(11) Rattner, B.A., et al. 1982. *Pest. Biochem. Physiol.* 18:132-138.

(12) Hoffman, D.J., et al. 1984. *Arch. Environ. Contam. Toxicol.* 13:15-27.

(13) Singh, H., et al. 1981. *Endrokrinologie.* 77(2):173-178.

(14) *17*

PCB's

(1) Ghirelli, R.P., et al. 1983.

(2) *9*

(3) Baumann, M., et al. 1983. *Arch. Environ. Contam. Toxicol.* 12:509-515.

(4) *29*

(5) Vos, J.G., et al. 1972. *Sci. Total Environ.* 1:289-302.

(6) Tilson, H.A., et al. 1990. *Neurotoxicol. Teratol.* 12:239-248.

(7) Sager, D.B., et al. 1987. *Bull. Environ. Contam. Toxicol.* 38:946-953.

(8) *31*

(9) *45*

(10) Ereitzer, J.F., et al. 1974. *Environ. Pollut.* 6:21-29.

(11) Haseltine, S.D. 1980. *Environ. Res.* 23:29-34.

pendimethalin

(1) Walker, A., et al. 1977. *Pesticide Science.* 8:359-365.

(2) *58*

(3) *20*

(4) U.S. Environmental Protection Agency. 1985. *Pesticide Fact Sheet. Pendimethalin.* Office of Pesticide Programs.

pentachlorobenzene

(1) Dunn, J.S., et al. 1979. *Toxicol. Appl. Pharmacol.* 48:425-433.

pentachlorophenol

(1) World Health Organization. 1987. Environmental Health Criteria *71-Pentachlorophenol.* International Programme on Chemical Safety.

(2) *30*

(3) *58*

(4) *2*

(5) *9*

(6) National Toxicology Program. 1989. *Toxicology and Carcinogenesis Studies of Two Pentachlorophenol Technical-Grade Mixtures (CAS No. 87-86-5) in B6C3F₁ Mice (Feed Studies).* NTP Technical Report Series No. 349.

(7) U.S. Environmental Protection Agency. 1987. *Final determination and notice of intent to cancel and deny applications for registration of pesticide products containing pentachlorophenol (including but not limited to its salt and esters) for non-wood uses.* Office of Pesticides Programs.

(8) *51*

(9) Holsapple, M.P., et al. 1987. *J. Toxicol. Environ. Health.* 20:229-239.

(10) Hattemer-Frey, H.A., et al. 1989. *Arch. Environ. Contam. Toxicol.* 18:482-489.

(11) Eisler, R. 1989. *Pentachlorophenol Hazards to Fish, Wildlife and Invertebrates: A Synoptic Review.* U.S. Fish and Wildlife Service. Biol. Rep. 85(1.17). 72 pp.

(12) Choudhury, H., et al. 1986. *Toxicol. Ind. Health.* 2(4):483-571.

(13) Stephenson, G.L., et al. 1991. *Arch. Environ. Contam. Toxicol.* 20:73-80.

(14) *1*

(15) Galt, D.E. 1988. *Can. Vet. J.* 29:65-67.

(16) Prescott, C.A., et al. 1982. *Am. J. Vet. Res.* 43(3):481-487.

(17) Little, E.E., et al. 1990. *Arch. Environ. Contam. Toxicol.* 19:380-385.

(18) Hodson, P.V., et al. 1981. *Aquatic Toxicology.* 112:113-127.

perchloroethylene
(1) U.S. Environmental Protection Agency. 1981. *Oxyfluorfen. Position Document. No. 1-2-3.* Office of Pesticide Programs.

perfluidone
(1) Mulisson, W.R., et al. 1979. *Herbicide Handbook.* 4th ed. Weed Science Society of America.
(2) Howell, S.M. 1975. *Perfluidone Persistence in Silty Clay Loam Soil Under Field Conditions in Texas.* Agrichemicals Technical Report. 3M Company. St. Paul, MN.
(3) *58*
(4) *5*
(5) *Acute Inhalation Toxicity of Perfluidone (MBR.8251).* 1975. Report No.224. Cannon Laboratiories, Inc. Reading, PA.
(6) *39. Perfluidone. 1985.*

permethrin
(1) *58*
(2) *3*
(3) Miyamoto, J. 1976. *Environ. Health Perspect.* 14:15-28.
(4) Qadri, S.S.H., et al. 1987. *J. Appl. Toxicol.* 7(6):367-371.
(5) Coats, J.R., et al. 1979. *Bull. Environ. Contam. Toxicol.* 23:250-255.
(6) Clark, J.R., et al. 1989. *Environ. Toxicol. Chem.* 8:393-401.
(7) *34*

phenol
(1) *3*
(2) Boutwell, R.K., et al. 1959. *Cancer. Res.* 19:413-424.
(3) *30*

phenothrin
(1) *9*
(2) *5*
(3) *39. Sumithrin. 1987.*

phorate
(1) *58*
(2) *8*
(3) Malhi, P.K., et al. 1987. *Mutat. Res.* 188:45-51.
(4) *32*
(5) *21*
(6) *1*

phosalone
(1) *32*
(20 *5*
(3) *21*
(4) *39. Phosalone. 1987.*
(5) *1*

phosgene gas
(1) National Institute for Occupational Safety and Health. Sept. 1978. *Occupational Health Guidelines for Methylene Chloride.* U.S. Department of Health and Human Services.
(2) Cucinell, S.A. 1974. *Arch. Environ. Health.* 28(5):272-275.

phosmet
(1) *39. Phosmet. 1986.*
(2) *28*

phosphamidon
(1) *8*
(2) *58*

(3) Behera, B.C. and Bhunya, S.P. 1987. *Toxicol. Lett.* 37(3):269-277.
(4) Soni, I., et al. 1989. *Tetratogenesis Carcinog. Mutagen.* 9(4):253-257.
(5) Bhatnagar, P., et al. 1990. *Environ. Contam. Toxicol.* 45(4):590-597.
(6) Bhatanager, P., et al. 1986. *Bull. Environ. Contam. Toxicol.* 37(5):767-773.
(7) *32*
(8) *19*
(9) *30*

phosphine gas
(1) World Health Organization. 1988. *Environmental Health Criteria 73. Phosphine and Selected Metal Phosphides.* International Programme on Chemical Safety.
(2) *3*

photodieldrin
(1) *13*

photomirex
(1) *9*

phoxim
(1) *7*
(2) Gomez-Arroyo, S., et al. 1988. *Cytologia.* 53:627-634.
(3) *9*
(4) *21*
(5) *1*

picloram
(1) *58*
(2) *39. Picloram. 1988.*
(3) *14*
(4) Reuber, M.D. 1981. *J. Toxicol. Environ. Health.* 7:207-222.
(5) Blakley, P.M., et al. 1989. *J. Toxicol. Environ. Health.* 28:309-316.
(6) Gorzinski, S.J., et al. 1987. *J. Toxicol. Environ. Health.* 20:367-377.
(7) *19*
(8) *21*
(9) *1*

piperonyl butoxide
(1) *8*
(2) *28*
(3) Massey, T.C., et al. 1982. *Toxicology.* 25:187-200.
(4) Falk, H.L., et al. 1965. *Arch. Environ. Health.* 10:847-858.
(5) *14*
(6) *21*
(7) *30*

pirimiphos-ethyl
(1) *32*
(2) *58*
(3) Hanna, P.J., et al. 1975. *Mutat. Res.* 28:405-420.
(4) *1*
(5) Moscioni, A.D., et al. 1977. *Biochem. Pharmac.* 26:2251-2258.

pirimiphos methyl
(1) *58*
(2) Rajini, P.S., et al. 1988. *J. Environ. Sci. Health.* B23(2):127-144.
(3) *13*

polyoxyethyleneamine
(1)   Sawanda, Y., et al. 1988. *Lancet.* 1(8580):299.
(2)   Servizi, J.A., et al. 1987. *Bull. Environ. Contam. Toxicol.* 39:15-22.
(3)   Envirologic Data. 1984. *Risk Assessment of Glyphosate for Maine Department of Transportation and Maine Board of Pesticides Control.*

potassium azide
(1)   *20*
(2)   *21*

potassium bromide
(1)   *1*

potassium permanganate
(1)   *3*
(2)   *40. Potassium permanganate. 1985.*
(3)   U.S. Environmental Protection Agency. 1985. *Potassium permanganate Pesticide Fact Sheet.* Office of Pesticide Programs.

procymidone
(1)   *5*
(2)   *58*
(3)   *Pesticide and Toxic Chemical News.* 1990. SAP Procymidone Report Basically Supports EPA, NACA. 19(6):9.

prometon
(1)   Ciba-Geigy Corporation. *Technical Bulletin. Prometon.* (2) *7*
(3)   Ciba-Geigy Corporation. 1972. *Pramitol Toxicology Data.* Department of Industrial Medicine.
(4)   *20*
(5)   *58*
(6)   *5*
(7)   Hoffman, D.J., et al. 1984. *Arch. Environ. Contam. Toxicol.* 13:15-17.

prometryn
(1)   *20*
(2)   Ciba-Geigy Corporation. *Caparol 80W Herbicide Technical Bulletin.*
(3)   *40. Prometryn. 1987.*
(4)   *9*
(5)   *30*
(6)   *1*

pronamide
(1)   *20*
(2)   *58*
(3)   *39. Pronamide. 1986.*
(4)   U.S. Environmental Protection Agency. 1989. *Pronamide Health Advisory Summary.* Office of Drinking Water.
(5)   *5*
(6)   *1*

propachlor
(1)   U.S. Environmental Protection Agency. *Propachlor Pesticide Fact Sheet.* Office of Pesticide Programs.
(2)   *39. Propachlor. 1984.*

propanil
(1)   *20*

(2)   McMillian, D.C., et al. 1990. *Toxicol. Appl. Pharmacol.* 105:503-507.
(3)   McMillian, D.C., et al. 1991. *Toxicol. Appl. Pharmacol.* 110:70-78.
(4)   *58*
(5)   Call, D.J., et al. 1983. *Environ. Contam. Toxicol.* 12:175-182.
(6)   Hoffman, D.J., et al. 1984. *Arch. Environ. Contam. Toxicol.* 13:15-27.
(7)   *1*

propargite
(1)   *39. Propargite. 1986.*
(2)   *58*
(3)   *1*

propazine
(1)   *39. Propazine. 1988.*
(2)   *1*

propham
(1)   *39. Propham. 1987.*
(2)   *20*
(3)   *32*
(4)   Georgian, L., et al. 1985. *Mutat. Res.* 147:296.
(5)   *21*
(6)   *1*

propionic acid
(1)   *3*
(2)   *9*

propoxur
(1)   *4*
(2)   *5*
(3)   U.S. Environmental Protection Agency. 1991. *List of Chemicals Evaluated for Carcinogenic Potential.* Health Effects Division, Office of Pesticide Programs.
(4)   Backus, B.T. 1985. U.S. Environmental Protection Agency Memorandum. *Baygon Oncogenicity, Mutagenicity, and Metabolite Studies.* Office of Pesticide Programs. pp. 13-14.
(5)   Cid, M.G., et al. 1990. *Mutat. Res.* 232:45-48.
(6)   Desi, I., et al. 1974. *Toxicol. Appl. Pharmacol.* 27:465-476.
(7)   *32*
(8)   *21*
(9)   *30*
(10) California Department of Agriculture. 1978. Report on Environmental Assessment of Pesticide Regulatory Programs. Vol. 2. State Component. p. 3.2-20.
(11) *1*
(12) *34*

pyrethrum
(1)   *22*
(2)   *5*
(3)   Mrak, Emil M. 1969. *Report of the Secretary's Commision on Pesticides and their Relationship to Environmental Health.* U.S. Department of Health, Education, and Welfare. p. 544.
(4)   *3*
(5)   *19*
(6)   *30*

red squill
(1)   *3*
resmethrin
(1)   U.S. Environmental Protection Agency. 1988. *Resmethrin Pesticide Fact Sheet.* Office of Pesticide Programs.
(2)   Miyamoto, J. 1976. *Environ. Health Perspect.* 4:15-28.
(3)   Doherty, J.D., et al. 1986. *Comp. Biochem. Physiol. [C].* 84C (2):373-379.
(4)   World Health Organization. 1989. Environmental Health Criteria 92: *Resmethrins-Resmethrin, Bioresmethrins, Cisresmethrins.* International Programme on Chemical Safety.
(5)   Dyball, R.E.J. 1982. *Pesticide Biochem. Physiol.* 17:42-47.
(6)   *19*
ronnel
(1)   *3*
(2)   *32*
(3)   *1*
(4)   *19*
(5)   *21*
(6)   Sobti, R.C., et al.1982. *Mut. Res.* 102:89-102.
rotenone
(1)   39. Rotenone. 1988.
(2)   *13*
(3)   *58*
(4)   Abdo, K.M., et al. 1988. *Drug Chem. Toxicol.* 11(3):225-234.
(5)   Gosalvez, M., et al. 1977. *Br. J. Cancer.* 36:243-253.
(6)   Khera, K.S., et al. 1982. *J. Toxicol. Environ. Health.* 10:111-119.
(7)   Spencer, F., et al. 1982. *Bull. Environ. Contam. Toxicol.* 28:360-368.
(8)   Gosalvez, M. 1983. *Life Sciences.* 32:809-816.
(9)   *19*
(10) *21*
(11) *30*
ryania
(10   *3*
(2)   *21*
(3)   *1*
sabadilla
(1)   *13*
(2)   *1*
safrole
(1)   *3*
(2)   Lu, L.J.W., et al. 1986. *Cancer Res.* 46:3046-3054.
(3)   *45*
siduron
(1)   *20*
(2)   Belasco, I.J., et al. 1969. *J. Agric. Food Chem.* 17:1004-1007.
(3)   *1*
silica aerogel
(1)   FMC Corporation. 1970. *Drione Toxicity.* Niagara Chemical Division. Middleport, NY.
(2)   *13*
(3)   *34*
silvex
(1)   *30*

(2)   *14*
(3)   Braun, A.G., et al. 1983. *J. Toxicol. Environ. Health.* 11:275-286.
(4)   Silvex (2,4,5-TP). 1988. *Rev. Environ. Contam. Toxicol.* 104:195-202.
(5)   *19*
(6)   *1*
simazine
(1)   *23*
(2)   *23*
(3)   *7*
(4)   *20*
(5)   U.S. Environmental Protection Agency. 1989. *Health Advisory Summary-Simazine.* Office of Water.
(6)   *30*
(7)   *1*
soap
(1)   Safer, Inc. *Product Guide.* Newton, MA.
(2)   Osborne, L.S. 1984. *J. Econ. Entomol.* 77:734-737.
sodium arsenate (I)
(1)   Seidenberg, J.M., et al. 1986. *Teratogenesis Carcinog. Mutagen.* 6:361-374.
(2)   Ferm, V. H., et al. 1968. *J. Reprod. Fertil.* 17:199-201.
(3)   Fowler, B.A. (Ed.). 1983. *Biological and Environmental Effects of Arsenic.* Elsevier Science Publishers B.V. p. 246-247.
sodium arsenate (II)
(1)   Zanzoni, F., et al. 1980. *Arch. Dermatol. Res.* 267:91-95.
(2)   Nakamuro, K., et al. 1981. *Mutat. Res.* 88:73-80.
sodium arsenate (III)
(1)   Fowler, B.A. 1983. *Biological and Environmental Effects of Arsenic.* Elsevier Publishers. B.V. p. 73.
(2)   *9*
(3)   *2*
(4)   Lee, Te-Chang, et al. 1985. *Carcinogenesis.* 6(10):1421-1426.
(5)   De Flora, S., et al. 1984. *Mutat. Res.* 133:161-198.
(6)   Hood, R.D., et al. 1972. *Teratology.* 6:235-238.
(7)   Willhite, C.C. 1981. *Exp. Mol. Path.* 34:145-158.
(8)   *19*
(9)   *30*
sodium arsenite
(1)   Fowler, B.A. 1983. *Biological and Environmental Effects of Arsenic.* Elsevier Publishers. p. 73.
(2)   *9*
(3)   *2*
(4)   Lee, T. et al. 1985. *Carcinogenesis.* 6(10):1421-1426.
(5)   Larramendy, M.L. et al. 1981. *Environ. Mutagen.* 3:597-606.
(6)   Hood, R.D. et al. 1972. *Teratology.* 6:235-238.
(7)   Willhite, C.C. 1981. *Mol. Pathol.* 34:145-158.
(8)   *19*
(9)   *30*
sodium azide
(1)   *20*
(2)   *5*
(3)   Jones, J.A., et al. 1980. *Mutat. Res.* 77:293-299.
(4)   *21*

sodium chlorite
(1) *3*
(2) *9*
(3) *30*
sodium fluoroacetate
(1) *5*
(2) *2*
(3) *40*
(4) *14*
(5) *24*
(6) Parkin, P.J., et al. 1977. *N. Z. Med. J.* 85:93-99.
(7) *9*
sodium hypochlorite
(1) U.S. Environmental Protection Agency. 1986. *Sodium and Calcium Hypochlorite Pesticide Fact Sheet.* Office of Pesticide Programs.
sodium omadine
(1) *28*
(2) *39. Sodium Omadine.* 1985.
sodium salt
(1) *20*
stoddard solvents
(1) *28*
streptomycin
(1) *22*
(2) *39. Streptomycin and Streptomycin Sulfate.* 1988.
strychnine
(1) *40. Strychnine.* 1980.
(2) *19*
(3) *Pesticide Information Manual.* 1966. Northeastern Regional Pesticide Coordinators. Cooperative Extension Services.
sulfone analogue of fenthion
(1) *33*
(2) *14*
sulfoTEPP
(1) *58*
(2) *5*
(3) *31*
sulfoxide analogue of fenthion
(1) *33*
sulfur
(1) *30*
(2) *39. Sulfur.* 1982.
sulfuryl fluoride
(1) *58*
(2) *13*
(3) Windholz, M., et al. 1983. *The Merck Index: An Encyclopedia of Chemicals, Drugs, and Biologicals.* 8th Edition. Merck and Co., Inc. p. 1235.
sulprofos
(1) *32*
(2) *39. Sulprofos.* 1981.
2,4,5-T
(1) *17*
(2) *30*
(3) *5*
(4) *58*
(5) *7*

(6) Hardell, L. and M. Eriksson. 1988. *Cancer.* 62:652-656.
(7) *14*
(8) Lilienfeld, D.E. and M.A. Gallo. 1989. *Epidemiol. Rev.* 11:28-58.
(9) *47*
(10) *1*
(11) Sanderson, C.A. and L.J. Rogers. 1981. *Science.* 22:593-595.
(12) Hoffman, D.J. and P.H. Albers. 1984. *Arch. Environ. Contam. Toxicol.* 13:15-27.
TCAB
(1) Mcmillan, D.C., et al. 1988. *Fundam. Appl. Toxicol.* 11:429-439.
TCDD
(1) Kearney, P.C., et al. 1987. *Environ. Sci. Technol.* 6(12):1017-1019.
(2) Tschirley, F.H. 1986. *Sci. Am.* 254(2):29-35.
(3) National Institute for Occupational Safety and Health. 1984. *Current intelligence Bulletin 40: 2,3,7,8-Tetrachlorodibenzo-p-dioxin (TCDD, "dioxin").* U.S. Department of Health and Human Services. Centers for Disease Control.
(4) *29*
(5) Baughman, R. and M. Meselson. 1973. *Environ. Health Perspect.* 5:27-35.
(6) Hebert, C.D., et al. 1990. *Toxicol. Appl. Pharmacol.* 102:362-377.
(7) *47*
(8) Abbott, B.D. and L.S. Birnbaum. 1989. *Toxicol. Appl. Pharmacol.* 99:287-301.
(9) Neubert, D., et al. 1973. *Environ. Health Perspect.* 5:67-79.
(10) Nikolaidis, E., et al. 1988. *Toxicol. Appl. Pharmacol.* 92:315-323.
(11) Hakansson, H., et al. 1987. *J. Nutr.* 117(3):580-586.
(12) Hakansson, H., and U.G. Ahlborg. 1985. *J. Nutr.* 115:759-771.
(13) Thunberg, T., et al. 1980. *Arch. Toxicol.* 45:273-285.
(14) *19*
tebuthiuron
(1) *20*
(2) *3*
(3) *33*
tefluthrin
(1) *58*
(2) U.S. Environmental Protection Agency. 1989. *Tefluthrin Pesticide Fact Sheet.* Office of Pesticide Programs. p. 7.
temephos
(1) *39. Temephos.* 1981.
(2) *3*
(3) *9*
(4) *5*
(5) *19*
(6) *21*
(7) Kpekata., A.E. 1983. *Bull. Environ. Contam. Toxicol.* 31(1):120-124.
(8) *1*
TEPP
(1) *33*

(2)   30
(3)   1
terbacil
(1)   20
(2)   39. Terbacil. 1989.
(3)   1
terbucarb
(1)   22
(2)   5
(3)   2
(4)   1
terbufos
(1)   32
(2)   58
(3)   39. Terbufos. 1988.
(4)   U.S. Environmental Protection Agency. 1989. *Terbufos Health Advisory Summary.* Office of Water.
terbutryn
(1)   58
(2)   39. Terbutryn. 1986.
(3)   20
(4)   19
(5)   1
tetrachlorocatechol
(1)   Renner, G., et al. 1986. *Toxicol. Environ. Chem.* 11:37-50.
(2)   Schäfer, W., et al. 1988. *J. Agric. Food Chem.* 36:370-377.
tetrachloroethylene
(1)   Reichert, D. 1983. *Mutat. Res.* 123(3):411-429.
(2)   12
(3)   National Toxicology Program. 1986. *Toxicology and Carcinogenesis Studies of Tetrachloroethylene (Perchloroethylene) (CAS. No. 127-18-4) in F344/N Rats and B6C3F₁ Mice (Inhalation Studies).*NTP Technical Report 311.
(4)   National Institute for Occupational Safety and Health. 1978. *NIOSH Current Intelligence Bulletin 20-Tetrachloroethylene (Perchloroethylene).* DHEW Publication No. (NIOSH) 78-112.
(5)   World Health Organization. 1987. *IARC Monographs on the Evaluation of the Carcinogenic Risk of Chemicals to Humans.* Suppl. 7. International Agency for Research on Cancer. pp. 355-357.
(6)   Seeber, A. 1989. *Neurotoxicol. Teratol.* 11:579-583.
(7)   Kerster, H.W., et al. 1983. *Ecotoxicol. Environ. Safety.* 7:342-349.
tetrachlorvinphos
(1)   58
(2)   54
(3)   National Cancer Institute. 1978. *Bioassay of Tetrachlorvinphos for Possible Carcinogenicity.* Technical Report Series No. 33. National Institutes of Health.
(4)   19
(5)   1
tetradifon
(1)   World Health Organization. 1986. *Environmental Health Criteria 67. Tetradifon.* International Program on Chemical Safety.
(2)   58

(3)   33
(4)   19
(5)   21
(6)   1
(7)   12
tetrahydroquinone
(1)   Renner, G., et al. 1986. *Toxicol. Environ. Chem.* 11:37-50.
(2)   Witte, I., et al. 1985. *Mut. Res.* 145:71-75.
tetramethrin
(1)   58
(2)   3
(3)   33
tetramethrin (1R)-isomers
(1)   58
thiabendazole
(1)   23
(2)   5
(3)   33
(4)   Davidse, L.C. and W. Flach. 1978. *Biochimica et Biophysica Acta.* 543:82-90.
(5)   Khera, K.S. and C. Whaler. 1979. *J. Environ. Sci. Health.* B14(6):563-577.
(6)   Seiler, J.P. 1975. *Mutat. Res.* 32:151-168.
(7)   Wright, M.A. and A. Stringer. 1973. *Pesticide Science.* 4:431-432.
(8)   4
thiophanate ethyl
(1)   5
(2)   U.S. Environmental Protection Agency. 1986. *Thiophanate-ethyl Pesticide Fact Sheet.* Office of Pesticide Programs.
(3)   Hashimoto, Y., et al. 1970. *Pharmacometrics.* 4(1):5-21.
thiophanate methyl
(1)   Fleeker, J.R. 1974. *J. Agric. Food Chem.* 22(4):592-595.
(2)   5
(3)   Hashimoto, Y. 1972. *Toxicol. Appl. Pharmacol.* 23:606-615.
(4)   39. Thiophanate Methyl. 1986.
(5)   Stringer, A., et al. 1973. *Pesticide Science.* 4:165-170.
thiram
(1)   30
(2)   5
(3)   58
(4)   Fishbein, L. 1976. *J. Toxicol. Environ. Health.* 1:713-735.
(5)   Prasad, M.N., et al. 1987. *Food Chem. Toxicol.* 25(9):709-711.
(6)   Paschin, Y.U.V., et al. 1985. *Food Chem. Toxicol.* 23(3):373-375.
(7)   Mattiaschk, G., et al. 1973. *Arch. Toxikol.* 30:251-262.
(8)   46
(9)   Dalvid, R.R., et al. 1985. *Acta Pharmacol. Toxocol.* 58:38-42.
(10) Dalvi, R.R., et al. 1984. *J. Environ. Science Health.* B19(8&9):703-712.
(11) 1
(12) Edwards, H.M. 1987. *J. Nutr.* 117:964-969.
(13) Serio, R., et al. 1984. *Toxicol. Appl. Pharmacol.* 72(2):333-342.
(14) 9

toluene
(1)   3
(2)   Granholm, A.-C., et al. 1988. *Toxicol. Appl. Pharmacol.*
      96:296-304.
(3)   9
(4)   Rebert, C.S., et al. 1983. *Neurobehav. Toxicol. Teratol.*
      5:59-62.
(5)   Glowa, J.R. 1987. *NeuroToxicol.* 8(2):237-248.
(6)   Devlin, E.W., et al. 1985. *Arch. Environ. Contam. Toxicol.*
      14:595-603.
(7)   Kerster, H.W., et al. 1983. *Ecotoxicol. Environ. Safety.*
      7:342-349.

toxaphene
(1)   Cohen, D.B., et al. 1982. *Toxaphene. Water Quality and*
      Pesticides. Vol.2. Special Projects No. 82-45P. California
      State Water Resources Control. Borard. p. 17.
(2)   Eisler, R., et al. 1985. *Toxaphene Hazards to Fish,*
      Wildlife, and Invertebrates: A Synoptic Review. U.S. Fish
      and Wildlife Service. Biol. Rep. 85(1.4).
(3)   51
(4)   Steinel, H.H., et al. 1990. *Mutat.Res.* 230:29-33.
(5)   Rosen, M.B., et al. 1988. *Teratology.* 37(5):486.
(6)   Mohammed, A., et al. 1985. 1985. *Toxicol. Lett.* 24:137-
      143.
(7)   Allen, A.L., et al. 1983. *J. Toxical. Environ. Health.* 11:61-
      69.
(8)   Moorthy, K.S., et al. 1987. *J. Toxicol. Environ. Health.*
      20:249-259.
(9)   19
(10) 21
(11) 1
(12) 30

tralomethrin
(1)   5
(2)   58

triadimefon
(1)   58
(2)   5

triallate
(1)   32
(2)   58
(3)   U.S. Environmental Protection Agency. 1980. *Triallate.*
      Decision Document. Office of Pesticide Programs.
(4)   Sandhu, S.S., et al. 1984. *Mutat. Res.* 35:173-183.
(5)   4
(6)   1
(7)   9

triazophos
(1)   23
(2)   5
(3)   Velazquez, A., et al. 1990. *J. Toxicol. Environ. Health.*
      31:313-325.
(4)   58

tributyltin chloride complex
(1)   Eisler, R. 1989. *Tin Hazards to Fish, Wildlife, and*
      Invertebrates: A Synoptic Review. U.S. Fish and Wildlife
      Service. Biol. Rep. 85(1.15). 83pp.

tributyltin fluoride
(1)   Eisler, R. 1989. *Tin Hazards to Fish, Wildlife, and*
      Invertebrates: A Synoptic Review. U.S. Fish Wild. Serv.
      Biol. Rep. 85(1.15). 83pp.

tributyltin oxide
(1)   Piver, W.T. 1973. *Environ. Health Perspect.* Exp. Issue
      No.4:61-79.
(2)   Davis, A., et al. 1987. *Mutat Res.* 188:65-95.
(3)   Crofton, K.M.,et al. 1989. *Toxicol. Appl. Pharmacol.*
      97:113-123.
(4)   31
(5)   Eisler, R. 1989. *Tin Hazards to Fish, Wildlife, and*
      Invertebrates: A Synoptic Review. U.S. Fish and Wildlife.
      Service. Biol. Rep. 85(1.15). 83pp.

trichlorfon
(1)   Devine, J.M. 1973. *J. Agric. Food Chem.* 21(6):1095-1098.
(2)   5
(3)   58
(4)   54
(5)   Martson, L.V., et al. 1976. *Environ. Health Perspect.*
      13:121-125.
(7)   Batzinger, R.P., et al. 1977. *J. Pharmacol. Exp. Therap.*
      200(1):1-9.
(8)   Berge, G.N., et al. 1987. *Acta Vet. Scand.* 28:313-320.
(9)   Gibel, V.W. 1973. *Arch. Geschwulstforsch.* 41(4):311-328.
(10) World Health Organization. 1986. *Environmental Health*
      Criteria 63. Organophoshorous Insecticdes: A General
      Introduction. International Programme on Chemical Safety.
      p. 71.
(11) 31
(12) 21
(13) *Report of Environmental Assessment of Pesticide*
      *Programs.*1978. California Department of Food and
      Agriculture. Vol. 2. p. 3.2-19-3.2-21.
(14) 1
(15) deKergommeaux, D.J, et al. 1983. *Mutat. Res.* 124:69-84.

1,3,7-trichlorodibenzo-*p*-dioxin
(1)   Cochrane, W.P. et al. 1981. *J. Chromatogr.* 217:289-299.

trichloroethene
(1)   World Health Organization. 1987. *IARC Monographs on*
      the Evaluation of the Carcinogenic Risk of Chemicals to
      Humans. Suppl. 7.
(2)   Fukuda, K., et al. 1983. *Ind. Health.* 21:243-254.
(3)   Onfelt, A. 1987. *Mutat. Res.* 182:135-154.

trichlorodibenzo-*p*-dioxin
(1)   Palmer, F.H., et al. 1988. *Chlorinated dibenzo-p-dioxin*
      and dibenzofuran contamination in California From
      chlorophenol wood preservative uses. Report No. 88-
      5WQ. Division of Water Quality. State Water Resource
      Control Board.

trichloroisocyanuric acid
(1)   39

2,4,5-trichlorophenol
(1)   17
(2)   5
(3)   51

2,4,6-trichlorophenol
(1)   National Cancer Institute. 1979. *Bioassay of 2,4,6-*
      Trichlorophenol for possible Carcinogenicity (CAS No. 88-

06-2). *Technical Report Series No. 155.* National Institutes of Health.
(2) 42
(3) 51

triclopyr
(1) Solomon, K.R., et al. 1988. *J. Agric. Food Chem.* 36(6):1314-1318.
(2) 58
(3) 20
(4) Shearer, R. 1983. *Affidavit for Sierra Club Legal Defense Fund.* Juneau, Alaska.
(5) Shipp, A.M., et al. 1986. *Worst case analysis study on forest plantation herbicide use.* K.S. Crump & Company, Inc. Ruston, LA.
(6) Gersich, K.M., et al. 1984. *Bull. Environ. Contam. Toxicol.* 32:497-502.

trifluralin
(1) 20
(2) Grover, R., et al. 1988. *J. Environ. Qual.* 17(4):543-550.
(3) 7
(4) 2
(5) 39. Trifluralin. 1987.
(6) Donna, A., et al. 1981. *Pathologica.* 73:707-721.
(7) Mortelmans, K., et al. 1986. *Environ. Mutagen.* 8 (suppl. 7):1-119.
(8) Ghiazza, G., et al. 1984. *Bollettino-Societa Italiano Biologia Sperimentale.* 60:2149-2153.
(9) Beck, S.L. 1981. *Teratology.* 23:33-55.
(10) 21
(11) 30
(12) 1

trimethacarb
(1) 23
(2) 5
(3) 39. Trimethacarb. 1985.

trimethylarsine
(1) 28

O,S,S-trimethyl phosphorodithiote
(1) 14

O,O,S-trimethyl phosphorothioate
(1) Imamura, T., et al. 1983 *Chem. Biol. Interactions.* 45:53-64.

triphenyltin acetate
(1) 58
(2) 13
(3) 31
(4) 30

unsymmetrical 1,1-dimethylhydrazine
(1) 3
(2) 29
(3) Suter, W. et al. 1982. *Mutat. Res.* 97:1-18.
(4) 43
(5) Mori, H., et al. 1988. *Japn. J. Cancer Res.* 79:204-211.
(6) Snyder, R.D. et al. 1985. *Environ. Mutagen.* 7:267-279.

vamidothion
(1) 58
(2) 24

vernolate
(1) 20

(2) 58
(3) 21
(4) 1

vinclozolin
(1) 58
(2) 5

warfarin
(1) 36
(2) 14
(3) Howe, A.M. et al. 1989. *Teratology.* 40: 259-260.

xylene
(1) 18
(2) 13
(3) National Toxicology Program. 1986. *Toxicology and Carcinogenesis Studies of Xylenes (Mixed) (60% m-Xylene, 14% p-Xylene, 9% o-Xylene, AND 17% Ethylbenzene) in F344/N Rats and B6C3F$_1$ Mice (Gavage Studies).* Technical Report Series No. 327.
(4) Brasington, R.D., et al. 1991. *Arthritis Rheum.* 34(5):631-633.

zinc coposil
(1) 29

zinc phosphide
(1) 5
(2) 3
(3) *Zinc Phosphide Specimen Label.* Bell Laboratories, Inc.
(4) 19
(5) California Department of Food and Agriculture. 1985. *Zinc Phosphide Safety Information Series I-3.* Worker Health and Safety Branch.
(6) Fisheries and Wildlife Research. 1978. U.S. Fish and Wildlife Service.

zineb
(1) Nash, R.G. and M.L. Beall Jr. 1980. *J. Agric. Food Chem.* 28:322-330.
(2) 58
(3) 14
(4) 32
(5) Van Leeuwen, C.J., et al. 1986. *Aquatic Toxicol.* 9:129-145.
(6) Ghate, H.V. 1985. *Riv. Biol.* 78:288-291.
(7) Bristow, P.R. and A.Y. Shaw. 1981. *J. Amer. Soc. Hort. Sci.* 106(3):290-292.

ziram
(1) 30
(2) 5
(3) Gulati, D.K. et al. 1989. *Environ. Mol. Mutagen.* 13:133-193.
(4) 46
(5) McGregor, D.B. et al. 1988. *Environ. Mol. Mutagen.* 12:85-154.
(6) Giovani, E. et al. 1983. *Ecotoxicol. Environ. Safety.* 7:531-537.
(7) 11
(8) Cilievici, O., et al. 1983. *Morphol. Embryol.* 29(3):159-165.
(9) Enomoto, A., et al. 1989. *Toxicology.* 54:45-58.
(10) Serio, R., et al. 1984. *Toxicol. Appl. Pharmacol.* 72(2):333-342.

ESSENTIAL BACKGROUND REFERENCES

Brooks, P. 1972. *The house of life: Rachel Carson at work.* New York: Houghton Mifflin.

Carr, A., et al. 1991. *Chemical-free yard and garden.* New York: Rodale.

Carson, R. 1962. *Silent spring.* New York: Houghton Mifflin.

Colborn, T. E., et al. 1990. *Great Lakes great legacy?* Washington, DC: Conservation Foundation and Institute for Research on Public Policy.

Dahlsten, D. L., and R. Garcia. 1989. *Eradication of exotic pests.* New Haven, CT: Yale University Press.

Dover, M., and B. Croft. 1984. *Getting tough: Public policy and the management of pesticide resistance.* Washington, DC: World Resources Institute.

Dunlap, T. R. 1981. *DDT—Scientists, citizens, and public policy.* Princeton, NJ: Princeton University Press.

Epstein, S. S. 1979. *The politics of cancer.* New York: Anchor Press/Doubleday.

Flint, M. L., and R. van den Bosch. 1981. *Introduction to integrated pest management.* New York: Plenum.

Franklin, S. 1988. *Building a healthy lawn.* Pownal, VT: Storey Communications.

Gips, T. 1988. *Breaking the pesticides habit: Alternatives to 12 hazardous pesticides for sustainable agriculture,* 2d ed. Minneapolis: International Alliance for Sustainable Agriculture.

Graham, F. 1970. *Since silent spring.* New York: Houghton Mifflin.

Kimbrough, R. D., et al. 1989. *Clinical effects of environmental chemicals: A software approach to etiological diagnosis.* Washington, DC: Hemisphere.

Leslie, A. R., and R. L. Metcalf. 1989. *Integrated pest management for turfgrass and ornamentals.* Office of Pesticide Programs, U.S. Environmental Protection Agency.

Morgan, D. P. 1989. *Recognition and management of pesticide poisoning.* U.S. EPA doc. no. 540/9-88-01.

Mott, L., and K. Snyder. 1987. *Pesticide alert.* San Francisco: Sierra Club.

National Research Council. 1980. *Urban pest management.* Washington, DC: National Academy Press.

National Research Council. 1984. *Toxicity testing strategies to determine needs and priorities.* Washington, DC: National Academy Press.

National Research Council. 1986. *Pesticide resistance: Strategies and tactics for management.* Washington, DC: National Academy Press.

National Research Council. 1989. *Alternative agriculture.* Washington, DC: National Academy Press.

Olkowski, W., S. Daar, and H. Olkowski. 1991. *Common-sense pest control.* Newtown, CT: Taunton.

Rudd, R. 1964. *Pesticides and the living landscape.* University of Wisconsin Press. (Now from University Microfilms International.)

Schultz, W. 1989. *The chemical-free lawn.* New York: Rodale.

van den Bosch, R. 1978. *The pesticide conspiracy.* University of California Press.

U.S. Environmental Protection Agency. 1988. *The Federal Insecticide, Fungicide and Rodenticide Act as amended.* Office of Pesticide Programs.

U.S. Environmental Protection Agency. 1990. *National pesticide survey phase I report.* Offices of Water and Pesticide Programs.

U.S. General Accounting Office. 1990. *Lawn care pesticides: Risks remain uncertain while prohibited safety claims continue.* GAO/RCED-90-134.

U.S. General Accounting Office. 1990. *Alternative agriculture: Federal incentives and farmers' options.* GAO/PEMD-90-12.

U.S. General Accounting Office. 1986. *Nonagricultural pesticides: Risks and regulation.* GAO/RED-86-97.

U.S. General Accounting Office. 1986. *Pesticides: Need to enhance FDA's ability to protect the public from illegal residues.* GAO/RCED-97-7.

APPENDICES

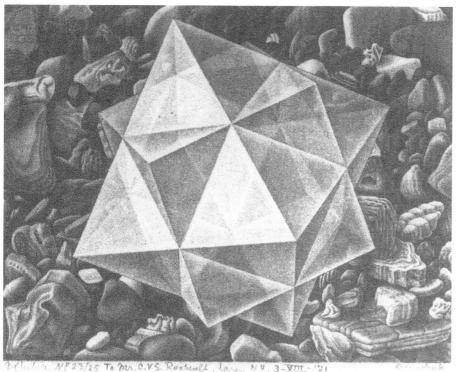

Using Pesticides

by Shirley A. Briggs

DETERMINING NEED, CHOICE, AND PRECAUTIONS

Whether a pest control problem involves large agricultural acreage or a home garden, the first step is deciding whether the pest has reached a level at which special control measures are needed. The following questions should be considered.

Is a Pesticide Necessary?

This can take study to learn whether the perceived pest is a real problem, and if so whether it has reached the point at which it can inflict economic damage or harm the appearance of landscape plants or injure desired animals. Are natural controls on hand to take care of the matter in a reasonable time?

Overuse of pesticides often comes from two causes: application on a routine basis to prevent damage in case a pest appears later, called prophylactic, or overreaction to something considered a pest that presents no real menace to a crop or the health or appearance of the environment. In their book *Urban Pest Management*, the National Academy of Sciences wonders whether the extreme aversion felt by many people to insects may have been increased by advertising for pest control firms. Beyond the waste of money and resources from these practices and the increase of toxic hazard, a serious consequence is the rapidly increasing resistance of pests to pesticides. This has caused increased amounts and toxicity of pesticides applied just to try to stay even with ever more resistance being developed. With destruction of normal biological controls of pests by this excessive application of pesticides, species that were never considered pests before have multiplied to serious pest status. This has happened with mites in several cases.

Which Pesticides Can Control the Pest Without Undue Damage to Nontarget Species or the Environment?

If emergency action is called for, careful choice of a pesticide can minimize unwanted side effects. If the only suitable product has a broad range of high toxicity, or is very persistent, try to find a method of application that will minimize the effect of these

traits. The best pesticide is very specific in its targets, has the minimum toxicity necessary, and has short persistence and low capacity to disperse widely in the environment.

The Netherlands government proposes reducing the volume of pesticides used by 50% by the year 2000. A recent report studying ways to achieve this goal lists several criteria for evaluating the environmental burden caused by pesticides. Aspects of pesticides that increase this burden include emission to air; leaching to ground and surface water; accumulation in the soil; damaging effects on soil life, on the water ecological system, and on human health.

What Method of Application is Most Effective and Least Harmful?

If the pesticide can be applied so that most of it reaches the target pest and is confined mainly to a limited area, unwanted side effects will be minimized. Aerial application, for example, may seem to cover a large area economically, but very little of the pesticide will actually reach the target, and most may drift long distances with unexpected effects.

What is Legal Pesticide Use?

FIFRA first regulated the use as well as the production and sale of pesticides in 1972. It is unlawful to disobey the provisions for use on the product label, and the producer may not claim any "safety" or "no toxic effects" or similar statements. Some state regulation is more strict than federal, at least in practice. The Federal Trade Commission has authority to rule against false and misleading advertising claims for pesticide products, whether by the producer or commercial applicator, but has rarely raised the issue. New York State has sued a lawn care firm successfully for their excessive advertising claims. The labels can tell you what the product contains, its immediate toxicity rating, and what precautions are required. It does not tell you whether the product has been adequately tested for the range of toxic effects that is now required for reregistration, a process that is not very far along.

If commercial applicators are hired to apply pesticides to your property, it is wise to be sure that their liability insurance covers any damage that may be done to your own or your neighbors' property and family.

STORING, TRANSPORTING, DISPOSAL

Labels on pesticide products contain "requirements and procedures for the transportation, storage, and disposal of the pesticide, any container of the pesticide, any rinsate containing the pesticide, or any other material used to contain or collect excess or spilled quantities of the pesticide" (FIFRA as amended, 1988, Sec. 19 (a)(B)). Though many of these conditions apply to large commercial or agricultural use, they must also be observed by any user.

Storage of pesticides may involve problems not contemplated on the labels: possible mixing with other chemicals, pesticides, or materials like fertilizers with unexpected results. Volatile substances can penetrate nearby containers or bins. Fertilizers stored with herbicides have absorbed enough active ingredient to kill plants to which they were applied. More drastic results can come from forcible mixing through severe wind storms, floods, or fire. Some otherwise low-combustible materials can combine to form highly flammable or explosive mixtures. In case of fire or explosion, very poisonous gases and liquids can be widely dispersed. Anyone storing a quantity of pesticides or similarly hazardous materials is advised to tell the nearest fire department of the types and amounts and exact locations of each so that firemen can be prepared to fight any fire with the best possible protection to themselves and others, including livestock and wildlife. Dust explosions can occur when pesticides in powdered form are dispersed in the air. Vapor explosions can come from mixing a pesticide with a combustible solvent and spraying in atomized form, especially in heated conditions.

Similar precautions pertain, of course, when quantities of pesticides are being transported. Very hazardous conditions can arise in case of truck or train accidents.

Disposal of pesticides and their containers requires compliance with federal and local rules for toxic waste disposal. Some communities have special collections for all such materials for transport to authorized toxic waste dumps. If there is such a dump nearby, the waste materials may be taken there directly.

Re-use of pesticide containers for anything else is strictly forbidden. It can be almost impossible to remove all traces of many pesticides. Using the same equipment to apply them, and then fertilizing or using another kind of pesticide can result in unanticipated effects.

Appendix Two

Alternatives to Pesticides; IPM

by Robert van den Bosch

Introduction

Alternative means of pest control are essential to any plan to reduce use of toxic chemical pesticides. Several books in our Reference lists explain these alternatives, referring to Integrated Pest Management and Sustainable Agriculture. Both of these concepts require intelligent observation, knowledge of pests and the plants and animals on which they prey, and a strategy geared to the species involved and their environmental setting. That is, they take more information, attention, and perhaps more work than the quick-fix attitude that has fostered the almost universal dependence on toxic chemical pesticides in recent decades. Their rewards are a large reduction in hazards to human and environmental health, lower costs, and a system that can improve the quality of soil, water, and air over the years.—SAB

INTEGRATED PEST MANAGEMENT (IPM)

The late Robert van den Bosch, chairman of the Division of Biological Control at the University of California, Berkeley, was an entomologist who played a key role in developing both the concept and the practical application of Integrated Pest Management. In his book, *The Pesticide Conspiracy,* he explained IPM concisely:

"What Is Integrated Control?"

Integrated control is simply rational pest control: the fitting together of information, decision-making criteria, methods, and materials with naturally occurring pest mortality into effective and redeeming pest-management systems.

Under integrated control, natural enemies, cultural practices, resistant crop and livestock varieties, microbial agents, genetic manipulation, messenger chemicals, and yes, even pesticides become mutually augmentative instead of individually operative or even antagonistic, as is often the case under prevailing practice (e.g., insecticides versus natural enemies). An integrated control program entails six basic elements: (1) man, (2) knowledge/information, (3) monitoring, (4) the setting of action levels, (5) methods, and (6) materials.

Man conceives the program and makes it work. *Knowledge* and *information* are used to develop a system and are vital in its day-to-day operation. *Monitoring* is the continuous assessment of the pest-resource system. *Action levels* are the pest densities at which control methods are invoked. *Methods* are the pathways of action taken to manipulate pest populations. *Materials* are the tools of manipulation. . . .

Integrated control systems are dynamic, involving continuous information gathering and evaluation, which in turn permit flexibility in decision-making, alteration of the pathways of action, and variation in the agents used. It is the pest-control adviser who gives integrated control its dynamism. By constantly "reading" the situation and invoking tactics and materials as conditions dictate, he acts as a surrogate insecticide, "killing" insects [or other pests] with knowledge and information as well as pesticides, pathogens, parasites, and predators. Integrated control's dynamism is a major factor that sets it off from conventional pest control. Thus, though the latter involves some of the same elements, it lacks dynamism in that it is essentially preprogrammed to the prophylactic or therapeutic use of pesticides. In other words, pesticides dominate the system and constitute its rigid backbone. Where a crop is involved, there is little or no on-going assessment of the crop ecosystem and the dynamic interplay of plant, pests, climate, and natural enemies. . . . Under the prevailing chemical control strategy, there is virtually no flexibility in decision-making, particularly as regards alternative pathways of action.

The game plan is set at the start and it is stubbornly followed.... It is the lazy man's approach, which characterizes so many aspects of modern life and for which society and the environment pay dearly....

The basic argument, then, is not against pesticides per se or chemical control as a tactic, but against chemical control *as our single-component pest management strategy and the biocide as its operational tool....* But along with this strategic change it is also vital that society insist on the establishment of standards that eliminate "biocides" as our chemical tools, and require, instead, safe, selective, and ecologically tenable pesticides.

[Current provisions of FIFRA direct EPA to conduct research in Integrated Pest Management, and develop strategy for applying it in various situations. They are also to build partnerships with industry, users, universities, and private organizations to spread word rapidly of improved pest management technologies. Some states have also adopted the principle that IPM is the goal of pest control policies.]

The Meaning of Carcinogenicity Testing

by William Lijinsky

Introduction

Dr. Lijinsky's testimony to a U.S. Senate committee summarizes the most important aspects of testing for cancer-inducing substances. He points out some key ways in which various formulas for presuming to rate this potential are not based on sound science, and clarifies the methods used and their reasons. When he submitted this statement he was director of the Chemical Carcinogenesis Program at the Frederick Cancer Research Center. He is now at the National Institutes for Environmental Health Sciences.—SAB

STATEMENT OF DR. WILLIAM LIJINSKY BEFORE THE COMMITTEE ON LABOR AND HUMAN RESOURCES, UNITED STATES SENATE, JUNE 6, 1989

Mr. Chairman and members of the committee, thank you for allowing me to present my views on the problem of regulation of pesticides in the food supply, an issue of great importance in protecting the public health. I am William Linjinsky and am providing my personal opinions, not those of the United States government nor those of my employer, Bionetics Research, Inc. I would like to address briefly the cautions that must be considered before changing the present conditions. These are based on more than 35 years of study of chemical carcinogenesis.

Cancer is probably the most fearful disease to people in the industrialized world, afflicting 1 person in 4 and killing 1 person in 6. This is a new pattern of the twentieth century, following the reduction through hygiene and better habits of the common infectious diseases. The experimental work that began early in this century revealed that chemical compounds of certain types possessed the property of inducing cancer in animals. Succeeding studies of several thousand chemicals showed that about 10% were carcinogenic, although this proportion is not very informative, since many compounds were tested because they were believed likely to induce tumors. Among them were many pesticides and other agricultural chemicals, and several of them were carcinogens.

Within the past 20 years carcinogenesis tests have become more precise and sophisticated, instead of hit-or-miss studies that were common before that. They were designed to provide reasonable certainty that the test substance posed little or no risk of cancer when humans were exposed to it. Dr. Umberto Saffiotti and I helped design those tests. Because of the expense of the tests, the substances chosen for testing in these systems (using many animals of two species at two dose levels) were not random. They were substances already suspect because of prior experiments, or because they resembled known carcinogens in structure, and because they had some economic importance. Fewer than half of those tested were carcinogenic, many were active only in one species (usually rats or mice), and some only in one sex. However, by definition, any substance that induced cancer in even one sex or one species was a carcinogen, and more likely to pose a cancer risk to humans than a substance that was not a carcinogen in those tests. Because of the well-documented differences between species in response to a particular carcinogen, it cannot be said with certainty that a substance that does not induce tumors in either rats or mice would not pose a carcinogenic risk to humans; it is an assumption we must make in the absence of any alternative means of testing for carcinogenic activity.

There have been many suggested short-term assays for detecting potential carcinogenic activity, most of them based on the unverified assumption that induction of tumors takes place by a mutagenic mechanism. The most commonly used are assays that produce colonies of mutated bacteria, such as that introduced by B.N. Ames of the University of California, Berkeley, and they have been useful.

However, in recent analyses of the results of testing in such assays compounds that had been examined in the long-term bioassays carried out by the National Cancer Institute and later by the National Toxicology Program, there was little more than 50% correlation between carcinogenicity and activity in the short term assays (Tennant et al. 1987; Zeiger 1987). Therefore, the abbreviated assays cannot be considered predictive of carcinogenic activity, and the chronic and expensive animal bioassays are all we have.

The causes of most human cancers are not known, but tobacco use is related to a considerable proportion, although we do not know which components of tobacco or tobacco smoke are responsible for tobacco related cancers. Diet has also been associated with certain patterns of cancer, but the data are imprecise and the carcinogens are unidentified. According to the epidemiologists Doll and Peto only about 4% of human cancer can be attributed to occupational or industrial exposures to carcinogens, which leave a large gap to be occupied by "unknown causes." Among these lie environmental carcinogens, including agricultural chemicals, food additives, food constituents and possibly carcinogens formed within the body (endogenous carcinogens). There are certainly some cancers that have a direct genetic origin, perhaps defective genes, but these seem to be small in number. The slowly accumulated evidence of epidemiology shows that cancer incidence has more to do with where people live than with their genetic background, as shown in the figures of migrant populations. For example the common cancers of second generation Japanese people in Hawaii resemble those of other Americans rather than those of people in Japan. The cancer patterns of black Americans resemble those of all Americans and differ from those of people in West Africa. This evidence from very large populations (much larger than the groups of animals on which we do our experiments) shows that exposure to carcinogenic agents is the main source of cancer in humans.

WHAT ANIMAL EXPERIMENTS TELL US

1. Well-conducted 2-year bioassays of substances in 2 species of rodents at high but nontoxic doses that give rise to statistically significant number of tumors compared with controls in at least one sex of one species identify them as carcinogens. Substances that do not give this result are assumed to be noncarcinogens.

2. Exposure of humans to substances identified as carcinogens are assumed to pose a carcinogenic risk to them. The magnitude of this risk cannot be calculated with present knowledge and estimates may be off by several orders of magnitude.

3. It is not possible to conduct plausible experiments that enable us to establish a safe threshold for exposure of people to any carcinogen revealed through experiments in animals. Few dose-response studies over a large range of doses have been carried out and in those, mainly with nitrosamines, there have been significant tumor responses even at the lowest doses used, which were lower than those to which workers in some occupations were exposed (nitrosodimethylamine, Peto et al. 1984; nitrosomorpholine, Lijinsky et al. 1988).

4. Comparing the carcinogenic effects of compounds in several species has indicated that there is no way to predict the potency in one species from that in another. However, a substance carcinogenic in rats, for example, is more likely than not to be carcinogenic in mice, and vice versa (Haseman and Huff 1987). It is consistent with protection of the public health to assume that man is as susceptible as the most sensitive species in which the substance has induced tumors.

5. There is no basis for assuming that the response of humans or animals to a carcinogen is related to the life span of the species. The little information relating to this subject that we have, for example the experiments of Schmähl in rats, chickens, cats, and snakes of very different life span, show that the time of appearance of the tumor is roughly related to the dose (Schmähl and Schorf 1984; Schmähl et al. 1978). Therefore, cancers usually appear late in the life of humans because they are exposed to low doses of carinogens over a long period, compared with our experimental animals.

6. Experiments in which animals are exposed to combinations of carcinogens show that the effects are usually additive, as is the case with increasing doses of a single carcinogen. Increasing doses of a carcinogen (or several) decreases the time of appearance of tumors and increases the incidence, and vice versa. Therefore the effects of low doses of a carcinogen, such as humans might experience, when repeated and combined with the effects of other carcinogens, accumulate over time (60 years or more, in the case of people exposed as children), and increase the probability that cancer will develop.

7. The long time that elapses between exposure to carcinogens and the appearance of tumors differentiates carcinogenesis from most other kinds of toxicity. Tumor cells implanted into

receptive animals grow very rapidly into tumors in a few weeks, whereas induction of tumor by chemicals always takes months. The process of transforming normal cells into tumor cells must have many stages, none of which is understood. Even the commonly accepted beginning of the process as a mutation (or several) in DNA of the nucleus is not a fact, since several substances that are not mutagens are quite effective carcinogens.

In view of the small amount of information about the mechanisms by which chemicals give rise to cancer (and the uncertainty about the relevance of that information), it is unwise to permit officials or experts to calculate tolerable or "safe" exposures for humans to carcinogens. All of us are fallible even when armed with sound information. Reliable information about carcinogens is limited almost to whether or not the substance is one.

The argument that has been advanced, but is not accepted by most investigators in this field, that human exposure to "natural" carcinogens in plants used as food is much more important than exposure to industrially produced carcinogenic pesticides, is not tenable. Carcinogens occurring naturally in plants (for example, aflatoxins) may not be ignored, but it is well known that vegetarians have a considerably lower cancer risk than do omnivorous humans. There is a danger in turning over custody of our health to those who profess more knowledge than they possess. My apprehensions about fellow scientists with such a mind-set are expressed in the following short letter which I recently submitted to the magazine *Science* in response to a much longer letter from Thomas Jukes, ridiculing our apprehensions about exposure to commercial agricultural chemicals that are carcinogens. *Science* would not publish my letter, but it speaks for itself.

May 12, 1989
The Editor, *Science:*

Thomas Jukes (Letters, 5 May, p. 515) has a point, that the assessment of risk in exposure to Alar (daminozide) by the National Resources Defense Council is arguable. What is not arguable is that unsymmetrical dimethylhydrazine (UDMH) formed from daminozide is a carcinogen, and daminozide is a commercial agricultural chemical leaving residues in apples. UDMH is weaker than the extremely potent dimethylnitrosamine in the species in which both were tested, but our limited knowledge of carcinogenesis does not permit us to estimate their relative potencies in humans. The Delaney Amendment requires that Alar be banned from human food because it has been shown by appropriate tests to cause cancer in man *or* animals. The exposure to Alar and UDMH might be small, but neither Dr.

Jukes nor anyone else knows how to establish a threshold for exposure to carcinogens, the effects of which accumulate over time with results that we cannot measure or estimate. Of course, exposure to "natural" carcinogens is also important in the process, whether we can do anything about them (e.g., natural radiation) or not. However, both Dr. Jukes and the Ames[1] article he quotes introduce highly questionable animal testing data with natural substances as "evidence" that commercial carcinogens can be ignored. For example, three tests of patulin have been published[2] and not one shows evidence of carcinogenicity, yet Jukes calls it a suspected carcinogen; there is no such doubt about Alar and UDMH. That Ames et al. rank one mushroom as equivalent in hazard to the aflatoxin in three peanut butter sandwiches does not make it so, when the hydrazine-in-mushroom test[3] is so dubious that it would be unacceptable as evidence of carcinogenicity if it were a synthetic chemical.

REFERENCES

Haseman, J.K., and J.E. Huff. 1987. Species correlation in long-term carcinogenicity studies. *Cancer Letter* 37:125–132.

Lijinsky, W., R.M. Kovatch, C.W. Riggs, and P.T. Walters. 1988. A dose-response carcinogenesis study of nitrosomorpholine in F344 rats. *Cancer Res* 48:2089–2095.

Peto, R., R. Gray, P. Brantom, and P. Grasso. 1984. Nitrosamine carcinogenesis in 5120 rodents: Chronic administration of 16 different concentrations of NDEA, NDMA, NPYR, and NPIP in the water of 4440 inbred rats, with parallel studies on NDEA alone of the effect of age of starting (3, 6, or 20 weeks) and of species (rats, mice or hamsters). In N-nitroso compounds: Occurrence, biological effects and relevance to human cancer, eds, I.K. O'Neill, R.C. VonBorstel, C.T. Miller, J. Long, and H. Bartsch, *IARC Scientific Publications* No. 57, 627–665.

Schmähl, D., M. Habs, and S. Ivankovic. 1978. Carcinogenesis of N-nitrosodiethylamine (DENA) in chickens and domestic cats. *Int. J. Cancer* 22:552–557.

Schmähl, D., and H.F. Scherf. 1984. Carcinogenic activity of N-nitrosodiethylamine in snakes (*Python reticulatus*, Schneider). In N-nitroso compounds: Occurrence, biological effects and relevance to human cancer, eds. I.K. O'Neil, R.C. VonBorstel, C.T. Miller, J. Long, and H. Bartsch. *IARC Scientific Publications* No. 57, 667–682.

Tennant, R.W., B.H. Margolin, M.D. Shelby, E. Zeiger., J. K. Haseman, J. Spalding, W. Caspary, M. Resnick, S. Stasiewicz, B. Anderson, and R. Minor. 1987. Prediction of chemical carcinogenicity from *in vitro* genetic toxicity assays. *Science* 236:933–941.

Zeiger, E. 1987. Carcinogenicity of mutagens: Predictive capability of the Salmonella mutagenesis assay for rodent carcinogenicity. *Cancer Res* 47:1287–1296.

[1]B.N. Ames, R. Magaw, and L.S. Gold. *Science 236,* 271 (1987).
[2]U.S. Public Health Service Publication No. 149, *Survey of Compounds Which Have Been Tested For Carcinogenic Activity,* 1961–1967 Volume, 1973 Volume, 1981–1982 Volume.
[3]B. Toth and J. Erickson. *Cancer Res. 46,* 4007, 1986.

Pesticides and Our Environment

by Charles R. Walker

Introduction

Effects of pesticides beyond the immediate purpose and location of the pestiferous annoyance or pestilent threat were best explained by Rachel Carson in *Silent Spring*. As the number, kinds, and amount have increased since 1962, so also have our problems in realizing their total impact. We must balance our incomplete knowledge of the action of these products and our far more imperfect understanding of the intricate mesh of interactions of the global natural world. The issues involved and a glimpse of the range of considerations are shown in an article written by Charles R. Walker in 1981 for a symposium sponsored by the American Society for Testing and Materials. The proceedings were edited by E. W. Schafer and C. R. Walker as *Vertebrate Pest Control and Management Materials*, ASTM Special Technical Publication 752. We reprint this chapter with permission of the author and ASTM. In 1981, C. R. Walker was senior scientist in the U.S. Fish and Wildlife Service; he is now retired.

To translate clues given on the pesticide charts into information needed to pursue the kinds of analysis explained by Walker, consider the forms of life we know to be affected, and such properties as water and oil solubility. Water soluble materials are dispersed through plants and the bodies of animals, while usually filtering down through soil into groundwater, streams, and lakes into estuaries and oceans. If they break down into transformation products before they have time to travel the whole way, consider the nature of these transformation products known and their persistence. Not all transformation products have been discovered, nor their toxicity tested, so we must allow for the travel of unknowns far in space and time from the application of the original product. Oil soluble materials are apt to collect in the fatty tissues of animals, from the tiniest forms to the largest. In soil organisms, they can travel through the soil also, often to considerable distances. Larger forms may travel much farther and faster. In either case, bioaccumulation can occur, as larger forms eat smaller ones, and the pesticide continues to build up in fat until they themselves are eaten or depletion of fat reserves releases the total amount into their bloodstream and so poisons their whole bodies. Remember the robins that eat the earthworms that are able to survive large amounts of accumulated pesticide in their bodies; a robin can get a fatal dose from 11 worms. Migrating birds may suffer during the effort of migration as their reserves are drawn down, or they bring their burden of poison into distant habitats.

Volatile pesticides contribute to the dispersal by air to near and far places. In trying to account for high levels of some persistent pesticides in the Great Lakes, when no such amounts were known to have been used in their watersheds, scientists found that these toxic materials had gotten into the air by evaporation, been carried long distances, and brought back to earth and water by rain. A striking example occurred several years ago when an especially volatile form of 2,4-D that was illegal in the state of Washington was found to have killed vineyards there. It was traced to application in Oregon, where it was legal. Though ground wind was going from north to south at the time, a higher inversion layer formed under an upper air current going north. The herbicide rose to the inversion layer, was brought north at almost full strength, and dropped to earth in Washington when the inversion layer dispersed. Complex research into meteorology and air transport went into solution of the mystery.—SAB

REDUCING POTENTIAL HAZARDS OF PESTICIDES TO FISH, WILDLIFE, AND HABITAT, By C. R. Walker

Recently, a worldwide effort has been made to bring together experts on environmental effects of pesticides and to seek a consensus of scientific opinion on harmonization of criteria for testing and registration of pesticides. The foremost examples are the efforts of the United Nations Food and Agriculture Organization regarding pesticides and the Organization for Economic Cooperative Development regarding chemicals in general.

The risks to the environment of a pesticide are

dependent on many factors. They include its toxic properties, persistence and mobility in the environment, the amount applied, formulation, the method and time of application, and the extent and pattern of its use. The overall effect of the pesticide also depends on the stage in the development of the nontarget organisms involved, the feeding habits of these species, and the extent to which toxic residues or metabolic compounds may accumulate or be concentrated in successive species in food chains. The hazards to wildlife may also be accentuated if the animals in the treated area are subject to some external stress, for example, by a lack of food or by adverse weather prevailing at the time.

Some pesticide effects on wildlife may be too complex, subtle, or delayed to be detected by ordinary routine testing in the laboratory or the field. In many cases, the pathway of a pesticide may be impossible to follow as it moves through the environment under an infinite variety of conditions under which the pesticide may be used in practice. Nevertheless, experience has shown that in many cases predictions can be made of the probable effects of a compound on the environment from consideration of certain basic studies [1–4].

Data are developed prior to registration of a pesticide to allow a reasonable judgment of safe use in relation to both human health and the ecosystem. Equally important is the environmental behavior of the product when applied according to the recommendations for use. These data are essentially predictive and intended to describe those characteristics of the product relevant to the environment. They should be sufficiently comprehensive to enable a reasonable judgment of environmental behavior of the products for the uses proposed. The actual test program has to be decided according to the products' characteristics and conditions of use. A good example of this is the use of copper sulfate in low or high pH and hard or soft water.

PESTICIDE CONCENTRATIONS

Influence of Use Patterns and Physical-Chemical Properties

A pesticide's use patterns can greatly influence its potential environmental impact. Very toxic or persistent compounds may not cause environmental hazards if used in ways which do not expose nontarget organisms. Conversely, compounds of low toxicity or low persistence can cause damage if applied too frequently or if poorly distributed so that they cause "hot spots" or become localized over the dosage area.

Methods of application differ greatly and are closely allied to the type of formulation. They range through granular types, dust, fogs, and a variety of sprays (inverts, biverts, and a variety of ultralow-volume, low-volume, and high-volume sprays). Usually, localized treatments are most often done with ground equipment whereas the broad-scale treatments are usually done with aircraft. Generally treatments with ground equipment minimize drift to non-target areas [1].

The site of application influences the environmental effects. A pesticide applied directly to the soil may influence soil fertility by its effects on soil organisms. Since most pesticides are applied to cultivated crops, contamination of nontarget organisms is more likely to occur in fields than in forest systems. A forest canopy can screen out most of an aerial spray, and, except for relatively water-soluble pesticides, the spray will be leached into water following precipitation. However, more fat-soluble and persistent pesticides have some mobility and are seldom confined to the site of application and therefore are potentially hazardous when transported by means of organic matter into aquatic ecosystems [5, 6]. Pesticides readily adsorbed onto soil are also actively transported along with sediments in runoff and are especially hazardous to benthic organisms [1, 7-9]. The more volatile pesticides are also subject to drift and may pose inhalation hazard to nontarget terrestrial organisms. Aerial application has a greater drift potential than ground-based application, and dust formulations are particularly prone to drift. While drift may be used deliberately to treat target pests, the potential hazard to nearby areas, particularly water, must always be considered by providing adequate "buffer strips."

The scale of use can greatly influence the impact of a pesticide on the environment. Small localized uses, even of very toxic chemicals, seldom cause widespread or long-lasting damage. On the other hand, very common use of a slightly toxic pesticide may have quite large effects. Frequent and widespread use can result in the development of resistance to the pesticide by the target organisms. Higher pesticide application rates may result in greater ecological damage and aggravate the resource impacts by contamination of edible tissues with residues. The size of the area to be treated can be the most important consideration of all since the recolonization of populations is restricted to the rate of immigration from the perimeter [10]. On an areal basis, the percentage of the population or habitat perturbation is directly related to the dysfunction of ecosystem processes and the maintenance of the species composition of the community within the ecosystem.

The amount and frequency of application of the pesticide clearly influences its environmental effects. Continuous or repeated applications of pesticides, for example, can seriously affect the populations of saprophagous invertebrates and macroorganisms that are important in the maintenance and improvement of soil structure and fertility through their role in decomposition, mineralization, and sequestration of organic matter [11, 12]. Since the relationship between dose and effect is usually logarithmic, any way in which either exposure or amount can be reduced will minimize the hazard [8]. Short-term benefits of fertilizers cannot compensate for long-term effects on soil-structure and fertility [13].

TIME AND NATURE OF EXPOSURE

The time of application is important in minimizing hazards to nontarget organisms. For example, harm to bees can be avoided by not spraying flowering crops or by using the less toxic formulation at the time during the day that bees are not working. Avian species are especially vulnerable during the nesting season. Both the young and adults of insectivorous birds and mammals are specifically exposed to insecticides.

Geographic locality can be of prime importance in influencing the extent of environmental effects. This is particularly true for areas with wide climatic differences. A pesticide which may be very persistent in a temperate situation may be much less so under moist and warm tropical conditions. This underlines the danger of uncritical extrapolation of data between temperate and tropical areas.

Formulation affects the availability of the pesticide. For example, the residue of a wettable powder may be much more toxic than that of an emulsifiable concentrate to an animal walking over the residue, while granular formulations applied to the soil minimize exposure to aerial species. The use of a pesticide as a seed-dressing or bait may present a direct hazard to birds and small mammals. The aquatic environment is particularly susceptible to adverse effects, and direct application to water (pest control in rice paddies; aquatic weed control) requires careful consideration. The physical chemical properties of pesticides and the nature of exposure can be altered by the formulation. For example, inhalation and drift can be minimized with granular formulations, invert emulsion sprays, sticker-spreader combinations, and encapsulated or controlled-release innovations. The use of a slow-release formulation (such as Casoron 6-10) results in a "persistent" pesticide that may constitute an efficacious exposure to target organisms while minimizing hazards to others. The pesticide antimycin has been incorporated into sand granules with a Carbowax resin in the Fintrol 5 formulation. This formulation allows release of antimycin in the first 1.6 m (5 ft) of the water column. This action allows selective toxic action to pest species in surface waters while minimizing the effects on deeper cold-water game fish species. On the other hand,

the highly toxic insecticides formulated in slow-release capsules or granules can directly poison birds that ingest them. These same formulations of certain organophosphate insecticides can make persistent-type exposures out of normally short-lived pesticides.

Chemical degradation data are important in predicting environmental hazards. Three main pathways exist: chemical degradation affected by chemical agents, biological degradation affected by living organisms as intermediaries, and photochemical degradation that utilizes light energy [8]. The rate of degradation affects the residual activity of the pesticide and impacts directly on its toxicity to nontarget organisms.

Knowledge of the rates and pathways of degradation of a compound in various systems and organisms is needed before a preliminary hazard assessment can be made of both the parent compound and the degradation products. An assessment of the expected environmental distribution gives an estimate of the level of exposure for the organisms that inhabit the various environmental compartments. This assessment will indicate which test systems and organisms should be used to evaluate the potential hazards.

In most cases, a knowledge of the use pattern of a pesticide (site and mode of application), together with information on its physical and chemical properties, and the type of nontarget organisms involved, enables a good estimate to be made of the likely environmental distribution of the compound and its products in the various ecological compartments.

In some cases further studies are needed to complement the information. For example, if laboratory soil testing indicated that the compound is easily leached, one should then check its leaching behavior in field circumstances. Again, where its behavior is abnormally sensitive to the soil type, some amplification of information for different soil types is required. Cases can also arise where soil residues consist primarily of a degradation product, and residue data on leaching products are needed.

Knowing whether the vegetation, soil, or water will receive the main burden of the pesticide is important. In some cases, the distribution pattern over the compartments might be calculated on the basis of the physical characteristics: vapor pressure, molecular weight, and solubility [2, 3, 7, 14–16]. According to Neely, no additional degradation studies are necessary if the calculated amount of the chemical reaching the soil is less than 4 percent or that reaching the water is less than 2 percent of the applied dose [17, 18]. However, those chemicals that persist readily bioaccumulate in tissues of fish, shellfish, waterfowl, and other economically important non-target organisms require monitoring for residues in field studies.

For bioconcentration, the suggestion of Metcalf and Sandborn [18] that pesticides with water solubilities of 50 ppm are unlikely to be biologically magnified might be applicable. The potential for bioaccumulation can also be estimated by the n-octanol/water partition coefficient [2, 4, 13, 15, 17–22]. When this coefficient is greater than 1000, a risk of accumulation exists, and laboratory testing for bioaccumulation should be conducted on fish [1]. Chemical stability and microbiological degradability, as well as volatilization characteristic, are more important in relation to the environmental compartments which are most likely to be contaminated. If vegetation and soil are to receive nearly all the pesticide used, an estimation could be made of the volatilization rate. This could be done on the basis of such physical data as vapor pressure and molecular weight, together with such climatological characteristics as temperature, hours of sunshine, humidity, and wind [8].

However, if the chemical is readily adsorbed onto soil or ionized, has surface-active properties or is bound to protein, then consideration must be given to conducting tests that properly characterize the environmental fate of the chemicals.

BIOLOGICAL AND ENVIRONMENTAL CONSIDERATIONS

The precise nature of the biological activity of pesticides should be determined in field tests under environmental conditions and the route of exposure associated with actual use patterns. Testing must consider the physical and chemical properties; mode of action; rate kinetics of uptake, metabolism, and excretion; and the behavior of organisms. The improper selection of the test species, of the appropriate test system in relation to the route of exposure, or of the prescription of testing conditions (humidity, water chemistry, temperature, and so on) may seriously compromise the usefulness of the data in hazard evaluation. Thus, some latitude in scientific judgment must be exercised at this level of testing to give adequate consideration to vital functions of organisms (growth, development, reproduction, and behavior) and ecosystem processes (structural and functional). Several examples illustrate the importance of biological considerations and their relevance to each environmental compartment and to the pesticide's mode of action.

Atmospheric-Terrestrial Compartment

Those pesticides that are volatile (that is, having high vapor pressure) and that have a low affinity for organic matter, water, or soil may not require further

testing. However, persistent substances that have high lipid solubility and sufficient density must be considered for testing against aerial and terrestrial forms of life such as honey bees, plants, and appropriate birds and mammals. For example, predatory species tend to be more vulnerable to pesticides than their prey; thus insecticide use should be restricted during the nesting season when young birds require a high-protein diet of insects. Organisms with food sources likely to be contaminated, such as seed or plant consumers, may be particularly at risk during planting time if seed is treated with toxic chemicals.

Water-Sediment Compartment

Water-soluble chemicals that are highly volatile or that undergo rapid degradation, such as by hydrolsis or photolysis, may not sustain toxic concentration in water and need only be tested minimally unless the degradation products are more toxic, less volatile, and more persistent than the parent compound [1, 8, 23]. However, once the chemical enters the aquatic system or is adsorbed in the sediment or organic substrate, such degradation phenomena as photolysis are reduced, and persistence becomes the essential criterion for testing. Pesticides thus adsorbed or retained in the soil-sediment substrate should be tested on appropriate representatives of the benthic community. Ionizable substances and those with surface-active properties that have an affinity for suspended organic matter or aquatic organisms in the limnetic zone or littoral zone should be tested accordingly. Most herbicides have low vapor pressures; some highly volatile substances such as xylene or acrolein are applied directly to water as for aquatic weed control. These compounds may require testing for safety to aquatic nontarget organisms to determine the fate and persistence of residues. Except for some insecticides, few pesticides applied to terrestrial crops reach important water bodies at levels sufficient to give rise to environmental concern. However, if physicochemical data indicate the mobility of a compound, then field studies should be undertaken to delineate the transport kinetics and conditions of exposure. If the pesticide is likely to be transported to or applied directly to a water body, then further studies are also needed to examine the effects on organisms living in the water bodies concerned.

Terrestrial Soil Compartment

A pesticide with low affinity for soil, water, or organic matter would not require testing unless it was deliberately injected or fixed by formulation into the soil. Highly volatile soil fumigants usually have temporary effects and leave little residue. Compounds that are tightly bound onto soil but desorbed onto organic matter would require tests on soil organisms and plant life. If these pesticides also have a high lipid solubility, bioaccumulation may be a serious concern in secondary poisoning of terrestrial organisms, and particularly of top predators. Long-term effects may be important where the organism is known to play an important function in maintaining the agricultural or wildlife environment. For example, in the terrestrial environment, various organisms contribute to the decomposition of crop residues, an important soil function. This raises the concept of studies of functional processes in soil rather than of specific organisms. In fact, experience has shown it to be very rare for pesticides used as recommended to give rise to long-term effects on important soil functions [23].

However, if the primary studies do show that respiration may be drastically reduced, or major effects can be expected in the soil biota, it may be valuable to conduct studies on the mineralization of organic debris. Again, deleterious effects shown in laboratory studies with the earthworm may also need to be followed up in the field to evaluate the effects on soil texture and structure [8, 12, 24].

Mode of Action

The spectrum of activity of the pesticide that has been demonstrated to be a herbicide, an insecticide, a fungicide, or a microbial agent automatically establishes conditions and levels for testing and hazard evaluation. A herbicide intended for use in the terrestrial environment should be first tested on appropriate representative nontarget species indicated to be important food for wildlife, and those herbicides that may be transported to or directly applied to the aquatic environment should be tested on algae and macrophytes in addition to the food chain organisms and fish.

Insecticides logically should be extensively tested on nontarget insects (such as honeybees, where pollination and commercial apiaries are important) and soil invertebrates (such as annelids) particularly where the formulation may concentrate the pesticide in one specific environmental compartment [8]. The next level of testing should include those species of wildlife dependent upon these organisms as a food supply (such as birds and insectivorous mammals) [25–27]. Insecticides that are transported by runoff are obvious candidates for tests on aquatic invertebrates and fish.

Microbial agents and fungicides may adversely impact on material transfer and sequestering (especially the reduction of organic matter and nitrogen or sulfur transformation), and thus the first-level testing priority should include appropriate tests such as the specific bacterial growth inhibition and function tests for processing organic matter and mineralization of soils [23]. A second level of conditional testing is needed when exposure to leaf litter occurs [28].

Predicides, avicides, and rodenticides require that priority be given to the toxicity to nontarget species, and especially if there is potential for secondary poisoning of birds and mammals when persistent chemicals are used (such as organochlorine pesticides, thallium salts, sodium fluoroacetate) [25, 29–32]. Special tests must be tailored to the mode of action of other control agents, and in some instances the use of special formulations, baits, and dispensing devices (locked bait stations) may even preclude the necessity for further testing.

Aquatic pesticides, including those used for the specific purpose of controlling aquatic weeds, parasites, snails, noxious insects, or disease vectors, require rigorous testing to ensure safety to nontarget species and evidence that residues will not compromise the resources used. Waterfowl, amphibians, and aquatic reptiles may also be subjected to the effects of bioaccumulated residues of persistent pesticides that are highly fat soluble. If these persist in a compartment of the environment for an appreciable time (due either to slow degradation or to frequent application) or if the compound's basic toxicological properties suggest the possibility of longer term effects, data additional to those on short-term toxicity may be needed. Tests on reproduction, neurotoxicity, and behavioral effects may be needed to determine long-term effects on both invertebrates and vertebrates, including fish, birds, mammals, and other forms dependent on aquatic food chains [25–27, 29–31].

Interpretation of Laboratory and Field Experiments

The extrapolation of data obtained in the laboratory to field situations is a fundamental problem. Usually only acute and chronic or semichronic toxicity data of a compound are known and in relation to a limited number of test species. For several reasons, the outcome of these tests is of restricted value for the prediction of hazards of an actual field application. One cannot be sure that the most sensitive nontarget organism or the ecologically most relevant criterion for effect have been used. A very important problem is the absence of data on secondary effects on the ecosystem. As a result of the mutual interrelationships between ecosystem components, a primary effect will always cause secondary effects as well. Sometimes the ecosystem will cushion these effects; but this is not always so, and a major disturbance may be the result.

The toxicity data available for the different classes of organisms tested may then be used to estimate the effect of the likely exposure on comparable local species. However, the environmental conditions also must be comparable to those under which the toxicity data were developed. Experience has shown that extrapolation from one species to another in a given class or family is a valid concept when we have knowledge of the mode of action and the pharmacodynamics of the chemical [8]. Toxicological extrapolation of data can be made with greater confidence where variations in the data within a given class are small [5, 9, 25–27, 33].

The extrapolation problem, then, is to make the best possible estimate of an acceptable level. This will always imply the application of arbitrary safety factors. The magnitude of these should be made dependent on the risk one is willing to accept in the situation (for example, some mortality of fish may be acceptable in one situation but not in another).

The ability to assess a certain hazard will always be linked with the quality and amount of data on the toxicology of the compound, that is, the level of testing. There are certain additional studies that can be carried out in the laboratory to study such issues as reproductive effects in a species such as *Daphnia*. But where the laboratory data give rise to serious doubts and where the product is intentionally applied to the water body, observations on fish and large invertebrates should be confirmed in field observation studies. The special value of these studies is that recovery of affected populations can only be studied in natural conditions. Moreover, the suspended matter and other factors characteristic of natural waters which cannot readily be reproduced in the laboratory can have a profound effect on the results. However, to ensure adequate protection of this environment, laboratory tests are usually run under "worse-case conditions," that is, with no suspended matter that would tie up pesticides. On the other hand, ecosystem types of investigation have to be carried out in order to collect data of a more predictive power. This could be done in a number of selected sites by means of pilot studies or outdoor model ecosystems.

By applying these considerations, it will become readily apparent if any classes of wildlife are likely to be at risk when the product is used as recommended. Moreover, modifications in the proposed uses can be developed to minimize the risk of these effects occurring in practice. Clearly a product with high avian toxicity could lead to difficulties if it were likely to contaminate food for the species in question. Similarly, compounds of high fish toxicity would have to be used in a way which prevented exposure to fish and other nontarget organisms.

A further issue may arise from the primary data regarding the possibility of bioaccumulation, particularly in economically important fish, shellfish, birds, and mammals. Such data as the partition coefficient, laboratory animal metabolism studies, and the way the product is to be used usually give a clear indication of the bioaccumulation potential [28, 34]. In cases where this may occur, separate follow-up residue monitoring studies may be needed to determine the extent of occurrence and the toxicological significance of multiple exposure to the parent compound, metabolites, and transformation products [5, 8, 9, 35]. With this appraisal, the proposed uses can be reviewed and any recommended modifications made to avoid unacceptable environmental effects. It may be possible to minimize exposure of vulnerable elements of the environment by altering the timing of the application, the formulation, the phasing of the dosage (such as split applications), and the application techniques.

RELATIONSHIP OF REGULATION OF PESTICIDES AND HAZARD

An important consideration in the registration of pesticides relates directly to inadvertent, unconscious, or accidental abuse associated with "approved" usage. These abuses can arise from lack of training among applicators, pesticide labels in a language foreign to the user, and inappropriate directions for use among other factors. Value systems in regard to hazard assessment factors may not even be commensurable, while enforcement, monitoring, and evaluation of efficacy and abuse may be absent. A pesticide which may be correctly and effectively used in temperate climates by skilled and experienced managers and applications for the reasons just mentioned may be a serious threat to humans and the environment in other hands. Thus, pesticide registration practices should take into account the potential for such abuse. Extrapolation of data from one geographic, edaphic, or climatic area to another can be extremely misleading in the evaluation of the efficacy and safety of the pesticide use. This is also true when toxicological data gathered on one plant or animal species is extrapolated to another organism or environmental situation. This may be most critical in the hazard evaluation of aquatic environments. Data on salmoid fish, a very sensitive test species from cool-temperate climates, need to be related to species in warm-temperate and near-tropic climates in a systematic manner. Certain biological and chemical principles in chemodynamics and toxicology remain inviolate between these different ecosystems. The critical point is that all usage experience must feed back into the regulatory-evaluative loops, whether by laboratory model ecosystems, simulated outdoor model ecosystems, or actual field testing [6, 11, 29, 30, 35, 36]. Although these studies are useful in the final hazard assessment stages, there are limitations (such as geographic, edaphic, climatic, and biologic variables). Understanding the extremes in these variables aids in validating or supporting efficacy and toxicology data generated on single species under laboratory conditions intended to best simulate natural environmental situations.

Probably no area needs site-specific information more than the physical-chemical and biodegradation measurements. Primarily, they reveal the actual response of typical target soils, waters, and microbiota. Overemphasis on certain properties as predictors (such as the partition coefficient for bioaccumulation) leads to large gaps: for protein-binding materials and for covalent reacting chemicals. Soil thin-layer chromatography and shake flask sorption and degradation measurements can reveal much about the anticipated distribution and biodegradation and can be performed relatively cheaply in the laboratory under controlled conditions. The estimates of environmental concentration are usually derived by models which assume achievement of some steady-state behavior of the chemical under a given environmental loading (usage pattern, frequency, volume). These fail to take into account the rates of phenomena which are treated as equilibrium processes (such as sorption, cation exchange, and bioaccumulation).

Media and biotic flux may create situations considerably different from the project assumptions used in the model, so that much less danger may be inherent in a given use, or the impact of that use might be very seriously underestimated. The matter extends to sorption phenomena. High humidity, for example, drives more water onto adsorption sites competing with the chemical. If these adsorption sites are important in surface-catalysis of hydrolysis, then the overall fate of the chemical will depend upon the effects of temperature and humidity relationships with all these phenomena (hydrolysis, binding, and volatilization) and is not easily predicted without temperature-dependent data for each phenomenon [8, 14, 15, 17, 18, 28]. These data (including such factors of a local environment as air exchange rates and the type of adsorptive surface) are not likely to be available. Efficacy can thus be greatly reduced or hazard significantly affected.

Reduction of the hazard of pesticide use can be accomplished by adequate training of pest control operators, the use of better formulations and application methods, wise timing of the application, and utilization of the safest pesticides or alternative pest control methods and cultural practices (integrated pest management) [1–3, 7, 14, 15, 23]. However, such precautions must be universally accepted and applied over broad geographic areas to be effective in minimizing losses. What one pest control operator does or does not do affects both the effectiveness of the pest control and the protection of fish, wildlife, and habitat.

The optimization of application techniques certainly is an important area for exploration, especially where integrated pest management is being introduced. The use of various controlled-release materials already has shown great promise, particularly in terms of the degree to which the product can be tailored to the conditions of use in the field. Furthermore, controlled-release materials show great promise for aspects of environmental and worker protection. However, the overall approach of integrated pest control is a much more sophisticated approach than many agriculturists want to employ. More research is needed in basic ecology and applied pest dynamics (including relationships of various predators, pathogens, and parasites) than many agriculturists may be able to support.

Conversion of large areas to agriculture (and especially monoculture) is the most significant ecological event in a region. This, in and of itself, creates "brittle" ecosystems of extreme simplicity and vulnerability. Thus caution and proper use of controlled and statistically well-designed experiments are needed for both toxicological evaluation and efficacy studies as a part of the hazard assessment scheme; the use should not provide risks which are not commensurate with the gains. The regulatory process should select out pesticides which a manager might use (and potentially abuse) simply because it looks good for a situation. This is particularly important when pesticides are applied to large areas where cumulative effects occur and the spreading population of plants and animals cannot compensate for the injury. The hazards may vary with each type of habitat, and consideration must be given to the sensitivity of organisms, the most vulnerable life stage, the weakest link in the food chain, the habitat requirement, and other limiting factors to the maintenance of growth, reproduction, health, and behavior. The bioaccumulation of residues may also compromise the wholesomeness and value of fish and wildlife for both recreational and nutritional uses. To reduce the hazard of pesticides to fish, wildlife, and habitat, we must (1) reduce adverse pesticide concentrations; (2) limit the critical time of exposure to residues; and (3) prevent contact with susceptible organisms, ecological components processes, or functions.

REFERENCES

[1] "Criteria and Rationale for Decision Making in Aquatic Hazard Evaluation. Aquatic Hazards of Pesticides Task Group of the American Institute of Biological Sciences," in *Estimating the Hazard of Chemical Substances to Aquatic Life*, ASTM STP 657, American Society for Testing and Materials, Philadelphia, 1978, pp. 241–273.

[2] Branson, D. R., "Predicting the fate of Chemicals in the Aquatic Environment from Laboratory Data," in *Estimating the Hazard of Chemical Substances to Aquatic Life*, ASTM

STP 657, American Society for Testing and Materials, Philadelphia, 1978, pp. 55–70.

[3] Branson, D. R. and Blau, G. E., "Predicting a Bioconcentration of Potential of Organic Chemicals in Fish from Partition Coefficients," in *Structure-Activity Correlations in Studies of Toxicity and Bioconcentration with Aquatic Organisms.* G. D. Veith and D. E. Konnoasewich, Eds., International Joint Commission, Great Lakes Research Advisory Board, 1975, pp. 99–117.

[4] Carlson, R. M., Kopperman, H. L., Caple, R., and Carlson, R. E., "Structure-Activity Relationships Applied," in *Structure-Activity Correlations in Studies of Toxicity and Bioconcentration with Aquatic Organisms,* International Joint Commission, Great Lakes Research Board, 1975, pp. 57–72.

[5] Walker, Charles R., "Aquatic Uses of Herbicides and Problems Associated with Forestry and Water Resource Management," Addendum to the Symposium on the Use of·Herbicides in Forestry, Environmental Protection Agency. Washington, D.C., 1978, pp. 55–60.

[6] Woodwell, G. M., "Effects of Pollution on the Structure and Physiology of Ecosystems," *Science,* Vol. 168, 1970, pp. 429–433.

[7] Baughman, G. L. and Lassiter, R. R., "Prediction of Environmental Pollutant Concentration," in *Estimating the Hazard of Chemical Substances to Aquatic Life, ASTM STP 657,* American Society for Testing and Materials, Philadelphia, 1978, pp. 35–54.

[8] Stern, A. M. and Walker, C. R., "Hazard Assessment of Toxic Substances: Environmental Fate Testing of Organic Chemicals an Ecological Effects Testing," in *Estimating the Hazard of Chemical Substances to Aquatic Life, ASTM STP 657,* American Society for Testing and Materials, Philadelphia, 1978, pp. 81–131.

[9] Walker, C. R., "Toxic Substances from Non-Point Sources Related to Water Management and Forestry Practices and Their Impact on Fish and Wildlife," in *Proceedings "208" Symposium: Non-Point Sources of Pollution from Forested Land.* G. M. Aubertin, Ed., Southern Illinois University, Carbondale, Ill., 1977, pp. 171–212.

[10] Dortman, R. J. and Koeman, J. H., "Recommendations on Guidelines for Standardized Methodologies in Predicting Environmental Toxicology. Expert Consultation on Harmonization of Environmental Criteria for Registration of Pesticides," Working Paper PE 79/3, United Nations Food and Agriculture Organization, Rome, 1979.

[11] European Inland Fisheries Advisory Commission "Report on Fish Toxicity Testing Procedures," EIFAC Tech. Paper No. 24(EIFAC/T24), United Nation Food and Agriculture Organization, Rome, 1975.

[12] Helling, C. S., Kearney, P. C., and Alexander, M., "Behavior of Pesticides in Soil," *Advances in Agronomy,* Vol. 23, 1971, pp. 147–240.

[13] Zitko, V. and Carson, W. G., "Avoidance of Organic Solvents and Substituted Phenols by Juvenile Atlantic Salmon," M.S. Rept. No. 1327, Fisheries Research Board of Canada, 1974.

[14] Moolenaar, R. J., "Environmental Impact of Chemicals," in *Chemicals, Human Health and the Environment, A Collection of Dow Scientific Papers.* Vol. 1, Dow Chemical, Midland, Mich., 1974, pp. 2–6.

[15] Neely, W. B., Branson, D. R., and Blau, G. E., "Predicting Bioconcentration Potential," *Environmental Science and Technology.* 1974, p. 1113.

[16] Hueck, H. J., "Principals of Testing for Potential Biotic Environmental Effects of Chemicals," TNO p 78/2-1978-01-17, Division of Technology for Society, Delft, Netherlands, 1978.

[17] Neely, W. B., "Mathematical Models to Describe the Time/Space Distribution of Chemicals in a Water System," in *Structure-Activity Correlations in Studies of Toxicity and Bioconcentration with Aquatic Organisms,* G. D. Veith and D. E. Konasewich, Eds., International Joint Commission, Great Lakes Research Advisory Board, 1975, pp. 261–269.

[18] Neely, W. B., "An Integrated Approach to Assessing the Potential Impact of Organic Chemicals in the Environment," Step Sequence Group 5.35/SSG-3-C. Organization for Economic and Cooperative Development Chemicals Testing Programme, Paris, France, 1978.

[19] Metcalf, R. L. and Sanborn, J. R., "Pesticides and Environmental Quality in Illinois," *Bulletin of Illinois Natural History Survey,* Vol. 31, 1975, pp. 381–436.

[20] Kimerle, R. A., Gledhill, W. E., and Levinskas, G. L., "Environmental Assessment of New Materials," in *Estimating the Hazard of Chemical Substances to Aquatic Life, ASTM STP 657,* American Society for Testing and Materials, Philadelphia, 1978, pp. 132–146.

[21] Martin, Y. C., "A Comparison of Methods for the Analysis of the Relationship Between Chemical and Biological Properties of Chemicals," in *Structure-Activity Correlations in Studies of Toxicity and Bioconcentration with Aquatic Organisms.* G. D. Veith and D. E. Konasewich, Eds., International Joint Commission. Great Lakes Research Advisory Board, 1975, pp. 143–150.

[22] Zitko, V., "Structure-Activity Relationships in Fish Toxicology," in *Structure-Activity Correlations in Studies of Toxicity and Bioconcentration with Aquatic Organisms.* G. D. Veith and D. E. Konasewich, Eds., International Joint Commission, Great Lakes Research Advisory Board, 1975, pp. 7–24.

[23] Stephenson, M. E., "An Approach to the Identification of Organic Compounds Hazardous to the Environment and Human Health," *Ecotoxicology and Environmental Safety.* Vol. 1, 1977, pp. 39–48.

[24] Edwards, C. A. and Thompson, A. R., "Pesticide and Soil Fauna," *Residue Reviews,* Vol. 45, 1973, pp. 1–80.

[25] Tucker, R. K., "An Avian Toxicological Testing Sequence," *Western Pharmacology Society Proceedings,* Vol. 19, 1976, pp. 408–411.

[26] Tucker, R. K. and Haegele, M. A., "Comparative Acute Oral Toxicity of Pesticides to Six Species of Birds," *Toxicology and Applied Pharmacology,* Vol. 20, 1971, pp. 57–65.

[27] Tucker, R. K. and Leitzke, J. S., "Comparative Toxicology of Insecticides for Vertebrate Wildlife and Fish," *Pharmacology and Therapeutics,* Vol. 6, 1979, pp. 167–220.

[28] Gillett, J. W. and Gile, J. D., "Pesticide Fate in Terrestrial Laboratory Ecosystems, *International Journal of Environmental Studies,* Vol. 10, 1976, pp. 15–22.

[29] Gilbertson, M., "Methods of Testing Effects of Chemicals on Birds," Regulations, Codes, and Protocols Report EPS 1-EC-75-1. Environment Canada, Environmental Contaminants Control Branch, 1975.

[30] Gilbertson, M., "Methods for Testing Effects of Chemicals on Mink Reproduction," Regulations, Codes, and Protocols Report EPS 1-EC-75-2, Environment Canada, Environmental Contaminants Control Branch, 1975.

[31] Haegele, M. A. and Tucker, R. K., "Effects of 15 Common Environmental Pollutants on Eggshell Thickness in Mallards and Coturnix," *Bulletin of Environmental Contamination and Toxicology,* Vol. 11, No. 1, 1974, pp. 98–102.

[32] Hill, E. F., Heath, R. G., Spann, J. W., and Williams, L. D., "Lethal Dietary Toxicities of Environmental Pollutants to Birds," Special Scientific Report-Wild, No. 191, U.S. Fish and Wildlife Service, Washington, D.C., 1975.

[33] Kenaga, E. E., "Factors to Be Considered in the Evaluation of the Toxicity of Pesticides to Birds in Their Environment," in *Environmental Quality and Safety. Global Aspects of Chemistry, Toxicology and Technology as Applied to the Environment.* Vol. II, Academic Press, New York, 1973, pp. 68–181.

[34] Metcalf, R. L. Sangha, G. K., and Kapoor, I. P., "Model Ecosystems for Evaluations of Pesticide Biodegradability and Ecological Magnification," *Environmental Science and Technology,* Vol. 5, 1971, pp. 709–713.

[35] Wedemeyer, Gary and Yasutake, Tosh, "Standard Methods for the Clinical Assessment of the Effects of Environmental Stress on Fish Health," Technical Paper No. 89, U.S. Fish and Wildlife Service, Washington, D.C., 1977.

[36] Whitworth, W. R. and Lane, T. H., "Effects of Toxicants on Community Metabolism in Pools," *Limnology and Oceanography,* Vol. 14, 1969, pp. 53–58.

Appendix Five

Environmental and Economic Impacts of Pesticide Use

by David Pimentel et al.

Introduction

For several years, Dr. David Pimentel and his associates at Cornell University have written articles calculating the total costs of pesticide use—to the whole environment, within the limits of converting nonmonetary values to dollar estimates. The latest and most extensive of these will appear in a book entitled *The Pesticide Question: Environment, Economics and Ethics,* edited by Dr. Pimentel and Professor Hugh Lehman of Guelph University. The following brief summary most importantly shows the range of aspects taken into consideration, and the difficulty of making comparable and precise evaluations, even on an annual basis.

The opening paragraphs of the full text make these points: Despite the widespread use of pesticides in the United States, pests (principally insects, plant pathogens, and weeds) destroy 37 percent of all potential food and fiber crops. Estimates are that losses to pests would increase 10 percent if no pesticides were used at all; specific crop losses would range from zero to nearly 100 percent. . . .

Although pesticides are generally profitable, their use does not always decrease crop losses. For example, even with the 10-fold increase in insecticide use in the United States from 1945 to 1989, total crop losses from insect damage have nearly doubled from 7 percent to 13 percent. This rise in crop losses to insects is, in part, caused by changes in agricultural practices. For instance, the replacement of corn-crop rotations with the continuous production of corn on about half of the original hectarage has resulted in a nearly 4-fold increase in corn losses to insects despite approximately 1000-fold increase in insecticide use in corn production. (Reference to Pimentel's 1991 *Handbook on Pest Management in Agriculture*.)

Most benefits of pesticides are based only on direct crop returns. Such assessments do not include the indirect environmental and economic costs associated with pesticides. To facilitate the development and implementation of a balanced, sound policy of pesticide use, these costs must be examined.—SAB

AN ASSESSMENT OF THE ENVIRONMENTAL AND ECONOMIC IMPACTS OF PESTICIDE USE

By D. Pimentel, H. Acquay, M. Biltonen, P. Rice, M. Silva, J. Nelson, V. Lipner, S. Giordano, A. Horowitz, and M. D'Amore

Pesticides have contributed to the impressive productivity of U.S. agriculture, with an estimated $16 billion in crops saved each year by the $4 billion U.S. investment in pesticide controls. This benefit, however, does not include the cost of serious human health and environmental problems associated with pesticide use.

Pesticides are applied by hand, by large mechanical sprayers as well as by aircraft. Although efforts are made to contain sprays to the target crop-area, the pesticides reach adjacent vegetation, wildlife, soil, and water. In this way the impact of pesticides is felt far beyond the designated target area.

Pesticides adversely affect the health of humans when they are exposed to them. Based on survey data, a recent World Health Organization report estimated there are 1 million human pesticide poisonings each year in the world with about 20,000 deaths worldwide. In the United States, pesticide poisonings are reported to total a minimum of 67,000, with the number of accidental fatalities estimated to be 27 each year.

In addition, several thousand domestic animals are poisoned by pesticides each year. Dogs and cats are the most commonly poisoned animals because they usually wander freely about the home and farm and therefore have greater opportunity to come into contact with pesticides than other domesticated animals. Central records on domestic animal poisonings are not kept, making an overall economic assessment extremely difficult.

In cultivated and wild areas, naturally present predators and parasites help keep pest species in check. When pesticides destroy both pest and beneficial natural enemies, frequently other pests present reach outbreak levels. For example, in cotton and apple crops, pesticide destruction of natural enemies results in the outbreaks of numerous pests, including: *cotton*—cotton bollworm, tobacco budworm, cotton aphid, and cotton loopers; and *apple*—European red mite, red-banded leafroller, San Jose scale, rosy apple aphid. The costs of the additional pesticide applications required to control these pests, plus the increased crop losses they cause, are estimated to be about $520 million per year.

Another vital group of insects that pesticides frequently kill are honeybees and wild bees, essential for the annual pollination of about $30 billion in fruits and vegetables. The losses incurred with the destruction of honey bees and loss of pollination each year are conservatively estimated to be $320 million.

Another serious and costly side effect of heavy pesticide use has been the development of pesticide resistance in pest populations of insects, plant pathogens, and weeds. At present some 900 species exhibit resistance to commonly applied pesticides. When resistance occurs, farmers must increase pesticide applications to save their crops. Even so, crop losses frequently are higher than normal. This resistance problem is estimated to cost the nation $1.4 billion each year, in increased costs of pesticides and reduced crop yields.

Basically pesticides are applied to protect crops from pests and to increase yields, yet at times the crops themselves are damaged by pesticide treatments. This occurs when the recommended dosages suppress crop growth, development, and yield; pesticides drift from the targeted crop to damage adjacent valuable crops; and, residual herbicides either prevent chemical-sensitive crops from being planted in rotation or inhibit the growth of crops that are planted. In addition, excessive pesticide residues accumulate on crops, necessitating the destruction of the harvest. When crop seizures and insurance costs are added to the direct costs of crop losses caused by pesticides, the total yearly loss in the United States is conservatively estimated to be nearly $1 billion.

Table 1. Total Estimated Environmental and Social Costs from Pesticides in the United States (in $million/yr)

Public health impacts	787
Domestic animal deaths and contamination	30
Loss of natural enemies	520
Cost of pesticide resistance	1400
Honey bee and pollination losses	320
Crop losses	942
Fishery losses	24
Bird losses	2100
Groundwater contamination	1800
Government regulations to prevent damage	200
Total	8123

Ground and surface waters frequently are contaminated by applied pesticides. Estimates are that nearly one-half of the ground and well water in the United States is or has the potential to become contaminated with pesticides. To adequately monitor this contamination, an estimated $1.3 billion would need to be spent each year. However, pesticides are not monitored, nor have steps been taken to prevent the widespread contamination of U.S. water resources.

Pesticides wash into streams and lakes where they cause substantial fishery losses. Thus, high pesticide concentrations in water directly kill fish; low dosages primarily kill small fish fry. Also, pesticides eliminate aquatic insects and other small invertebrates, which are food for fish.

Birds, mammals, and other wildlife also are killed by pesticides. The full extent of wildlife destruction is difficult to determine because these animals are often hidden from view, camouflaged, highly mobile, and live in protected habitats. Based on the available data, U.S. bird losses associated with pesticide use represent an estimated loss of about $2.1 billion per year. No estimate can be made of mammal losses because of a lack of data.

The known costs of human and animal health hazards, plus the costs of diverse environmental impacts associated with U.S. pesticide use total approximately $8 billion each year (Table 1). Thus, based on a strictly cost/benefit basis, pesticide use remains beneficial. Decisions about future pesticide use need to be based not only on the benefits, but consider carefully the risks they create. Perhaps in this way an equitable balance can be achieved. Now is the time for individuals to make known to political leaders the value they place on human health and, in the broader sense, the integrity of their natural environment.

U.S. Federal Regulation of Pesticides, 1910–1988

by Shirley A. Briggs

1910 The first U.S. federal pesticide act was passed to protect farmers from fraudulent, ineffectual, and misbranded products. The law was placed under the administration of the U.S. Department of Agriculture. It was based on post-market control; if fraud was discovered the product was taken off the market. This came at the end of a sequence of consumer protective legislation begun about 1900 to protect the public from medical quackery and gross abuses in food processing.

1947 The Federal Insecticide, Fungicide and Rodenticide Act (FIFRA) replaced the 1910 law. Administration of the law continued under the Department of Agriculture (USDA). The new law was passed in response to the need for regulation of the large number of synthetic organic pesticides, which were coming onto the market in a flood of new products, many of them developed during World War II. FIFRA, like the 1910 act, was designed to protect the farmer from material that might either be dangerous or not effective. No controls over the *use* of pesticides were included. Registration of the label with USDA was required before sale was permitted. The law required that the product be "safe" when used as directed on the label. A loophole provision for "protest registration" allowed a manufacturer to market a product even if USDA refused to register it, and 17 pesticides were marketed this way. FIFRA was originally a labelling act, providing no sanctions against misuse of the pesticide, no authority for immediate stop-sale orders against dangerous pesticides, and only feeble penalties for companies selling pesticides in violation of safety criteria.

1947–1969 The Department of Agriculture's record in enforcing the law reflected its bias as an enthusiastic promoter of pesticides. The Pesticide Regulation Division (PRD) of USDA was not inclined to ecological thinking. It had the power to recall dangerous products from 1947 on, but did not use this power until 1967. It did not set up a formal procedure for recall of products until May 5, 1969, just before Government Accounting Office hearings into PRD operations were scheduled. Only once in 22 years did PRD start criminal proceedings against violators of FIFRA, though in that period pesticide acci-

dents to people rose to 50,000 a year. They looked into 60 accidents to people, and twice as many to farm animals. They approved pesticides over the objections of the Food and Drug Administration and the Public Health Service, including mercury-treated seeds, alkylmercury compounds, use of no-pest strips around infants, the aged, and in food areas, and products that caused cancer in test animals.

1957 The Pesticide Research Act provided funds for research in the Department of the Interior. Since 1946, the Fish and Wildlife Service had done critical research of the effects of pesticides in the environment, with inadequate staff and funds. With expanded research after this act, they continued to find more effects on fish, birds, and mammals, and gave us the first warnings of the overall hazards of pesticides.

1962 Publication of Rachel Carson's *Silent Spring* began a series of events that were to culminate in 1970 with creation of the Environmental Protection Agency and complete restructuring of pesticide regulation. Pesticide review and control boards and committees were set up at state and federal levels. Reports from the President's Science Advisory Committee under Kennedy and Johnson made forceful recommendations: the 1963 report *Use of Pesticides* and the 1965 *Restoring the Quality of Our Environment.*

1964 An amendment to FIFRA closed the "protest registration" loophole.

1968 The Environmental Defense Fund filed suit against both USDA and the Department of Health, Education, and Welfare (HEW) on grounds that pesticides were being registered contrary to FIFRA, products that when used as directed were not safe for humans.

1969 Secretary of HEW Finch commissioned a study of pesticides and their relation to environmental health. The report concluded that the Department of the Interior and HEW be given greater roles in pesticide registration. Much basic research on pesticide hazards, especially long-term, was first made public in this *Report of the Secretary's Commission on Pesticides and Their Relationship to Environmental Health.* By 1969 several scientific reports firmly implicated DDT as a cause of cancerous growth in test animals. In November, Secretary Finch asked the Departments of Agriculture and Interior to join him in "phasing out" DDT within two years.

1970 The Environmental Protection Agency was established on December 2. Offices from other federal agencies were incorporated in it, including the Pesticide Registration Division from USDA.

1971 William D. Ruckelshaus, administrator of EPA, canceled almost all uses of DDT on June 14, announcing a virtual ban in the U.S. effective December 31, 1972.

1972 The Federal Environmental Pesticide Control Act (FEPCA) was adopted. It first recognized the quality of the environment as an important value to be protected, permitted regulation of the *use* of pesticides, and directed their classification as either general or restricted use pesticides. The latter could only be used by persons certified or working under a certified applicator. Qualifications for certification were set. More streamlined cancellation procedures were set forth, and the principle was established that in a contested case it is the responsibility of the manufacturer/registrant to prove a product's safety, instead of putting the burden of proof of hazard on EPA, citizen's groups, and the public. The act required reauthorization after three years.

FEPCA extended federal control over pesticides marketed within a state as well as in interstate commerce. EPA was given more authority to expand research and monitoring, to find ways to reduce dangers of exposure to pesticide residues. New civil and criminal penalties were established for violations. FEPCA is an act to amend FIFRA, and is thus incorporated into FIFRA.

1972–1975 EPA assumed principal authority over pesticides, with some human health aspects under the Food and Drug Administration and HEW. USDA remained committed to promotion of pesticides in the interests of large agricultural enterprises. It continued to rely on massive, costly chemical pesticide campaigns to "eradicate" pests like the fire ant or the boll weevil, regardless of the growing evidence of the ineffectiveness, economic waste, and environmental damage.

EPA took action against other organochlorine pesticides, by the slow, costly cancellation process, with hearings before an administrative law judge. Aldrin and Dieldrin were suspended April 5, 1975, after four years of proceedings. Mirex hearings began in 1973, and it was cancelled December 1, 1977. Chlordane/heptachlor were suspended in August 1975, while cancellation proceedings continued, but instead of a final, legal judgment, EPA made an agreement with the producer, Velsicol, to phase out most used by December 31, 1980, except for subsurface treatment for termites. (It was later found that these treatments did not stay in place as had been thought.) At that time EPA had to compensate producers for any remaining stocks of a cancelled product, which could exceed the whole EPA pesticide budget. Endrin was cancelled for most uses July 17, 1979. Many mercury pesticides were cancelled between 1970 and 1978.

Reregistration of pesticides to classify them as either general or restricted use now required testing for long-term health and environmental effects as well as immediate toxicity. Progress was slow, while the necessary tests were determined.

Standards for certification of applicators were set, with which state programs for actually certifying must comply.

New requirements for experimental use permits were defined, rules were developed for production and disposal of pesticides, and EPA studied ways to make the cost/benefit analyses required by the new system.

1975 FIFRA was reauthorized in an amended version, until 1977, with some weakening. A new Science Advisory Panel for pesticides was set up that must review all proposed actions, as must the secretary of agriculture and the Senate and House Agriculture Committees, who retained jurisdiction over EPA regulation of pesticides. In the certification program, EPA was forbidden to require that states make taking or passing a test part of the training and certification. States may do so if they wish, however.

The Science Advisory Panel for pesticides (FIFRA-SAP) has members chosen from lists from the National Science Foundation and the National Academy of Sciences.

Between reauthorizations, congressional committees exercise oversight to see whether agencies are carrying out the law properly. In 1976, the Senate Judiciary Committee studied EPA's administration of FIFRA. They were able to obtain three data files from EPA's archives, which had been kept safe from any public inspection by the "trade secret" claim of the producers, despite EPA statements that toxicology information should not come under this rule. Study of the files for methoxychlor, captan, and ferbam showed that decisions had been made on old data that was clearly inadequate, misinterpreted, and sometimes clearly contradicted the claims of the manufacturers concurred by EPA. This gave impetus to further amend FIFRA when it next came up for reauthorization.

1978 FIFRA was reauthorized for one year, and some changes were made. Reregistration was proceeding so slowly that reclassifying as general or restricted use, and thus involving certified applicators, had made little headway. EPA proposed a new "conditional registration" by which a new product could be put on the market before testing, which would only need to be done by the time old products being reregistered had to provide all data. The excuse for this was EPA's claim that they had all of the commercial products to reregister, 35,000 products, and each would take years, so testing could not be finished for decades in the future. This was pure invention because the testing is done by active ingredients, of which there were perhaps 500 used enough to cover most of the commercial formulations based on them, and in fact most would be accounted for under about 150 active ingredients. But Congress was fooled and permitted the new bypass of careful scrutiny to be made.

The trade secrets issue was tackled, with EPA insisting, and Congress agreeing, that data on toxicity, and environmental behavior and fate of pesticides, are not trade secrets. Explicitly, FIFRA states, "All information concerning the objectives, methodology, or significance of any test or experiment performed on or with a registered or previously registered pesticide or its separate ingredients, impurities, or degradation products, and any information concerning the effects of such pesticide on any organisms or the behavior of such pesticide in the environment, including, but not limited to, data on safety to fish and wildlife, humans and other mammals, plants, animals, and soil, and studies on persistence, translocation and fate in the environment, and metabolism, shall be available for disclosure to the public."

Some time passed before this clause went into effect, since Monsanto challenged the 1978 amendments in court. Finally, in 1981, the Supreme Court ruled in favor of public disclosure of this information. EPA still requires that inquiries go through the Freedom of Information Act procedure, however, so considerable time can elapse.

The requirement that the efficacy of a pesticide be determined, the principal reason for the 1910 law, was made incidental: the administrator can waive any requirements for such testing for any pesticide.

EPA was also authorized to delegate enforcement of FIFRA to the states, which was done, removing centralized, consistent administration of this function.

1980 A General Accounting Office report criticized the poor planning and failure to set priorities that had made reregistration so slow. They also brought up the matter of inadequate testing. Falsified test results had been revealed, especially in the case of the Industrial Biotest Laboratories, under investigation since 1977. At least 200 pesticides were registered on the basis of tests shown to be astonishingly faulty. EPA claimed no authority to take off the market any pesticides known to have been registered on false data, and they began the process of reviewing the suspect tests, and asking to have many redone. This left the dubious pesticides on the market.

The 10 years between 1978 and 1988 saw no changes in FIFRA, though the opposing groups battled constantly, one side trying to strengthen the law to give effective protection to environmental and human health, the other determined to improve the economic benefits to producers and major users of pesticides in the existing rules. Bills might pass one house of Congress only to be defeated in the other. Meanwhile, the legal authority of FIFRA was maintained by just voting to keep the old law in place, from year to year. Finally, a compromise was reached in which each side postponed some cherished issues in order to get what was called a "core" bill through. The resulting amendments produced what was called "FIFRA Lite," but it made some important reforms.

Reregistration of pesticides marketed before the expanded testing required under the 1972 amendments was to be speeded, with revised procedures and deadlines. Although almost all of the pre-1984 pesticides still had to be reviewed, another goal was set, this time to have the job done by 1997. Specific stages and procedures were ordained: EPA would have to publish lists of active ingredients subject to reregistration and ask their registrants (i.e., producers) whether they wanted to seek reregistration. These lists were to be issued in four stages over 10 months. Second, registrants must respond to this request, and for each one must agree to fill the existing data gaps required to fit EPA's current test requirements, on a schedule of firm deadlines, and agree to pay the first installment of a new reregistration fee.

The fee for reregistration, or regular registration for new products, has two purposes: to help pay the high costs to EPA of the process, and to make registrants think seriously about whether they want to continue with a product that may no longer be profitable, or whose toxic profile revealed by more tests may result in restrictions. EPA estimates that the total costs over the nine years will be about $250 million. Perhaps half of this will continue to come from EPA's regular budget, which has been inadequate over the years. Considering the high profits enjoyed by most pesticide producers, an initial fee of $150,000 for a pesticide used on food or feed crops, which require the widest range of tests, is not out of proportion. For pesticides aimed at uses not deemed so hazardous, the fee ranges from $50,000 to $150,000, depending in part on whether EPA has already done some of the job by issuing a Registration Standard spelling out what is known and what must be added. These fees are paid over the three phases of the reregistration process. Once a pesticide is reregistered it becomes liable for the second kind of fee, which is levied on all of the individual commercial formulations, the trade name products on the market that use the registered active ingredient. This varies somewhat by the number of such products per company, being $425 each for the first 50 and $100 for the rest up to 200. The maximum maintenance fees for a company puts a cap on this to some extent: for up to 50 products it is $20,000, and for any number above that, $35,000. These fees apply during the nine-year period. After that, fees that were set by regulation before passage of these amendments would come back into effect.

Already, the fee payments have had a major impact on the number of active ingredients subject to reregistration. There used to be about 1400, the present estimate is around 650. The change in the total commercial formulation products is even more impressive. Where the figure used to be put at about 50,000, EPA estimates now range from 21,000 to 24,000. (These are from different parts of the agency.) Only about 600 active ingredients were in enough use to matter, so EPA is getting down to reality.

A major curb on EPA ability to cancel or suspend dangerously toxic pesticides has been the previous requirement that they pay the full value of any stores still in existence with the producer and on the market. This amount was often more than EPA's annual budget for the whole pesticide regulation program. They also had to take care of storage and disposal of these now surplus products. The new law relieves them of these obligations and provides funds for the few still to be recompensed, certain "end-users" who bought the products unaware of the uncertainty.

Before, there had been no restraint on a producer who built the inventory up to the moment of EPA action, and the impossibility for EPA to pay all the costs resulted in decisions to let the existing stocks be used up before the decree went into effect. Other categories of users, formulators, dealers, and distributors are no longer compensated, and the few end-user payments come from the Judgment Fund of the U.S. Treasury.

Criminal penalties are increased for registrants or other pesticide producers who knowingly violate the law. Submitting false test data, violating cancellation and suspension orders, or failing to submit required records, or to allow plant and record inspection, are now unlawful, as one would have assumed they should have been. In spite of attempts to pass an amendment clearly authorizing EPA to take a pesticide off the market if it is found to have been registered on fraudulent data, this has not yet been accomplished.

EPA's authority to inspect storage of pesticides was increased, to enforce better standards. The Scientific Advisory Panel (FIFRA-SAP) was made permanent instead of having to be reauthorized every five years. And the time Congress is given to review final regulations was shortened to a flat 60 days.

The amendment process over the years has included other issues, mainly of concern to the producers, involving compensation between companies that do the original testing, and others that want to register a similar product later, and protection of everything possible under the trade secret provisions.

PRESENT STATUS OF FIFRA

FIFRA is still not comparable to the other laws for control of pollution in the United States. As spelled out by The National Coalition Against Misuse of Pesticides, these follow certain principles, including no one has the right to pollute the environment of all of us; harmful contaminants should be reduced to the lowest practical levels; rulings of regulatory agencies should have precedence over the polluters, as long as they are not arbitrary or capricious—that is, the government should not have to be on the defensive in pleading its case; penalties for breaking the rules should be large enough, and quickly enforced, to encourage voluntary compliance; prevention of contamination is cheaper and easier than trying to clean it up; and pollutants should be tracked and managed throughout their life cycle, from production, marketing, use, and fate in the environment, to achieve comprehensive environmental protection.

FIFRA satisfies none of these. It has always been a law to manage the marketing of pesticides, with whatever concessions to public health concerns it has had to accommodate. This reflects its congressional jurisdiction. The other environmental legislation has come through the Environment and Public Works Committees of the House and Senate, and they continue their oversight over their administration. FIFRA has always been under the agriculture committees, reflecting their concept of the welfare of agriculture, which has included an assumption that continued use of pesticides in the current mode is essential. Manufacturers of products important to agriculture have a large influence. If it were possible to move the jurisdiction to the environment committees, all the issues that affect pesticide use could have a broader forum.

Advocates of pesticide control in the general public interest still have a number of issues for future amendments of FIFRA. One is the elimination of the concept of "cost/benefit" balancing. In the EPA formulas, if the money to be made from use of a pesticide overbalances the risks to health and environment, it is still registered. The benefits are short-term, the risks often continuing for years after application, and the equations that have been devised for figuring this are artificial and may include unsupportable assumptions. These are like the rules EPA has been led to concoct for ruling on carcinogens, with the unsupported notion of lifetime exposures, and extrapolations between species. Clean Air and Clean Water acts have no cost/benefit balancing.

Efforts to get around even the current permissive regulations continue: EPA administration and the producers long for a no-effect level (NOEL) to be found for everything. (No-effect on what? We still know little about total reactions to these poisons.) Now they are sure that they can assign a "negligible risk" level to toxic contaminants, whether regulated by EPA or the Food and Drug Administration. This is part of a long campaign to somehow get rid of the one scientifically sound rule we have for regulating carcinogens, the Delaney Clause, which directs that no known carcinogen be deliberately added to food. It has not been applied to pesticides that get into food by the slightly indirect method of use during the growing and storage of food, but this would be desirable. Instead, they want to set up more formulas to claim that what is being used does not cause too much risk to the "average person."

EPA also operates as though FIFRA had a clause that requires the federal government to be sure that for every possible pest, occurring under any kind of agricultural practice, there be available a toxic chemical remedy, no matter how harmful it may be. If this insistence on keeping unjustified pesticides on hand were removed, farmers would have more incentive to try other integrated pest management systems.

FIFRA was passed to set basic national rules, a floor under pollution. But recent administrations, and some court rulings, have tried to make it the ceiling, not allowing states or local communities to be more strict. The Supreme Court, in June 1991, ruled against this, finding no such intent in the original FIFRA. Federal law does not preempt state law, nor can states forbid localities from being more careful than the state. Commercial interests are sure to contest this.

Efficacy data should be required again, for all pesticides. This can help prevent overuse for no purpose, and make it possible to detect growing pest resistance to pesticides. Use of hazardous pesticides for purely cosmetic purposes, to make sound produce look prettier, should be discouraged. And the lip service to promoting Integrated Pest Management, now in FIFRA, could be more enthusiastically followed through.

Concern about the widespread contamination of groundwater by pesticides, something that cannot be rectified for decades, if ever, has also sparked suggestions of including that in FIFRA as more of a criteria for regulating the toxic chemicals prone to leaching or running off into surface and groundwater.

More inert ingredients should be tested and regulated, following EPA's late start. The public should have the right to know the health and safety data for all components of a product.

The contest between the proponents of health and environment and those intent on more use of toxic chemical pesticides will go on. It has been a sometimes dramatic, often tedious and frustrating business for both sides. More public attention to the process can improve the quality of the debate.

SILENT SPRING

PUNCH, February 12, 1964

Printed and bound by CPI Group (UK) Ltd, Croydon, CR0 4YY

23/10/2024

01778230-0019